Electrodynamical Reduction of Carbon Dioxide

Energy and Environment Series

Editor-in-chief:
Heinz Frei, Lawrence Berkeley National Laboratory, USA

Series editors:
Nigel Brandon OBE FREng, Imperial College London, UK
Roberto Rinaldi, *Imperial College London, UK*
Vivian Wing-Wah Yam, *University of Hong Kong, Hong Kong*

Titles in the series:

1: Thermochemical Conversion of Biomass to Liquid Fuels and Chemicals
2: Innovations in Fuel Cell Technologies
3: Energy Crops
4: Chemical and Biochemical Catalysis for Next Generation Biofuels
5: Molecular Solar Fuels
6: Catalysts for Alcohol-Fuelled Direct Oxidation Fuel Cells
7: Solid Oxide Fuel Cells: From Materials to System Modeling
8: Solar Energy Conversion: Dynamics of Interfacial Electron and Excitation Transfer
9: Photoelectrochemical Water Splitting: Materials, Processes and Architectures
10: Biological Conversion of Biomass for Fuels and Chemicals: Explorations from Natural Utilization Systems
11: Advanced Concepts in Photovoltaics
12: Materials Challenges: Inorganic Photovoltaic Solar Energy
13: Catalytic Hydrogenation for Biomass Valorization
14: Photocatalysis: Fundamentals and Perspectives
15: Photocatalysis: Applications
16: Unconventional Thin Film Photovoltaics
17: Thermoelectric Materials and Devices
18: X-Ray Free Electron Lasers: Applications in Materials, Chemistry and Biology
19: Lignin Valorization: Emerging Approaches
20: Advances in Photoelectrochemical Water Splitting: Theory, Experiment and Systems Analysis
21: Electrochemical Reduction of Carbon Dioxide: Overcoming the Limitations of Photosynthesis

How to obtain future titles on publication:
A standing order plan is available for this series. A standing order will bring delivery of each new volume immediately on publication.

For further information please contact:
Book Sales Department, Royal Society of Chemistry, Thomas Graham House, Science Park, Milton Road, Cambridge, CB4 0WF, UK
Telephone: +44 (0)1223 420066, Fax: +44 (0)1223 420247
Email: booksales@rsc.org
Visit our website at www.rsc.org/books

Electrochemical Reduction of Carbon Dioxide
Overcoming the Limitations of Photosynthesis

Edited by

Frank Marken
University of Bath, UK
Email: fm202@bath.ac.uk

and

David Fermin
University of Bristol, UK
Email: david.fermin@bristol.ac.uk

ROYAL SOCIETY
OF **CHEMISTRY**

Energy and Environment Series No. 21

Print ISBN: 978-1-78262-042-6
PDF ISBN: 978-1-78262-380-9
EPUB ISBN: 978-1-78801-452-6
ISSN: 2044-0774

A catalogue record for this book is available from the British Library

The Royal Society of Chemistry is a charity, registered in England and Wales, Number 207890, and a company incorporated in England by Royal Charter (Registered No. RC000524), registered office: Burlington House, Piccadilly, London W1J 0BA, UK, Telephone: +44 (0) 207 4378 6556.

For further information see our web site at www.rsc.org

Printed in the United Kingdom by CPI Group (UK) Ltd, Croydon, CR0 4YY, UK

Contents

Energy and Environment Series No. 21
Electrochemical Reduction of Carbon Dioxide: Overcoming the Limitations of Photosynthesis
Edited by Frank Marken and David Fermin

Published by the Royal Society of Chemistry, www.rsc.org

CHAPTER 1

Introduction to the Eletrochemical and Photo-electrochemical Reduction of CO_2

DAVID J. FERMIN*[a] AND FRANK MARKEN*[b]

[a] School of Chemistry, University of Bristol, Bristol BS8 1TS, UK;
[b] Department of Chemistry, University of Bath, Bath BA2 7AY, UK
*Email: david.fermin@bristol.ac.uk; f.marken@bath.ac.uk

1.1 Introduction to the (Photo-)Electrochemical Reduction of CO_2

The scale of the challenge in carbon dioxide conversion is enormous, with 35 gigaton of anthropogenic CO_2 generated by fossil fuel combustion/consumption every year. The average percentage of CO_2 in the atmosphere had risen to 403.38 ppm by September 2017 according to the data released by the National Oceanic and Atmospheric Administration of the US Department of Commerce,[1] with serious effects expected on global warming, climate patterns, sea levels, biodiversity, food production, human displacement, and so forth. There is a global consensus that immediate action must be taken to halt the relentless increase in atmospheric carbon dioxide levels, with the Global Apollo programme being a prime example of international initiatives being undertaken.[2] In addition to the ever growing number of

Energy and Environment Series No. 21
Electrochemical Reduction of Carbon Dioxide: Overcoming the Limitations of Photosynthesis
Edited by Frank Marken and David Fermin

Published by the Royal Society of Chemistry, www.rsc.org

installations of renewable energy sources in different areas of the planet, there are great opportunities in "recycling"[3,4] carbon dioxide with biochars,[5] by absorption,[6] or by electrolysis powered by photovoltaic, wind, tidal, hydro-electric, salinity gradient energy, or the so-called "blue energy" systems.[7] Although CO_2 sequestration has also been proposed,[8] mimicking natural photosynthesis represents the most attractive but also scientifically challenging avenue.

In the context of artificial photosynthesis, links can be established between catalytic,[9] electrochemical, and photo-electrochemical conversion of atmospheric CO_2, as schematically illustrated in Figure 1.1. Taking inspiration from nature, integrating functional units at the nanoscale capable of (i) capture solar light, (ii) CO_2 accumulation, and (iii) selective reduction to products can lead to new technologies that can be widely deployed for local mitigation of carbon emission. There are also integration approaches for "semi-artificial photosynthesis", for example employing PS1 and PS2 apparatus extracted from cells and immobilised onto electrode surfaces. Although synergistic effects can be envisaged from such "nano-integrated" systems, such level of structural complexity currently rarely achieves high efficiency or stability. The electro-reduction of CO_2 is a tough problem so why combine this with the added complexity of an integrated solar cell? Why not just use conventional solar electricity without integration? So far, there are no technological developments that enable answering this question unambiguously. However, there is a large community of scientists, including those contributing to this book, that identifies a direct and selective path of CO_2 to a valuable carbon structure as one of the grand challenges in the path towards a low carbon economy.

This book primarily focuses on electrochemical conversion of CO_2, establishing correlation between the nature of the catalysts and the electrode potential. Figure 1.1 sketches how the potential bias (input energy) can be provided by solar energy either *via* photovoltaic solar cells,[10] or by direct photo-generation of carriers at the electrocatalytic site.[11] These two approaches, illustrated by Figures 1.1A and B, respectively, represent two different levels of nano-integration. The third level (Figure 1.1C) involves the integration of CO_2 capture moieties directly to the catalysts active site in a combined unit. Depending on the catalyst properties, a variety of products can be generated directly (*in situ*), or by a separate process exploiting, for example, the formose reaction[12,13] to build up carbohydrates from formaldehyde. A large number of approaches has been presented in the literature for generating higher added value compounds based on various electrochemical reactors,[14] homogeneous catalysts[15,16] or inorganic photosynthetic sites.[17] Independently of the approach used for providing the driving force for CO_2 reduction, investigating the properties of the catalytic sites by (spectro)-electrochemical techniques provides extremely valuable mechanistic information. Some of the most recent developments in spectro-electrochemical tools are reviewed in Chapters 8–10.

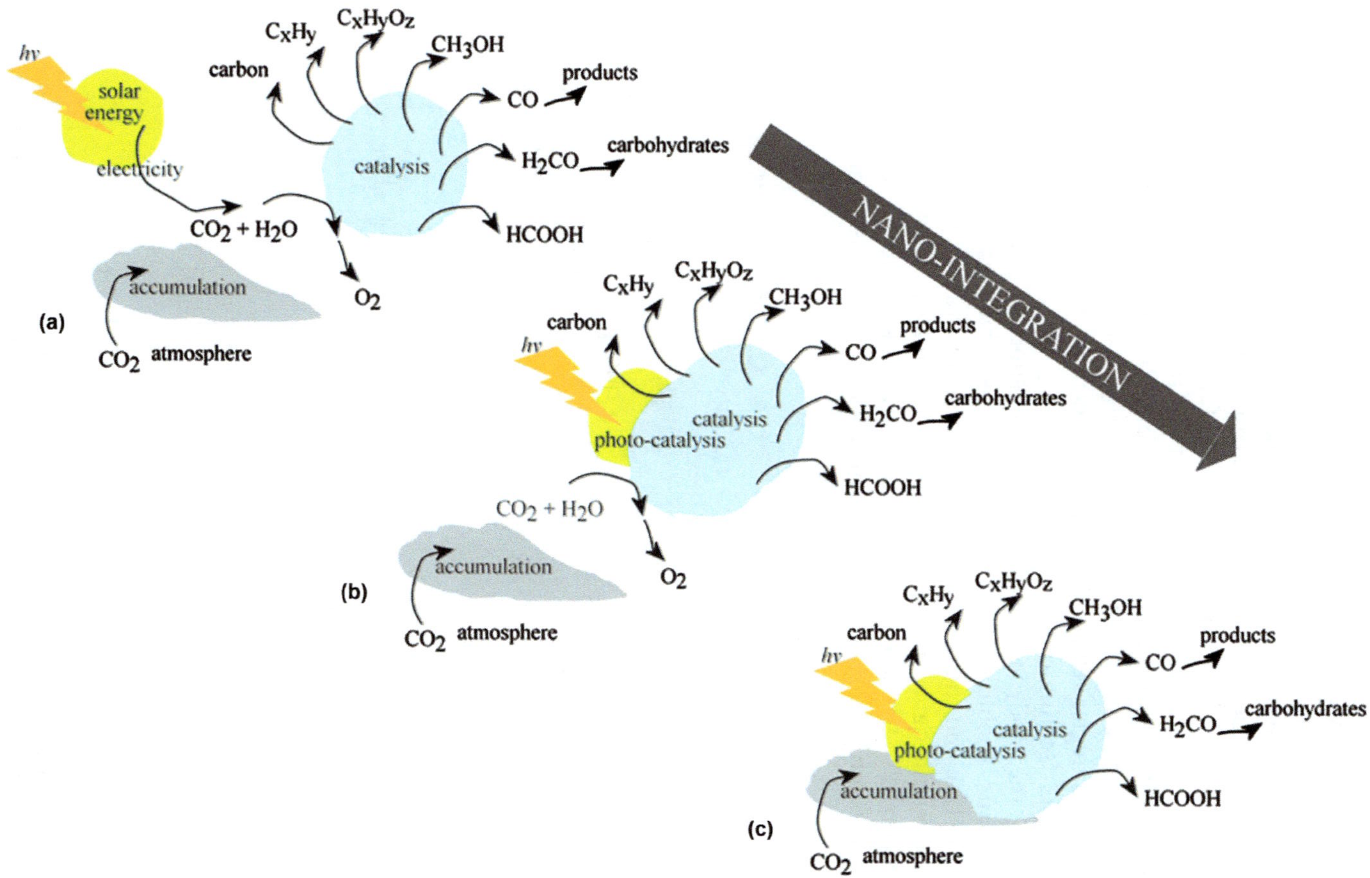

Figure 1.1 Schematic depiction of carbon dioxide accumulation, energy harvesting, and catalytic reduction of CO_2 into products at three levels of nano-integrated artificial photosynthesis.

Numerous reviews[18–20] and authoritative books[21–23] have been published on the topic of CO_2 reduction and electrocatalysis at electrode surfaces. The high symmetry of the carbon dioxide molecule has often been cited as a key contributor to the activation barrier, with a price in energy often paid by applying a high overpotential (excess potential with respect to the thermodynamic reduction potential). More recent studies, based on the so-called "scaling relations" formalism, point to the fact that the binding energy of intermediate species in multi-electron transfer reactions correlate in a linear fashion, resulting in large overpotentials for a variety of metals.[24–26] Azofra and Sun elaborate on this point in Chapter 6. At the fundamental level, a number of strategies reviewed in this book focus on breaking the challenge posed by this "scaling relation".

From the thermodynamic point of view, the reduction of CO_2 to give useful hydrocarbons should occur under mild conditions, as summarised by eqn (1.1)–(1.5), and the corresponding Pourbaix diagrams[27] (see Figure 1.2). Thermodynamic data for organic media linking to aqueous media have also been reported.[28] Reduction to elemental carbon, and indeed nano-carbon products, is only accessible in molten salts and at high temperature.[29]

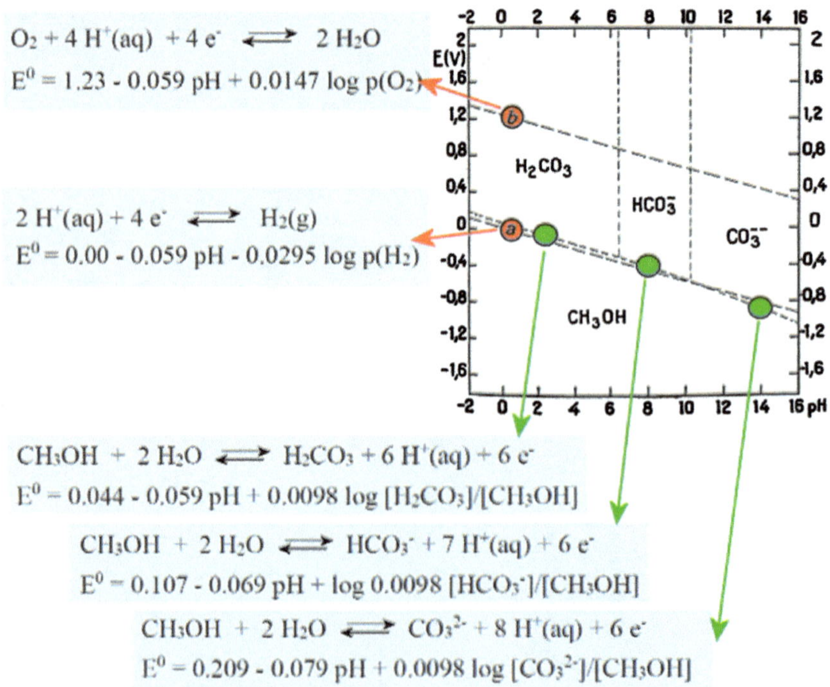

Figure 1.2 Pourbaix diagram[30] for methanol in an aqueous carbonate buffer environment. Indicated in *a* and *b* are the water reduction and oxidation processes. The reversible potential for methanol oxidation is dependent on the pH and solution composition.

The reversible potential for the CO_2/CO redox system (eqn (1.1)) has been elegantly confirmed by experiment (at pH 7) with enzyme-laden electrodes.[30]

$$CO + H_2O \rightleftarrows CO_2 + 2\ H^+(aq) + 2\ e^-$$

$$E^0 = -0.103 - 0.059\ \mathrm{pH} + 0.0295\ \log\ (pCO_2/pCO) \quad (1.1)$$

$$CH_4 + 2\ H_2O \rightleftarrows CO_2 + 8\ H^+ + 8\ e^-$$

$$E^0 = 0.169 - 0.059\ \mathrm{pH} + 0.0074\ \log\ (pCO_2/pCH_4) \quad (1.2)$$

$$C(\mathrm{graphite}) + 2\ H_2O \rightleftarrows CO_2 + 4\ H^+(aq) + 4\ e^-$$

$$E^0 = 0.207 - 0.059\ \mathrm{pH} + 0.0148\ \log\ (pCO_2) \quad (1.3)$$

$$CH_3OH(\mathrm{methanol}) + H_2O \rightleftarrows CO_2 + 6\ H^+(aq) + 6\ e^-$$

$$E^0 = -0.016 - 0.059\ \mathrm{pH} \quad (1.4)$$

$$C_6H_{12}O_6(\beta\text{-glucose}) + 6\ H_2O \rightleftarrows 6\ CO_2 + 24\ H^+(aq) + 24\ e^-$$

$$E^0 = 0.014 - 0.059\ \mathrm{pH} \quad (1.5)$$

As mentioned previously, photoexcitation is the most attractive approach to generating highly energetic electrons capable of driving CO_2 reduction, as exemplified by the scheme in Figure 1.3, integrating light absorbing units

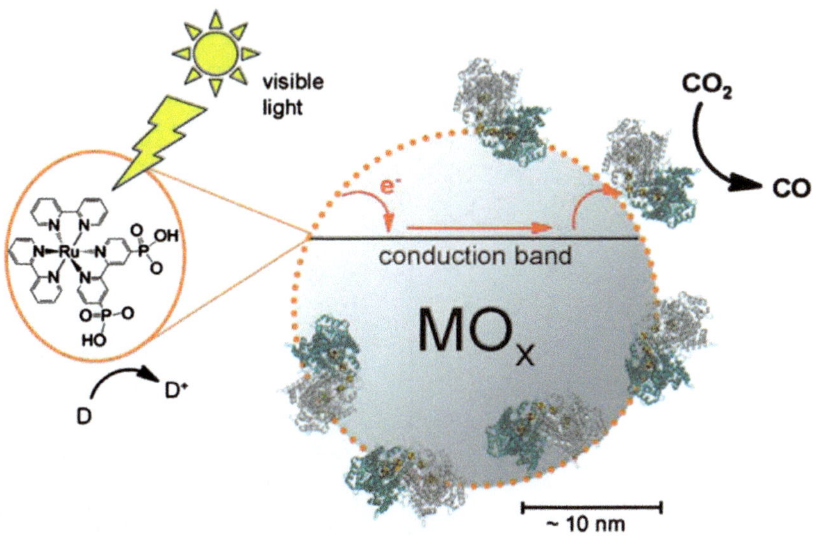

Figure 1.3 Schematic description of a model system for the photo-reduction of CO_2 to CO using visible light based on a metal oxide nanoparticle functionalised with the enzyme carbon monoxide dehydrogenase (CODH) and sensitised to visible light using a ruthenium bipyridyl photosensitiser. Using the light intensity and illuminated facial area of the reaction vessel an average quantum yield of approximately 0.07% is estimated. Reproduced from ref. 31 with permission from The Royal Society of Chemistry.

(ruthenium bipyridyl photosensitiser), a metal oxide as electron accepting moiety, and carbon monoxide dehydrogenase.[31]

A key component in the overall process is the photo-absorber, which locally generates charge carriers. In addition to molecular dyes (*e.g.* ruthenium bipyridyl), other materials such as quantum dots, oxide nanoparticles,[32,33] carbon dots, and carbon based heterostructures.[34–36] With regards to catalytic centres, biological systems,[37] with metabolically engineered microorganisms,[38] and aided by synthetic biology, have been generating a tremendous amount of interest.[39,40] Other prime examples of nanoscale hybrid integrated systems involve immobilising "PS1 and PS2" centres extracted from cells,[41–43] or dehydrogenase enzymes[44] onto electrodes. Risbridger and Anderson review some of these strategies in Chapter 2 of this book.

Selectivity in the catalytic process can offer major advantages. Direct and selective catalytic transformations have been achieved (without additional catalyst) at semiconductor materials such as $BiVO_4$, for example, leading to high efficiency in the formation of (nearly) single products for methanol[45] (see Figure 1.4) or ethanol.[46] Also the formation of relatively complex molecules such as lactate,[47] or dimethylcarbonate in a single process has been reported.[48]

Another dynamic area of research involves the development of electrochemical reactors for continuous flow CO_2 photo-reduction.[49] Conventional

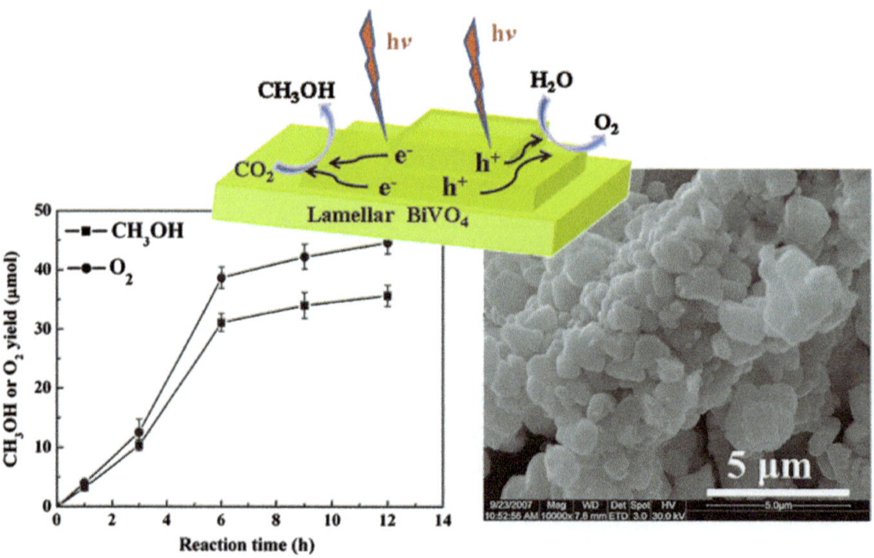

Figure 1.4 Schematic drawing of photo-induced CO_2 reduction on lamella $BiVO_4$. Also shown are a typical methanol and oxygen photosynthesis curve and the electron optical image of lamella $BiVO_4$.
Reprinted from *Catalysis Communications*, **28**, J. Mao, T. Peng, X. Zhang, K. Li and L. Zan, Selective methanol production from photocatalytic reduction of CO_2 on $BiVO_4$ under visible light irradiation, 38–41,[45] Copyright 2012, with permission from Elsevier.

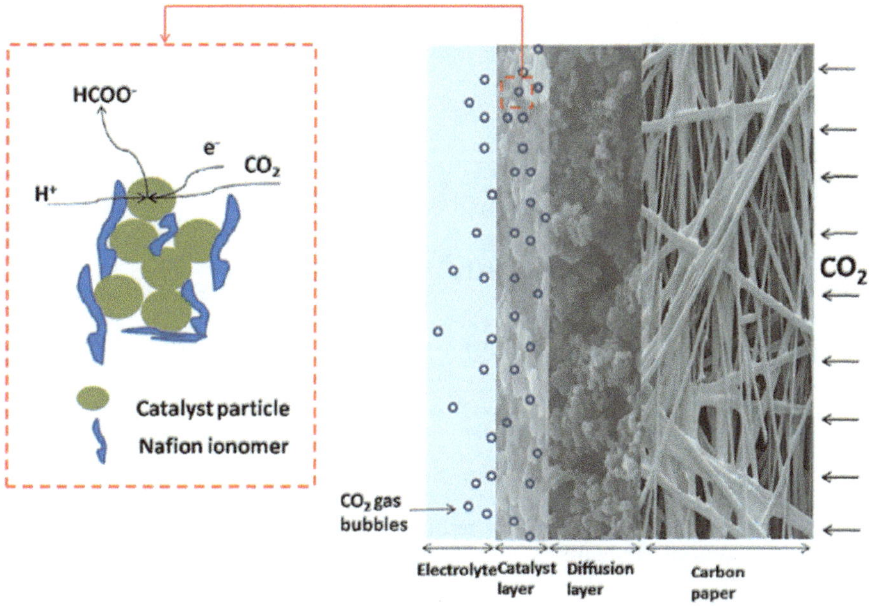

Figure 1.5 Schematic drawing of a gas diffusion electrode membrane assembly to allow CO_2 to enter *via* a gas diffusion layer to effectively contact the catalyst layer.
Reprinted from *Journal of Power Sources*, **258**, J. Wu, P. P. Sharma, B. H. Harris and X.-D. Zhou, Electrochemical reduction of carbon dioxide: IV dependence of the Faradaic efficiency and current density on the microstructure and thickness of tin electrode, 189–194,[55] Copyright 2014, with permission from Elsevier.

studies consist of catalytic electrodes placed in a reactor[50] in contact with a CO_2 saturated electrolyte solution, which could be either aqueous, organic, or ionic liquid media.[51] The solubility of CO_2 in these systems is a crucial factor, particularly in aqueous media. The low CO_2 solubility in water can be somewhat alleviated under high pressure.[52] Centi *et al.* demonstrated an inverted or "driven" fuel cell reactor design with direct gas feed and a gas diffusion electrode, as illustrated in Figure 1.5, leading to non-Fischer–Tropsch product distribution on platinum catalysts.[53] Membrane reactor designs[54] with an effective triple phase boundary reaction zone can be employed to overcome solubility limitations.

1.2 New Catalysts for the (Photo-)Electrochemical Reduction of CO_2

Manipulating the activity of catalysts can be generally described in terms of (i) increasing the number of active sites, or (ii) the intrinsic activity of the material.[56] The first approach is based on the concept of nano-structuring,[57] as exemplified by the work of Broekman and co-workers employing the

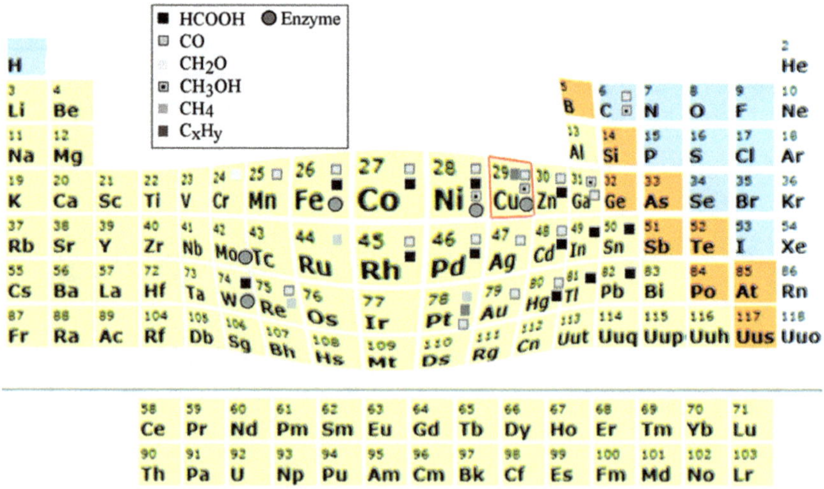

Figure 1.6 Schematic overview of CO_2 reduction catalytic activity across the periodic system of elements.[63]

so-called Cu sponges.[58] However, entirely new catalysts and catalyst architectures are also needed. Intrinsic activity of late transition metals has been extensively investigated (see Figure 1.6) from the experimental[59–62] and theoretical points of view.

There are two distinct groups of catalysts with (i) CO forming metals (Cu, Au, Ag, Zn, Pd, Ga, Ni, and Pt) and (ii) formate forming metals (Pb, Hg, In, Sn, Cd, and Tl). Copper takes a special place with its capacity of generating a wide range of CO_2 reduction products and promoting carbon–carbon coupling. Although this classification relates to main generated products, there is a vast difference in the overpotentials required to drive the CO_2 reduction at these metals. Our understanding of the CO_2 reactivity at model surfaces has been significantly improved by detailed studies at single crystal electrodes, as reviewed in Chapter 4 of this book. New insights are possible employing sophisticated *in situ* spectroscopic techniques, as highlighted in Chapter 10. On the other hand, a lot less is understood about materials featuring highly correlated electrons, such as metal oxides.[64]

The intrinsic activity of catalysts can also be modulated by alloying,[65] or by introducing metastable catalytic materials.[66] CO_2 reduction studies at ultrathin Pd layers grown on Au nanoparticles have provided interesting insights into the effect of the so-called electronic and strain effects.[67] Montes de Oca *et al.* have shown that decreasing the thickness of the Pd shells from 10 to 1 nm leads to an increase in the effective strain of the Pd lattice from less than 1 to 3.5%.[68] As illustrated in Figure 1.7, the reduction of CO_2 leads to the generation of mainly CO at the strain Pd shells, while relaxed shells produce HCOOH, CH_4, and C_2H_6.[67,69] *In situ* FTIR studies showed that the CO coverage and binding to Pd is strongly decreased at the thin Pd shells,

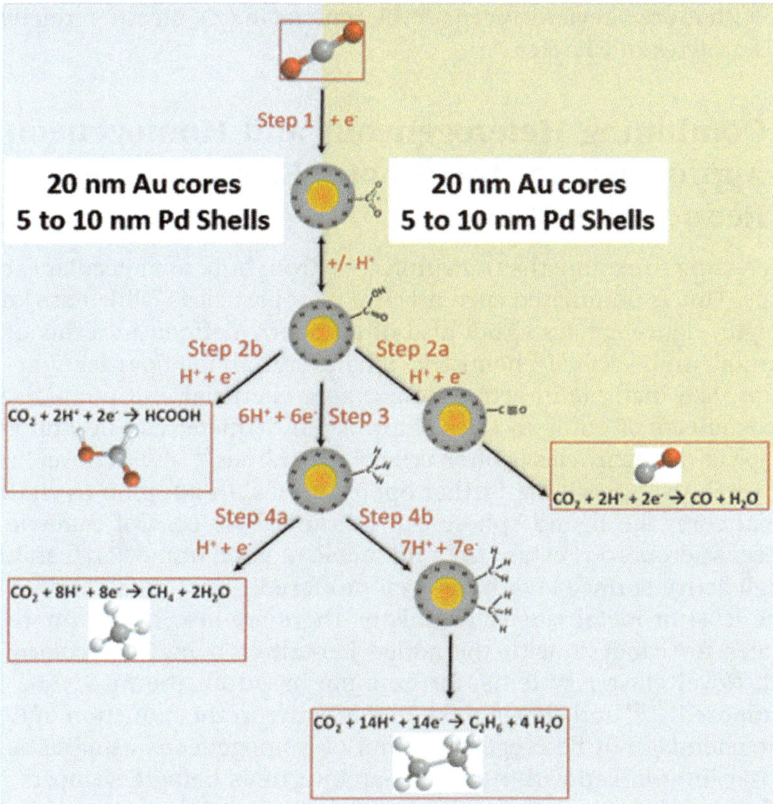

Figure 1.7 Production distribution of the CO_2 reduction products at Pd shells between 1 and 10 nm at 20 nm Au nanocores.[67]

which was rationalised in terms of an upward shift of the d-band centre due to lattice strain as estimated by DFT.[70] This strategy has been further extended with a large variety of core-shells and alloyed nanostructures.[71,72]

Transition-metal cations, including rare earth elements, play a key role in the activity of molecular catalysts.[73,74] The reactivity of these systems is also strongly affected by the ligand structure. For example, the catalytic reduction of CO_2 to CO at zinc porphyrinato complexes occurs through binding to the ligand rather than the metal centre.[75] Homogeneous catalysts based on coordination compounds will be discussed in Chapter 5.

Finally, the electrolyte salt (in particular the electrolyte cation) can be an important factor[76] as shown, for example, by the role of the alkali cation reported by Cuesta and coworkers.[77] Alkali cations have also been shown to affect the product distribution in CO_2 reduction.[78] Halide anions have been observed to improve the efficiency of CO_2 electro-reduction at copper surfaces.[79] Very intriguing are the effects of electrolyte cations in ionic liquid media, which strongly affect the structure of the double layer as well as enabling stabilisation of reaction intermediates such as $CO_2^{\bullet -}$ radical anions.[80] Zhao *et al.* reviews recent developments in CO_2 electro-reduction in ionic electrolytes in Chapter 7.

1.3 Combining Heterogeneous and Homogeneous Approaches for the (Photo-)Electrochemical Reduction of CO_2

It is interesting to explore the transition zone from bulk to molecular catalyst materials. This is dominated currently by "nanoparticles", which are known to be highly active catalysts, but also often poorly defined in terms of size distribution and surface chemistry. Single crystal nanoparticles are the exception. The increase in active surface area is crucial, but often also the electronic effects of going to sizes below 5 nm diameter can be important. The shape of nanoparticles is often crucial in catalysis[81,82] with novel "nanoframe" catalysts[83] providing further opportunities. In addition to changing the metal core, the ligand sphere can be tuned[84] to control catalytic performance. Microporous catalysts[85] and metal–organic framework catalysts[86] with high active surface area have been proposed.

At the level of metal cluster chemistry, there are also highly interesting candidates for catalysis with the added benefit of being molecularly well defined. Novel cluster systems, for example based on the Au_{25} core, have been proposed[87–89] and shown to be highly active in the reduction of CO_2 to CO. The chemistry of heterogeneous and of homogeneous catalysis is now ever closer inter-linked with new nano-architectures being developed.

Confining homogeneous catalyst to interfaces offers new approaches combining the better understanding/tunability of molecular systems with the practicality of heterogeneous systems immobilised at electrode surfaces. Although methods based on conducting polymers and immobilised coordination polymers have been successfully developed in the past, there is a new emphasis on the architecture at nano-scale. A recent review by Reisner *et al.*[90] highlights the developments and promise in this field. Work by Domen and coworkers[91] demonstrated the principle of combining a Ru(II) dinuclear complex used for CO_2 reduction and a Ag-loaded TaON (Ag/TaON) semiconductor and light absorber. Active substrate materials such as Cu_2O are combined with surface immobilised Re metal complexes[92] to fundamentally change the pathway of the catalytic reaction (Figure 1.8). Similarly, the immobilisation of a molecular Mn catalyst on TiO_2[93] has been reported to allow stable photo-electro-reduction of CO_2 to CO in conjunction with a photo-anode generating oxygen.

1.4 Summary and Chapter Overview

A brief overview of current developments and the state-of-the-art in electrochemical reduction of carbon dioxide has been attempted. The existing breadth of catalysis materials and processes is impossible to fully cover and

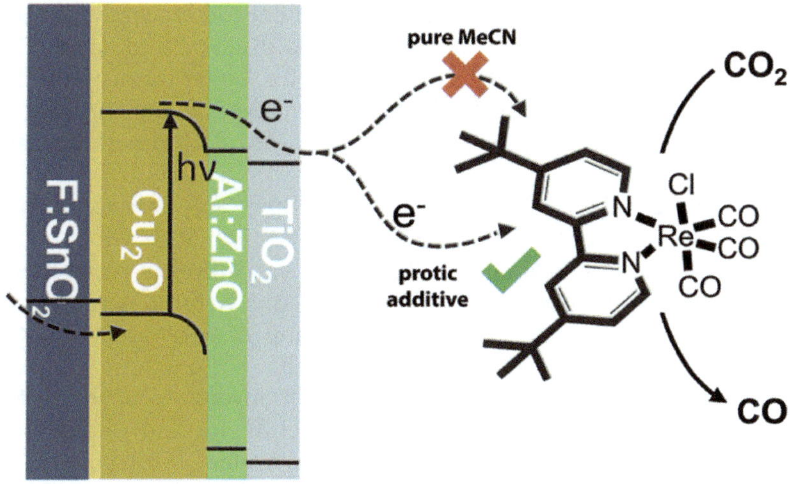

Figure 1.8 Schematic drawing of a nano-integrated structure for light absorption into Cu_2O, a unidirectional electron injection into an immobilised Rhenium catalyst, and formation of CO from CO_2. The hole is conducted into the external circuit.
Reproduced from ref. 92 with permission from The Royal Society of Chemistry.

the development appears to be rapid. New computational theory and the new opportunity of computational artificial intelligence will open up further avenues for the sunlight-driven conversion of atmospheric CO_2 to useful fuels and products.

This book offers a broad and up-to-date perspective on topics including pure CO_2 reduction electrocatalysis, photo-electrocatalysis, the transition from homogeneous to heterogeneous catalyst systems, biological perspectives, *in situ* spectroscopy, and aspects of computational theory. There is a strong materials chemistry element in catalyst development and still a lot of adventure in catalyst discovery. The complexity of nano-integrated materials and technologies offer potential for economic rewards for the future, but also leads to frustratingly difficult research challenges based on finely controlled nano-architectures and well-understood reaction conditions.

Contributions to this book include in Chapter 2 "Bio-Inspired and Bio-Electrochemical Approaches in CO_2 Reduction Catalysis" by Thomas Risbridger and Ross Anderson with a focus on biological and bio-inspired processes for photo-electrochemical carbon dioxide reduction. This includes an overview of processes based in micro-organisms.

In Chapter 3 entitled "Copper Catalysts for the Electrochemical Reduction of Carbon Dioxide" by Hyung Mo Jeong, Boon Siang Yeo, and Youngkook Kwon copper is the focus. In this chapter reactivity trends and novel catalyst designs are critically assessed and an outlook on copper catalysis is provided.

Chapter 4 on "Single-Crystal Surfaces as Model Electrocatalysts for CO_2 Reduction" by Adam Kolodziej, Paramaconi Rodriguez, and Angel Cuesta introduces single crystal methods as a tool for better understanding of catalytic surface processes and for better linking experiments to theory and computational chemistry. This chapter focuses on platinum-group metals, copper, silver, and gold.

Chapter 5 entitled "Homogeneous $M(bpy)(CO)_3\times$ and Aromatic N-Heterocycle Catalysts for CO_2 Reduction" by Mitchell C. Groenenboom, Karthikeyan Saravanan, and John A. Keith offers an atomic scale perspective and view based on computational theory. Classic systems such as homogeneous metal complex catalysts and the group of N-heterocycle CO_2 reduction catalysts are investigated and explained.

In Chapter 6 "DFT Modelling Tools in CO_2 Conversion: Reaction Mechanisms Screening and Analysis" by Luis Miguel Azofra and Chenghua Sun the theory on heterogeneous catalyst system is added. Information is provided about "where", "how", and "why" *in silico* hypothesising and how to assess promising catalysts before the experimental testing of their catalytic performance.

Chapter 7 entitled "Electrocarboxylation in Ionic Liquids" by Shu-Feng Zhao, Michael D. Horne, Alan M. Bond, and Jie Zhang offers a perspective on CO_2 reduction in ionic liquid media. These media, often non-volatile and able to adsorb CO_2, offer a "game changer" in terms of room temperature electrocatalytic CO_2 reduction.

In Chapter 8 David E. Ryan and František Hartl introduce "IR Spectro-Electrochemistry and Group-6 α-diimine Catalysts of CO_2 Reduction" to demonstrate that new *in situ* reaction monitoring tools (in particular at low temperatures) can be highly powerful even for very complex homogeneous reaction pathways during carbon dioxide reduction reactions.

In Chapter 9 "Probing CO_2 Reduction Intermediates Employing *in situ* Spectroscopy and Spectrometry" by S. Pérez-Rodríguez, G. García, M. J. Lázaro, and Elena Pastor describe the state-of-the-art in monitoring reaction intermediates and catalytic pathways. A range of spectroscopy and spectrometry tools are reported and spectral properties of intermediates are linked to reactivity.

Finally, in Chapter 10 entitled "Surface-Selective and Time-Resolved Spectro-electrochemical Studies of CO_2 Reduction Mechanisms" by Alexander J. Cowan an overview is provided into surface-selective, surface-enhanced, and time-resolved forms of spectro-electrochemistry applied to mechanistic challenges in the field of CO_2 reduction.

References

1. Monthly Average Mauna Loa CO_2, www.esrl.noaa.gov/gmd/ccgg/trends/index.html. Reported 5th October 2017.
2. http://cep.lse.ac.uk/pubs/download/special/Global_Apollo_Programme_Report.pdf.

3. M. Mikkelsen, M. Jorgensen and F. C. Krebs, *Energy Environ. Sci.*, 2010, **3**, 43–81.
4. A. Navarrete, G. Centi, A. Bogaerts, A. Martin, A. York and G. D. Stefanidis, *Energy Technol.*, 2017, **5**, 796–811.
5. P. Lahijani, Z. A. Zainal, M. Mohammadi and A. R. Mohamed, *Renewable Sustainable Energy Rev.*, 2015, **41**, 615–632.
6. A. Al-Mamoori, A. Krishnamurthy, A. A. Rownaghi and F. Rezaei, *Energy Technol.*, 2017, **5**, 834–849.
7. N. Y. Yip, D. Brogioli, H. V. M. Hamelers and K. Nijmeijer, *Environ. Sci. Technol.*, 2016, **50**, 12072–12094.
8. J. Xie, N. Wei, L. Z. Wu, K. N. Zhang and M. Xu, *Rock Soil Mech.*, 2017, **38**, 181–188.
9. T. M. Su, Z. Z. Qin, H. B. Ji, Y. X. Jiang and G. Huang, *Environ. Chem. Lett.*, 2016, **14**, 99–112.
10. M. Schreier, L. Curvat, F. Giordano, L. Steier, A. Abate, S. M. Zakeeruddin, J. Luo, M. T. Mayer and M. Grätzel, *Nat. Commun.*, 2015, **6**, 7326–7329.
11. X. Zhou, R. Liu, K. Sun, Y. Chen, E. Verlage, S. A. Francis, N. S. Lewis and C. Xiang, *ACS Energy Lett.*, 2016, **1**, 764–770.
12. I. A. Shkrob, T. W. Marin, H. Y. He and P. Zapol, *J. Phys. Chem. C*, 2012, **116**, 9450–9460.
13. G. Harsch, M. Harsch, H. Bauer and W. Voelter, *Z. Naturforsch. Sec. B*, 1983, **38**, 1269–1280.
14. C. Ampelli, C. Genovese, M. Errahali, G. Gatti, L. Marchese, S. Perathoner and G. Centi, *J. Appl. Electrochem.*, 2015, **45**, 701–713.
15. J. Hawecker, J. M. Lehn and R. Ziessel, *Chem. Commun.*, 1983, 536–538.
16. E. E. Barton, D. M. Rampulla and A. B. Bocarsly, *J. Am. Chem. Soc.*, 2008, **130**, 6342–6348.
17. M. Anpo, H. Yamashita, Y. Ichihashi, Y. Fujii and M. Honda, *J. Phys. Chem. B*, 1997, **101**, 2632–2636.
18. L. Zhang, Z. J. Zhao and J. L. Gong, *Angew. Chem., Int. Ed.*, 2017, **56**, 11326–11353.
19. E. E. Benson, C. P. Kubiak, A. J. Sathrum and J. M. Smieja, *Chem. Soc. Rev.*, 2009, **38**, 89–99.
20. S. C. Roy, P. K. Varghese, M. Paulose and C. A. Grimes, *ACS Nano*, 2010, **4**, 1259–1278.
21. *Carbon Dioxide as Chemical Feedstock*, ed. M. Aresta, Wiley-VCH, Weinheim, 2010.
22. *Electrochemical and Electrocatalytic Reactions of Carbon Dioxide*, ed. B. P. Sullivan, K. Krist, and H. E. Guard, Elsevier, Amsterdam, 1993.
23. *Carbon Dioxide Fixation and Reduction in Biological and Model Systems*, ed. C. I. Branden and G. Schneider, Oxford University Press, Oxford, 1994.
24. Y. Li and Q. Sun, *Adv. Energy Mater.*, 2016, **6**, 1600463.
25. W. J. Durand, A. A. Peterson, F. Studt, F. Abild-Pedersen and J. K. Norskov, *Surf. Sci.*, 2011, **605**, 1354–1359.

26. A. A. Peterson, F. Abild-Pedersen, F. Studt, J. Rossmeisl and J. K. Norskov, *Energy Environ. Sci.*, 2010, **3**, 1311–1315.
27. M. Pourbaix, *Atlas of Electrochemical Equilibria in Aqueous Solutions*, NACE Interrnational, Cebelcor, Brussels, 1974.
28. M. L. Pegis, J. A. S. Roberts, D. J. Wasylenko, E. A. Mader, A. M. Appel and J. M. Mayer, *Inorg. Chem.*, 2015, **54**, 11883–11888.
29. A. Douglas and C. L. Pint, *ECS J. Solid State Sci. Technol.*, 2017, **6**, M3084–M3089.
30. V. C. C. Wang, S. W. Ragsdale and F. A. Armstrong, *ChemBioChem*, 2013, **14**, 1845–1851.
31. T. W. Woolerton, S. Sheard, E. Pierce, S. W. Ragsdale and F. A. Armstrong, *Energy Environ. Sci.*, 2011, **4**, 2393–2399.
32. T. Inoue, A. Fujishima, S. Konishi and K. Honda, *Nature*, 1979, **277**, 637–638.
33. T. Hisatomi, K. Takanabe and K. Domen, *Catal. Lett.*, 2015, **145**, 95–108.
34. J. B. Essner and G. A. Baker, *Environ. Sci.: Nano*, 2017, **4**, 1216–1263.
35. L. B. Jiang, X. Z. Yuan, Y. Pan, J. Liang, G. M. Zeng, Z. B. Wu and H. Wang, *Appl. Catal., B*, 2017, **217**, 388–406.
36. X. Chen, Q. Liu, Q. L. Wu, P. W. Du, J. Zhu, S. Y. Dai and S. F. Yang, *Adv. Funct. Mater.*, 2016, **26**, 1719–1728.
37. S. Bajracharya, S. Srikanth, G. Mohanakrishna, R. Zacharia, D. P. B. T. B. Strik and D. Pant, *J. Power Sources*, 2017, **356**, 256–273.
38. H. M. Woo, *Curr. Opin. Biotechnol.*, 2017, **45**, 1–7.
39. F. Y. Gong, Z. Cai and Y. Li, *Sci. China: Life Sci.*, 2016, **59**, 1106–1114.
40. N. V. D. Long, J. Lee, K. K. Koo, P. Luis and M. Lee, *Energies*, 2017, **10**, 473.
41. T. Kothe, S. Poller, F. Y. Zhao, P. Fortgang, M. Rogner, W. Schuhmann and N. Plumere, *Chem. – Eur. J.*, 2014, **20**, 11029–11034.
42. V. Hartmann, T. Kothe, S. Poller, E. El-Mohsnawy, M. M. Nowaczyk, N. Plumere, W. Schuhmann and M. Rogner, *Phys. Chem. Chem. Phys.*, 2014, **16**, 11936–11941.
43. M. Kato, J. Z. Zhang, N. Paul and E. Reisner, *Chem. Soc. Rev.*, 2014, **43**, 6485–6497.
44. T. Reda, C. M. Plugge, N. J. Abram and J. Hirst, *Proc. Natl. Acad. Sci. U. S. A.*, 2008, **105**, 10654–10658.
45. J. Mao, T. Y. Peng, X. H. Zhang, K. Li and L. Zan, *Catal. Commun.*, 2012, **28**, 38–41.
46. Y. Y. Liu, B. B. Huang, Y. Dai, X. Y. Zhang, X. Y. Qin, M. H. Jiang and M. H. Whangbo, *Catal. Commun.*, 2009, **11**, 210–213.
47. K. Ogura, H. Sugihara, J. Yano and M. Higasa, *J. Electrochem. Soc.*, 1994, **141**, 419–424.
48. Y. J. Zhou, S. J. Wang, M. Xiao, D. M. Han, Y. X. Lu and Y. Z. Meng, *RSC Adv.*, 2012, **2**, 6831–6837.
49. B. Endrodi, G. Bencsik, F. Darvas, R. Jones, K. Rajeshwa and C. Janaky, *Prog. Energy Combust. Sci.*, 2017, **62**, 133–154.

50. K. F. Li, X. Q. An, K. H. Park, M. Khraisheh and J. W. Tang, *Catal. Today*, 2014, **224**, 3–12.
51. H.-K. Lim, *Molecules*, 2017, **22**, 536.
52. X. Lu, D. Y. C. Leung, H. Z. Wang, M. K. H. Leung and J. Xuan, *ChemElectroChem*, 2014, **1**, 836–849.
53. G. Centi, S. Perathoner, G. Wine and M. Gangeri, *Green Chem.*, 2007, **9**, 671–678.
54. I. Merino-Garcia, E. Alvarez-Guerra, J. Albo and A. Irabien, *Chem. Eng. J.*, 2016, **305**, 104–120.
55. J. Wu, P. P. Sharma, B. H. Harris and X. Zhou, *J. Power Sources*, 2014, **258**, 189–194.
56. Z. W. Seh, J. Kibsgaard, C. F. Dickens, I. B. Chorkendorff, J. K. Norskov and J. F. Jaramillo, *Science*, 2017, **355**, 146–148.
57. H. Mistry, A. S. Varela, S. Kuhl, P. Strasser and B. R. Cuenya, *Nat. Rev. Mater.*, 2016, **1**, 16009.
58. A. Dutta, M. Rahaman, M. Mohos, A. Zanetti and P. Broekman, *ACS Catal.*, 2017, **7**, 5431–5437.
59. J. H. Montoya, L. C. Seitz, P. Chakthranont, A. Vojvodic, T. F. Jaramillo and J. K. Norskov, *Nat. Mater.*, 2017, **16**, 70–81.
60. Z.-L. Wang, C. Li and Y. Yamauchi, *Nano Today*, 2016, **11**, 373–391.
61. Q. Lu and F. Jiao, *Nano Energy*, 2016, **29**, 439–456.
62. K. P. Kuhl, T. Hatsukade, E. R. Cave, D. N. Abram, J. Kibsgaard and T. F. Jaramillo, *J. Am. Chem. Soc.*, 2014, **136**, 14107–14113.
63. J. L. Qiao, Y. Y. Liu, F. Hong and J. J. Zhang, *Chem. Soc. Rev.*, 2014, **43**, 631–645.
64. D. F. Gao, Y. Zhang, Z. W. Zhou, F. Cai, X. F. Zhao, W. G. Huang, Y. S. Li, J. F. Zhu, P. Liu, F. Yang, G. X. Wang and X. H. Bao, *J. Am. Chem. Soc.*, 2017, **139**, 5652–5655.
65. J. W. Vickers, D. Alfonso and D. R. Kauffman, *Energy Technol.*, 2017, **5**, 775–795.
66. K. Hashimoto, *Mater. Sci. Eng., A*, 1997, **226**, 891–899.
67. J. J. L. Humphrey, D. Plana, V. Celorrio, S. Sadasivan, R. P. Tooze, P. Rodríguez and D. J. Fermín, *ChemCatChem*, 2016, **8**, 952–960.
68. M. G. Montes de Oca, H. Kurnarakuru, D. Cherns and D. J. Fermin, *J. Phys. Chem. C*, 2011, **115**, 10489–10496.
69. D. Plana, J. Florez-Montano, V. Celorrio, E. Pastor and D. J. Fermin, *Chem. Commun.*, 2013, **49**, 10962–10964.
70. V. Celorrio, P. M. Quaino, E. Santos, J. Flórez-Montaño, J. J. L. Humphrey, O. Guillén-Villafuerte, D. Plana, M. J. Lázaro, E. Pastor and D. J. Fermín, *ACS Catal.*, 2017, **7**, 1673–1680.
71. D. Friebel, F. Mbuga, S. Rajasekaran, D. J. Miller, H. Ogasawara, R. Alonso-Mori, D. Sokaras, D. Nordlund, T. C. Weng and A. Nilsson, *J. Phys. Chem. C*, 2014, **118**, 7954–7961.
72. J. Monzo, Y. Malewski, R. Kortlever, F. J. Vidal-Iglesias, J. Solla-Gullon, M. T. M. Koper and P. Rodriguez, *J. Mater. Chem. A*, 2015, **3**, 23690–23698.

73. K. A. Grice, *Coord. Chem. Rev.*, 2017, **336**, 78–95.
74. H. Takeda, C. Cometto, O. Ishitani and M. Robert, *ACS Catal.*, 2017, 7, 70–88.
75. Y. S. Wu, J. B. Jiang, Z. Weng, M. Y. Wang, D. L. J. Broere, Y. R. Zhong, G. W. Brudvig, Z. X. Feng and H. L. Wang, *ACS Cent. Sci.*, 2017, **3**, 847–852.
76. P. P. Sharma and X. D. Zhou, *WIREs Interdiscipl. Rev. Energy Environ.*, 2017, **6**, e239.
77. O. Ayemoba and A. Cuesta, *ACS Appl. Mater. Interfaces*, 2017, **9**, 27377–27382.
78. J. Resasco, L. D. Chen, E. Clark, C. Tsai, C. Hahn, T. F. Jaramillo, K. Chan and A. T. Bell, *J. Am. Chem. Soc.*, 2017, **139**, 11277–11287.
79. D. F. Gao, F. Sohoken and B. R. Cuenya, *ACS Catal.*, 2017, **7**, 5112–5120.
80. S. F. Zhao, M. Horne, A. M. Bond and J. Zhang, *J. Phys. Chem. C*, 2016, **120**, 23989–24001.
81. W. Karim, S. A. Tschupp, J. Herranz, T. J. Schmidt, Y. Ekinci and J. A. van Bokhoven, *Chimia*, 2017, **71**, 160–169.
82. Y. Tang and W. L. Cheng, *Nanoscale*, 2015, 7, 16151–16164.
83. J. E. Macdonald, M. Bar Sadan, L. Houben, I. Popov and U. Banin, *Nat. Mater.*, 2010, **9**, 810–815.
84. Z. Cao, D. Kim, D. C. Hong, Y. Yu, J. Xu, S. Lin, X. D. Wen, E. M. Nichols, K. Jeong, J. A. Reimer, P. D. Yang and C. J. Chang, *J. Am. Chem. Soc.*, 2016, **138**, 8120–8125.
85. W. L. Li, M. Seredych, E. Rodriguez-Castellon and T. J. Bandosz, *ChemSusChem*, 2016, **9**, 606–616.
86. J. W. Maina, C. Pozo-Gonzalo, L. X. Kong, J. Schutz, M. Hill and L. F. Dumee, *Mater. Horiz.*, 2017, **4**, 345–361.
87. D. R. Kauffman, D. Alfonso, C. Matranga, H. F. Qian and R. C. Jin, *J. Am. Chem. Soc.*, 2012, **134**, 10237–10243.
88. D. R. Alfonso, D. Kauffman and C. Matranga, *J. Chem. Phys.*, 2016, **144**, 184705.
89. R. C. Jin, C. J. Zeng, M. Zhou and Y. X. Chen, *Chem. Rev.*, 2016, **116**, 10346–10413.
90. C. D. Windle and E. Reisner, *Chimia*, 2015, **69**, 435–441.
91. K. Sekizawa, K. Maeda, K. Domen, K. Koike and O. Ishitani, *J. Am. Chem. Soc.*, 2013, **135**, 4596–4599.
92. M. Schreier, P. Gao, M. T. Mayer, J. Luo, T. Moehl, M. K. Nazeeruddin, S. D. Tilley and M. Grätzel, *Energy Environ. Sci.*, 2015, **8**, 855–861.
93. T. E. Rosser, C. D. Windle and E. Reisner, *Angew. Chem., Int. Ed.*, 2016, **55**, 7388–7392.

CHAPTER 2

Bio-inspired and Bio-electrochemical Approaches in CO_2 Reduction Catalysis

THOMAS RISBRIDGER*[a] AND ROSS ANDERSON*[b]

[a] School of Chemistry, University of Bristol, Bristol BS8 1TH, UK;
[b] School of Biochemistry, University of Bristol, Bristol BS8 1TH, UK
*Email: Tom.Risbridger@ilika.com; Ross.Anderson@bristol.ac.uk

2.1 CO_2 Reduction: The Biological Example

Biological carbon dioxide reduction is a critical reaction for sustaining life on earth. CO_2 is the primary chemical building block for most complex and higher energy carbon compounds required for metabolism and growth, and nature has developed incredibly elegant mechanisms for its reduction and up-conversion into such materials.[1] Enzymes broadly termed carbon dioxide reductases (CO_2Red) efficiently catalyse the reduction of CO_2 into a variety of lower oxidation state carbon products at rates that enable the functioning of complex organisms.[2] Autotrophs utilise CO_2 as a substrate, powered by solar radiation in the case of phototrophs (Figure 2.1), or from reducing chemicals in the case of chemotrophs; such organisms are the fundamental chemical energy source for all non-autotrophic life-forms.[3]

CO_2Reds in these organisms function within a highly complex environment. Sophisticated electron transport chains supply reducing power, while regulatory mechanisms continually generate or break down enzymes in response to changing conditions.[3] CO_2 reduction itself is only one of the myriad chemical steps in the metabolic pathways involved in cell growth and

Energy and Environment Series No. 21
Electrochemical Reduction of Carbon Dioxide: Overcoming the Limitations of Photosynthesis
Edited by Frank Marken and David Fermin

Published by the Royal Society of Chemistry, www.rsc.org

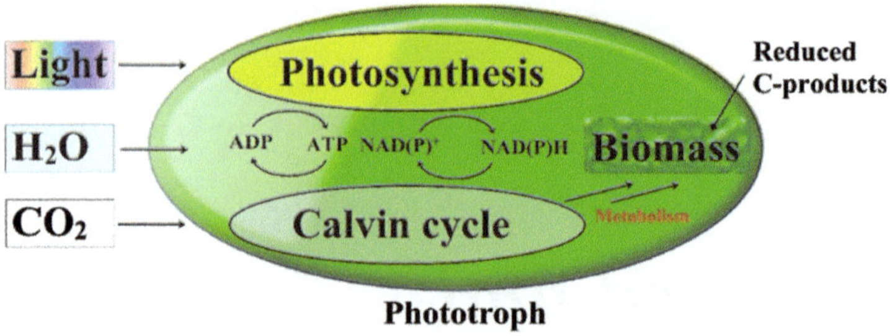

Figure 2.1 Representation of the natural CO_2 reduction process that occurs in phototrophs. Biomass, incorporating reduced carbon products, is produced *via* the Calvin cycle and further metabolism. These processes are powered by reduced species generated in photosynthesis.

biomass generation. These involve many different enzymes and cofactors,[4,5] and include mechanisms such as the Wood–Ljungdahl pathway and the Calvin cycle.[6]

The potential benefit of harnessing such organisms or some of their components for anthropogenic application has helped fuel the development of integrated bio-electrochemical CO_2 reducing devices. By incorporating a single type of enzyme into electrochemical cells, the efficient catalysis of specific reactions has been achieved. However this approach typically suffers from high isolation costs, low turnover and low stability to degradation.[7,8] Coupling whole organisms into an electrochemical cell offers improvements in stability as the organism regulates and maintains its internal components, and both short C-chain products and/or biomass can be produced using autotrophs, though with a lower energetic efficiency than single enzymes.[9,10] Natural catalysis also provides a highly optimised blueprint, which has inspired the design of non-biological catalysis.[2] Artificial photosynthesis is an example of a CO_2 reducing system, the design of which takes inspiration from biological CO_2 reduction in natural photosynthesis.[11,12]

Within this chapter, a range of bio-electrochemical and bio-inspired systems for CO_2 reduction to single carbon products are considered. Enzyme based approaches are covered first, followed by those incorporating microbes. Finally photoactive devices including those mimicking photosynthesis are discussed. An important distinction can be made between most bio-electrochemical systems based upon whether the catalyst interacts directly or indirectly with the electrode. In the direct case, DET between the immobilised bio-catalyst and electrode occurs. Here careful bio-catalyst selection is particularly important, as is the manner in which it is linked to the electrode. Systems with indirect interactions utilise some form of soluble reducing equivalent, which can be in the form of the NADH cofactor, acting either as hydride donor or terminal electron donor, a reduced species such as H_2 or formate, or a biologically compatible redox mediator such as methyl viologen (MV). The interaction, regeneration and stability of this transfer species must therefore be considered in addition to the catalyst itself.

2.2 Bio-electrocatalysis

2.2.1 Bio-electrochemical Cells and Product Detection

Several practical considerations should be considered when designing bio-electrochemical cells, which almost exclusively operate in solution.[9,13–16] The choice between a one- or two-compartment cell must take into account factors including product oxidation at non-selective anodes in one-compartment cells, while in two-compartment cells the effects of pH gradients and resistance across separating membranes must be considered.[17–19] In some cases the cell must be operated anaerobically, and minimising cell volume is often desirable to facilitate detection of products at low concentrations. Temperature control can be utilised to optimise rates. The reaction solution must be suitable for both the catalyst and any mediators, and selection of parameters such as pH can involve compromising between rate and stability, as well as affecting $CO_2/CO_3H^-/CO_3^{2-}$ solubility and equilibrium position. Electrochemically generated species should also be considered, for example at reasonable working potentials hypochlorite can be formed in the presence of Cl^-, which is detrimental to microbial species necessitating exclusion of Cl^- based salts in some systems.[20] In enzymatic cells, care must be taken to avoid build-up of inhibiting products.[21,22] In cells utilising DET, the immobilisation of the bio-catalyst is crucial, and several strategies including electrostatic[23] and covalent control[24] can be employed to optimise the binding orientation for maximum activity. Diagrammatic representations of mediated and DET bio-electrochemical cells including a list of set-up variables are shown in Figure 2.2.

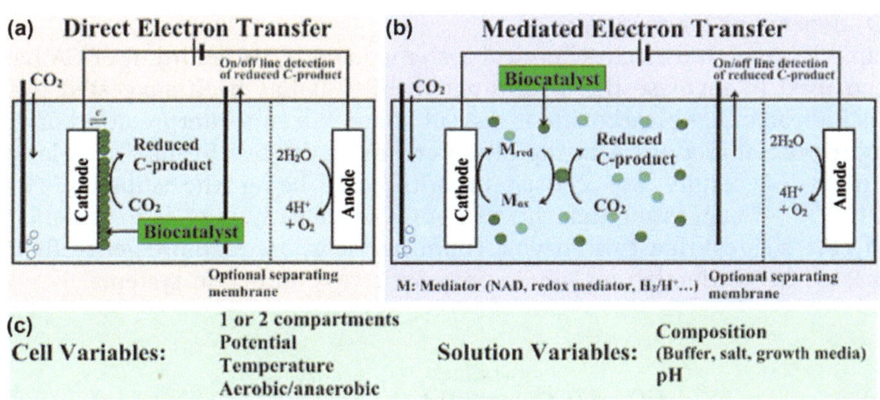

Figure 2.2 Bio-electrochemical cells involving (a) direct and (b) mediated electron transfer to the biocatalyst. The dashed line denotes an optional separating membrane, which is present in two compartment cells. Box (c) summarises controllable system variables.

CO_2 reduction product detection by online or offline methods should seek to eliminate contamination and can be achieved using several methods. Chromatographic methods including gas chromatography (GC) and high performance liquid chromatography (HPLC) have often been used to detect low molecular weight products from such systems,[20,23,25] as well as other techniques including spectroscopic enzymatic assays.[21]

2.2.2 Performance Metrics

A variety of performance metrics exist for these systems.[26] The overpotential and faradaic efficiency (FE) (or catalytic selectivity) indicate the energetic efficiency and selectivity, whilst the turnover frequency and turnover number give an indication of the activity and resilience of the catalytic material. Activity measurements should be supported by direct product detection.[27] The short and long term stability is important for determining the viability of the system.

2.2.3 Substrate

CO_2 has a mole fraction solubility in water of 6.14×10^{-4} (at 25 °C, 100 kPa partial pressure)[28] and slowly equilibrates with the hydrated bicarbonate (HCO_3^-) form ($pK_{a1} = 6.4$, $pK_{a2} = 10.3$) at physiological pH values, as summarised in Scheme 2.1.[29] Bicarbonate is the substrate for some carboxylases,[30,31] while the non-hydrated form is required for enzymes such as formate dehydrogenase (FDH) and carbon monoxide dehydrogenase (CODH).[32,33] Within organisms, zinc-containing carbonic anhydrase (CA) reversibly interconverts CO_2 with the hydrated form at incredibly fast rates (up to 10^6 s^{-1}),[34,35] and when present is able to quickly replenish CO_2 from the bicarbonate reservoir. This results in an overall increase in CO_2-dependent reaction rates.

In enzymatic bio-electrochemical systems using FDH, addition of CA has been used to increase the system activity.[36,37] It has been suggested that this indicates that bicarbonate is the substrate, but this interpretation often ignores the ability of CA to rapidly interconvert the two forms.[36,37] A large number of results are consistent with CO_2 being the substrate for FDH,[8,33,38,39] suggesting that the role of CA in these systems is to replenish CO_2. We suggest that this enzyme could be incorporated more generally to increase CO_2 reduction activity in enzymatic CO_2 reduction systems.

$$CO_2 + H_2O \underset{}{\overset{K_H = 1.7 \times 10^3}{\rightleftharpoons}} H_2CO_3 \quad \begin{matrix} k_{\text{CA}} \sim 10^{-3} - 10^{1}\ s^{-1} \\ k_{CA} \sim 10^{6}\ s^{-1} \end{matrix}$$

$$CO_2 + H_2CO_3 \overset{pKa = 6.4}{\rightleftharpoons} HCO_3^- \overset{pKa = 10.3}{\rightleftharpoons} CO_3^{2-}$$

Scheme 2.1 pH dependent CO_2 solution behaviour.[29,34]

2.2.4 NADH Regeneration

In a large number of biological CO_2 reduction mechanisms, the (light-sensitive) cofactor nicotinamide adenine dinucleotide (NADH) is of key importance as a hydride transfer agent, and a key challenge for generating stable turnover lies in the efficient regeneration of this species. Several methods currently exist for regenerating NADH, and have been reviewed elsewhere.[40–42] Effective metal-based heterogeneous catalysts have been developed,[43] but typically non-electrochemical approaches have utilised enzymes coupled with other reduced substrates, following a hydride transfer mechanism involving concerted transfer of two electrons and one proton. Dehydrogenases including FDH (formate oxidation),[42] glutamate dehydrogenase (glutamic acid oxidation),[44] phosphite dehydrogenase (phosphite oxidation)[45] and others have all been used successfully.[42] However although suitable for activity and mechanistic studies these require constant addition of non-interfering sacrificial substrates for continued catalytic performance, and add another layer of complexity to the system, which must also be optimised in terms of pH, temperature and concentration.[46]

An alternate method is to regenerate NADH ($E^{0\prime} = -0.32$ V *vs.* SHE, pH 7.0[47]) using an electrochemical system, either utilising DET, or a mediated or catalysed reducing process. This regeneration must be optimised for low overpotential, high activity and high specificity. Generally, it is found that electrochemical reduction of NADH follows two single electron transfer steps, as shown in Scheme 2.2. The radical intermediate formed by the initial one electron reduction of NAD^+ can either be further reduce to NADH *via* (slow) proton-coupled electron transfer, or dimerise to form NAD_2, which is inert towards enzymatic oxidation.[48] Regeneration systems must be specific towards the active NADH form, and activity tests should distinguish between these forms.[49,50]

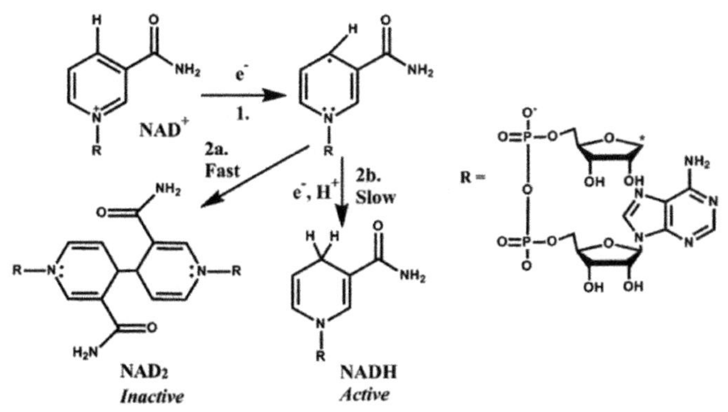

Scheme 2.2 Two-step electrochemical NADH regeneration mechanism proceeding *via* a reactive radical intermediate,[50] distinguishing between relatively fast radical dimerisation and slow proton coupled single electron reduction forming inactive and active NAD reduction products respectively.

NADH regeneration at metal electrodes through DET has typically required high overpotentials, resulting in the inactive NAD_2 dimer being favoured.[51,52] However, NADH selectivity up to 100% has been obtained on glassy carbon–Pt and glassy carbon–Ni electrodes where the metal provided a high concentration of 'active' adsorbed hydrogen,[52,53] thereby optimising the relatively slow proton-coupled electron transfer process. In-depth analysis of metal electrodes has suggested that the precise potential and the $M–H_{ads}$ bond strength are very important for optimising activity and selectivity.[54] A selectivity of 96% has been achieved on Ti electrodes at −0.8 V *vs.* SHE, which decreased as the potential was shifted negatively.[54] It may also be possible to hinder the dimerisation pathway by limiting the movement of NADH through immobilisation or encapsulation in a biocompatible supporting matrix.[55]

The regeneration overpotential can also be decreased by incorporation of catalysts or mediators into the system. Transition metal complexes based on Ru,[41,53,56] Co[56] and Rh[39] have been designed, and some of these are able to regenerate NADH at approximately −0.5 V *vs.* SHE, though not with 100% efficiency.[39] NAD analogues, which can be regenerated more efficiently and exhibit higher stability than NAD itself, have also been developed, though their interaction with enzymes has not yet been evaluated.[57,58]

2.3 Enzymatic CO_2 Electro-reduction

Enzymes are of interest as CO_2 reduction catalysts because they exhibit very high energy efficiency towards specific products. This is due to a combination of their ability to catalyse reactions very close to the thermodynamic (zero overpotential) limit[8,13] and their precise reaction pathways. Several enzymes in nature can act as CO_2 reductases, either directly reducing CO_2 to form small molecules, or by carrying out the first step in complex molecule bio-synthesis. These include FDH, CODH, ribulose-1,5-bisphosphate carboxylase oxygenase (RuBisCO) and carboxylase, which reduces carbonates.[6] Examples of FDHs and CODHs from several organisms catalyse the reaction in both directions, usually with a kinetic preference for formate or CO oxidation generating CO_2.[8] However, by biasing the enzyme chemically through substrate and cofactor concentrations, or electrochemically through substrate concentration and potential, the equilibrium can be shifted to favour CO_2 reduction.

The production of enzymes can be achieved biochemically *via* the expression or overexpression of native or recombinant enzymes by a host organism, followed by extraction and purification steps; an introduction to this field can be found in ref. 59 and 60. Enzyme-specific protocols are required, and ease of production varies significantly, strongly influencing the scale at which a target enzyme can be produced.

There are currently several significant challenges to the implementation of enzyme-based catalysts for medium- to large-scale industrial CO_2 reduction, which are linked to various combinations of poor stability, high O_2

sensitivity, high cost, production challenges and low rates, depending on the particular enzyme used. Although some success has been achieved in developing industrial non-reductive CO_2 conversion systems based on the highly active enzyme CA,[61] it is becoming increasingly accepted that enzymatic systems are not feasible for industrial CO_2 reduction.[8,62] However, electrochemical studies can be useful for understanding enzymatic mechanisms and informing the design of synthetic catalysts.

Energy requirements for the reduction of CO_2 to various different products are given as reduction potentials *vs.* NHE at pH 7, 25 °C in eqn (2.1)–(2.5).[63] The single electron reduction forming the bent radical anion $CO_2^{\bullet-}$ is particularly energetically challenging, and electrocatalysis *via* this intermediate often competes with the hydrogen evolution reaction (HER), resulting in low product specificity and limiting the FE. Catalysts that have a low activation energy requirement and are capable of multiple electron transfer pathways are desirable. In enzymatic CO_2 reduction mechanisms, two electron pathways are favoured. This is usually due to the presence of a bimetallic reaction centre with both a nucleophilic and electrophilic region,[64] or a cofactor-coupled mechanism involving hydride transfer. Electrochemical experiments have demonstrated that enzymes can function very close to the thermodynamic limit,[8,65,66] whilst also maintaining high specificity.

$$CO_2 + e^- \rightarrow CO_2^{\bullet-} \qquad E^0 = -1.90\ \text{V} \tag{2.1}$$

$$CO_2 + 2H^+ + 2e^- \rightarrow HCOOH \qquad E^0 = -0.61\ \text{V} \tag{2.2}$$

$$CO_2 + 2H^+ + 2e^- \rightarrow CO + H_2O \qquad E^0 = -0.53\ \text{V} \tag{2.3}$$

$$CO_2 + 6H^+ + 6e^- \rightarrow CH_3OH \qquad E^0 = -0.38\ \text{V} \tag{2.4}$$

$$CO_2 + 8H^+ + 8e^- \rightarrow CH_4 + H_2O \qquad E^0 = -0.24\ \text{V}. \tag{2.5}$$

Broadly speaking there are two classes of enzymatic bio-electrochemical (EzBEC) systems that involve either direct or mediated electron transfer, and both will be considered here.

2.3.1 Carbon Monoxide Generation

2.3.1.1 Carbon Monoxide

Carbon monoxide is an industrially significant molecule as it is readily converted into liquid fuels using the Fisher–Tropsch process, and is an important feedstock for carbonylation and in the production of a number of low molecular weight carbon compounds.[67,68] Industrially, CO is typically produced at high temperatures, for example by the Boudouard reaction.[69] Additionally, the two electron reduction of CO_2 to CO can be catalysed by a wide variety of metallic electrode materials, with control over the metal–CO adsorption strength being an important factor in controlling reaction rate.[70]

2.3.1.2 Carbon Monoxide Dehydrogenase

CODHs catalyse the reversible reduction of CO_2 to CO within a variety of organisms. Two classes exist, which are obtained from aerobic or anaerobic bacteria. Those from aerobic bacteria such as *Oligotropha carboxidovoran* are generally found to be air-stable and are characterised by an active site containing Mo and Cu.[71] CODHs from anaerobic bacteria, including *Moorella thermoacetica*[13] (also known as *Clostridium thermoaceticum*[72]), *Rhodospirillum rubrum*[73] and *Carboxydothermus hydrogenoformans* (*Ch*CODH),[65] have an active site containing Ni and Fe species. These are generally not air-stable, although CODH from *Methanothrix soehngenii* has demonstrated O_2 stability and CO oxidation activity.[74] *Ch*CODHs have shown turnover rates for CO oxidation of 39 000 s^{-1} at 70 °C in a pH 8.0,[75] and can also catalyse the reduction of CO_2 to CO, albeit typically with lower turnovers of up to 45 s^{-1}.[21,76] In autotrophic organisms, this reduction is the first step of the carbon fixation process *via* mechanisms such as the Wood–Ljungdahl pathway,[38] and CODHs from many organisms including *Moorella thermoacetica* are closely associated with acetyl-CoA synthase (ACS) at the level of quaternary structure (termed CODH/ACS).[13,76–78] ACS in these bifunctional enzymes couples CO with coenzyme A to form the acetyl-CoA metabolite.[79,80] Both CODH and CODH/ACS have been considered as bio-electrocatalysts for the CO_2 reduction reaction.[13,65]

2.3.1.3 Carbon Monoxide Dehydrogenase Structure

Structures of CODHs from each of the anaerobic bacteria mentioned above have been determined by X-ray crystallography, and exhibit a high degree of similarity.[81–84] The quaternary structure of the CODH subunit is that of a homodimer containing several metal clusters involved in electron transfer (Figure 2.3a), catalysis and possibly as structural components; these are termed the B, C and D clusters. Two $[Fe_4S_4]^{2+/1+}$ B-clusters are present, and are involved in electron transfer within the enzyme and with external mediators.[72] A distinct $[Fe_4S_4]$ D-cluster bridges the two monomeric units, though its role in structure or electron transfer is not yet fully understood.[78] The C-cluster is the active site for reversible CO_2 reduction to CO, and has been the subject of significant study.[85,86] This cluster consists of a distorted cubane $NiFe_3S_4$ cluster, with an attached unique Fe that does not form part of the distorted cubane (Figure 2.3b).[78,83,85] For a number of years, the importance of a bridging sulfide between Ni and unique Fe was strongly debated on account of its presence in the crystal structure of some CODHs. However, several works have since shown that this is not present in catalytically relevant crystal structures, and it is now generally accepted that this species is not important for catalysis.[78,85] CO_2 or CO substrates are thought to reach the active site *via* a hydrophobic channel,[87] while protons may be transported *via* a transfer network.[88] Interestingly, it has been proposed that

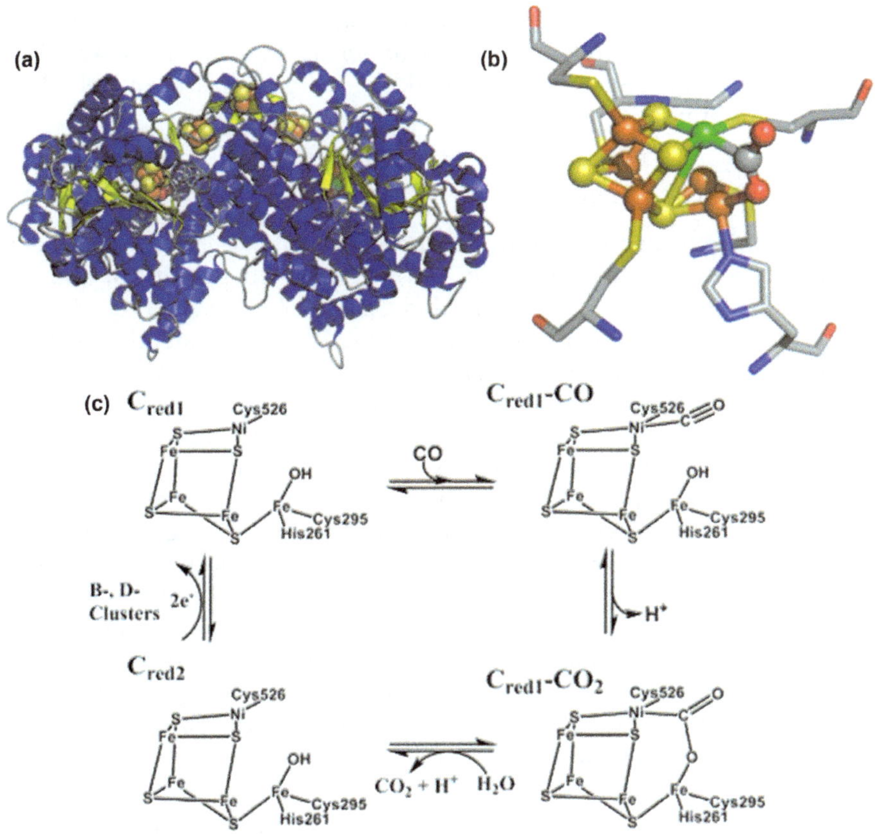

Figure 2.3 (a) Homodimeric structure of Ni/Fe–CODH highlighting metal containing B-, C- and D- clusters. (Protein representation created from PDB structure 3B5290 using PyMOL.[91]) (b) Ni/Fe C-cluster active site with bound CO_2 measured at ≤1.1 Å resolution,[83] (c) possible reversible CO oxidation mechanism at the C-cluster.[78,85,90,92]

this may exhibit proton translocation directionality, which changes dependent upon the oxidation state of the C-site.[89]

2.3.1.4 Carbon Monoxide Dehydrogenase/Acetyl-CoA Synthase Structure

In CODH/ACS, the close associations of the ACS results in an overall $\alpha_2\beta_2$ type structure, where the β units are very similar in structure and function to the mono-functional CODH enzymes, and the α units contains the A-cluster, which catalyses the conversion of CO into acetyl-CoA. In this case, the cluster is another Fe_4S_4 cluster bridged to Ni *via* cysteine, which is subsequently cysteine-bridged to a second Ni in active forms of the enzyme.[78,93] A hydrophobic channel is found between C- and A-sites, and it is thought that CO generated at the C-site travels to the A-site *via* this channel.[81,94]

The development of synthetic catalysts based upon the NiFe–CODH C-cluster has seen significant progress in terms of structure emulation, though functionality remains poor. Work in this area has been well summarised by Majumdar.[95]

2.3.1.5 CO/CO_2 Mechanism

Four electronically distinct forms of the active site C-cluster have been determined from EPR studies under different applied potentials, and are commonly designated C_{ox}, C_{red1}, C_{int}, and C_{red2}.[86,96,97] However, catalysis is thought to only involve the two reduced states.[78,85] In CO oxidation, carbon in CO and oxygen in water or hydroxide first bind to the nucleophilic Ni and electrophilic unique Fe sites of the hydroxylated C_{red1} state, respectively.[83,85] If water is bound, a shift of the CO carbon position may follow water deprotonation,[85,92] and the hydroxide species is then able to attack the carbon species accompanied by proton loss. Loss of CO_2 and attachment of water/hydroxide then occurs *via* a two electron transfer, generating the further reduced C_{red2} state. The C_{red1} state is regenerated by electron transfer through the enzyme (Figure 2.3c). This mechanism has been described in greater detail elsewhere.[78,85] Studies examining CO_2 reduction have shown that the distortion from linearity of Ni-bound CO_2 is very similar to that of ${CO_2}^{2-}$, and a surprisingly short Ni–C bond suggests significant backbonding is occurring. This strong metal to ligand charge transfer is thought to be important in activating the CO_2 reduction reaction.[83]

2.3.1.6 Bio-electrocatalysis: Mediated Electron Transfer

In bio-electrocatalytic systems the electronic interaction with the enzyme is key, and can occur by either mediated or direct electron transfer at an electrode surface. Mediated CO_2 reduction has been demonstrated for CODHs from several organisms including *Rr*CODH[21] (Figure 2.4a) and *Mt*CODH/ACS.[13,76,77] Several different mediators have been used to regenerate CODH during CO_2 reduction,[21,76,77] and, of these, the one electron mediator methyl viologen (MV^+) (Figure 2.4b) is generally found to be most suitable for high turnover.[21] In several studies, the presence of CO has been shown to inhibit CO_2 reduction[21,22] and consequently removal of CO from solution has been shown to increase ongoing activity.[21]

CO_2 reduction in systems incorporating electrochemical mediator regeneration generally occurs close to the thermodynamic potential (−0.51 V *vs.* SHE at pH 7[63]),[13,77] and functions best at slightly acid pHs.[13] The *Mt*CODH/ACS system developed by Shin *et al.* required an overpotential of approximately 100 mV for CO_2 reduction,[13] while when using *Ct*CODH/ACS CO_2 it was reduced at the thermodynamic limit.[77] However, turnover tends to be low, even in optimised systems, with rates of 0.19 s^{-1} per C-cluster observed in the *Mt*CODH/ACS system. This turnover further decreased significantly over 24 h, probably due to losses in enzyme activity.[13]

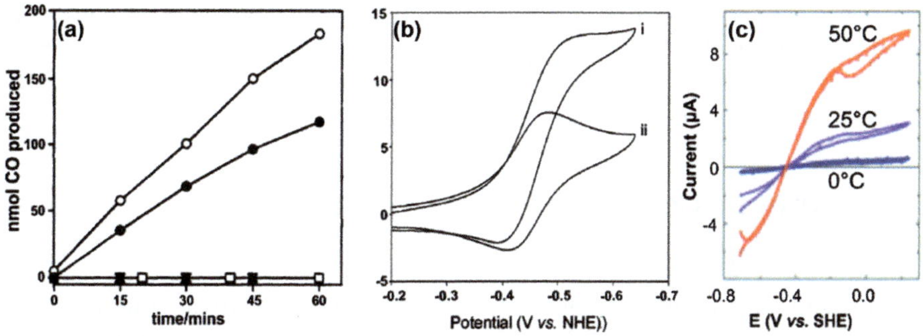

Figure 2.4 (a) GC analysis demonstrating $CO_2/NaHCO_3$ reduction to CO catalysed by different concentrations of *R. rubrum* CODH in the presence of reduced MV mediator (○ and ●). Reprinted (adapted) with permission from S. A. Ensign, *Biochemistry*, 1995, **34**, 5372–5381.[21] Copyright 1995 American Chemical Society. (b) Catalytic CO_2 reduction by CODH mediated by MV. Cyclic voltammograms showing MV response (i) with and (ii) without CODH (0.1 M phosphate buffer (pH 6–7), 1.0 mM MV^{2+}, 1 atm CO_2, 50 °C, glassy carbon electrode, 10 mV s^{-1}). Reprinted (adapted) with permission from W. Shin, S. Lee, J. Shin, S. Lee and Y. Kim, *J. Am. Chem. Soc.*, 2003, **125**, 14688–14689.[13] Copyright 2003 American Chemical Society. (c) DET voltammograms showing reversible CO_2 reduction and CO oxidation at surface immobilised ChCODH I in MES buffer (pH 6) containing 1 : 1 CO_2/CO at 0, 25 and 50 °C. Reprinted (adapted) with permission from A. Parkin, J. Seravalli, K. A. Vincent, S. W. Ragsdale and F. A. Armstrong, *J. Am. Chem. Soc.*, 2007, **129**, 10328–10329.[65] Copyright 2007 American Chemical Society.

2.3.1.7 Bio-electrocatalysis: Direct Electron Transfer

In 2007, Parkin *et al.* demonstrated that it was possible to perform DET measurements with mono-functional Ni-CODH I from the anaerobic thermophile *Carboxydothermus hydrogenoformans* (Figure 2.4c).[65] Reversible oxidation and reduction of CO and CO_2 were achieved at a formal potential of −0.5 V *vs.* SHE under anaerobic conditions at 50 °C and pH 6.7 for the enzyme immobilised onto a pyrolytic graphite edge plane electrode. Only very low overpotentials were required to drive the reaction in either direction, demonstrating the high energetic efficiency of the system. The set-up was found to be stable for several hours, and relative CO_2 reduction/CO oxidation activity was increased by decreasing pH. DET to the enzyme was thought to occur *via* the solvent-exposed Fe_4S_4 D-cluster. It has been suggested that the negative potential (−0.53 V *vs.* SHE)[98] of this cluster contributes to the high CO_2 reduction activity of this CODH.[78,99] Insights into CODH behaviour using DET electrochemical methods (often termed protein film electrochemistry, PFE) have included electrochemical determination of Michaelis constants. Furthermore, these have afforded several insights into the potential-dependent behaviour of inhibitors, which in turn leads to improved understanding of the different C-cluster states.[22] For example, it was shown that the ability of CO to inhibit CO_2 reduction becomes less significant at potentials more negative than −0.7 V *vs.* SHE, possibly suggesting that CO is not strongly bound to the C_{red2} state.[22,78] Discovery of this system has also led to the development of several photocatalytic CODH systems, as discussed in Section 2.5.2.2.[23,64,100,101]

2.3.2 Formate Generation

2.3.2.1 Formate

Formic acid is an industrially important molecule, with a global market value of approximately \$450 k in 2012.[102] It has large scale application in agriculture and leather production, and is used in a wide variety of other industries. With regard to energy vectors, it has recently been considered as an H_2 storage material, as it can readily be produced by CO_2 reduction involving addition of two protons, and then oxidised in a fuel cell.[103,104] It is also a precursor in the production of methanol, which has potential as a fuel.[105]

2.3.2.2 Formate Dehydrogenase

FDHs catalyse the reversible interconversion of formic acid and carbon dioxide. A variety of FDH structures exist, which are broadly classed as metal-dependent and metal-independent.[106] These variants show a range of different activities to $HCOO^-$ oxidation and CO_2 reduction, displaying a preferred reaction direction depending on their biochemical role.[6]

2.3.2.3 Metal-independent Formate Dehydrogenase: Structure and Properties

Metal-independent (MI) FDHs are found in methylotrophic yeast and aerobic bacteria[67] and generally share the same homodimeric α_2 substructure. They contain no metallic or other prosthetic groups, and have molecular weights between 70 and 100 kDa (Figure 2.5a). This class of FDH generally have desirable handling properties. Most are found to be air-stable,[107] can exhibit activity over a wide pH range,[25,108] and are relatively easy to produce in bulk.[25,108] Some, such as that from *Candida boidinii* (*Cb*FDH), are commercially available. However, most show a lower activity than metal-dependent FDHs,[25,108] and typically have a much higher activity towards formate oxidation than CO_2 reduction.[25] Several reviews of this class of FDH can be found elsewhere.[108,109]

2.3.2.4 Metal-independent Formate Dehydrogenase: CO_2/Formate Mechanism

Reversible interconversion between CO_2 and formate occurs by hydride transfer and is dependent upon NADH/NAD^+ (Figure 2.5b). The active site,

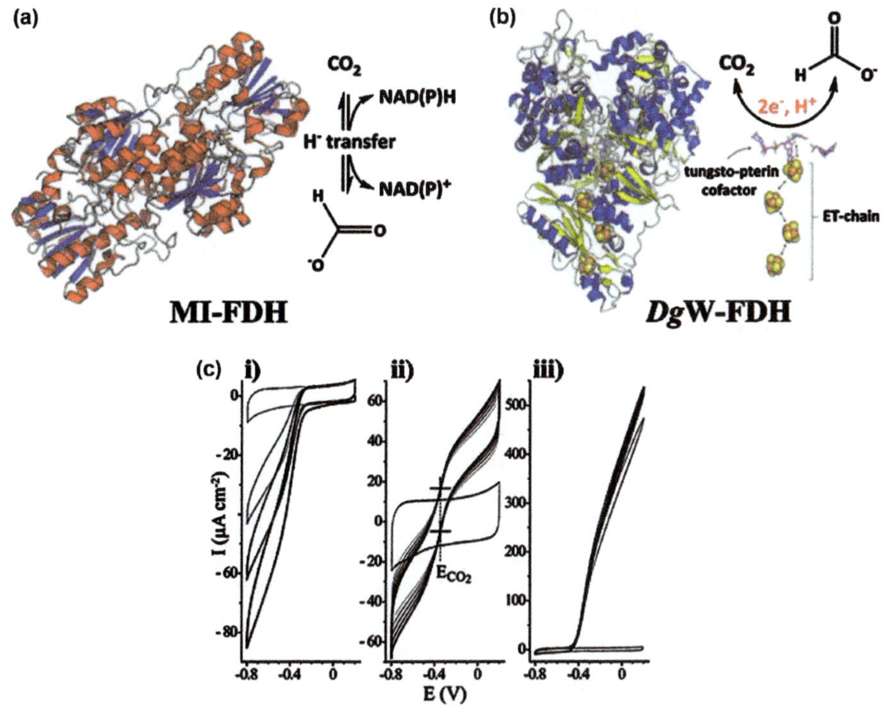

Figure 2.5 (a) Homodimeric MI-FDH structure and hydride-transfer mechanism (protein representation created from PDB structure 2NAD110 using PyMOL[91]), (b) DgW-FDH highlighting an FeS cluster electron transfer chain and a tungsto-pterin cofactor active site, (c) CO_2 reduction and formate oxidation at SfW-FDH1 immobilised onto a graphite electrode enabling enzyme regeneration *via* DET (black lines show first cycle; grey lines show subsequent cycles or background scans excluding substrate). Voltammograms carried out anaerobically at 37 °C in an MES buffer containing (i) 10 mM CO_2 (pH 5.9), (ii) 10 mM CO_2 : 10 mM formate (pH 6.4) and (iii) 10 mM formate (pH 7.8).
Reproduced with permission from ref. 8. Copyright (2008) National Academy of Sciences, USA.

composed of binding sites for both NADH/NAD^+ and CO_2/formate,[110] has been observed in both so-called open and closed conformations.[25,111] Upon NAD^+-binding, the open, NAD^+-unbound form changes to the more rigid closed NAD^+-bound form, which helps to fix the substrate and cofactor in an optimal position for hydride transfer.[107] In the closed form, two holes in the structure allow formate or CO_2 to reach the active site,[107] and slight changes in these structures between FDH variants have been linked to differences in CO_2 reduction activity.[67,107,111]

Formate oxidation has been more closely studied than CO_2 reduction for this enzyme class. During this reaction, direct two electron hydride transfer occurs between the C in formate and the C4 position in NAD^+,[67] and this transfer is the limiting step in the process.[112] Atomic-scale crystallographic measurements have helped to confirm that, when bound in the active site of FDH from *Moraxella sp. C-1*, NAD^+ adopts a so-called bipolar state in which the C4 has enhanced positive charge and electrophilicity, aiding oxidation. However, the effect of this upon the reverse CO_2 reduction reaction has not been examined to our knowledge.[111]

2.3.2.5 Metal-independent Formate Dehydrogenase: Bio-Electrocatalysis

As MI-FDH is relatively easy to obtain and has desirable handling properties, it has often been used in electrochemical CO_2 reduction systems, and a number of photo-electrochemical systems based upon it are also discussed in Section 2.5.2.1. Formate generating electrolysis cells, which use MI-FDH, typically incorporate commercial (*Cb*FDH) FDH and NADH in solution in a one- or two-compartment bio-electrochemical reactor. Indirect NADH regeneration has been achieved in such systems; in one case *via* the redox dye neutral red, itself regenerated at a graphite cathode,[14] and in another *via* a [Cp*Rh(bpy)Cl]Cl complex (where Cp* is pentamethylcyclopentadienyl and bpy is bipyridine), itself regenerated at a copper cathode.[39] At an applied potential of −1 V *vs.* Ag/AgCl, formate was generated in these systems at a rate of 0.07 and 0.002 $\mu mol\, min^{-1}\, mg^{-1}$ respectively, and was found to increase as the potential was shifted negatively, due to improvements in NADH regeneration. However, NADH concentrations greater than 0.3 mM result in FDH inhibition.[39] A mediated/DET intermediate system involving both viologen mediators and immobilised enzymes was fabricated by immobilising long chain viologen species and FDH from *Saccharomyces cerevisiae* onto FTO particles.[113] Formate was produced at a stable rate of 7.6 $\mu mol\, h^{-1}$ over the course of three hours at −0.55 V *vs.* Ag/AgCl, when a nine carbon, carboxylate-terminated viologen alkyl linker was used.[113]

*Cb*FDH can currently be obtained from commercial companies, including Sigma-Aldrich. For NADH-powered CO_2 reduction, the activity of this FDH has been shown to be around 1.6 $mU\, mg^{-1}$ ($1\ U = 1\ \mu mol_{NADH,ox}\, min^{-1}$).[25] At the time of writing, 500 U of commercial FDH with a formate oxidation activity of $10 \pm 5\ U\, mg^{-1}$ costs £633,[114] making the FDH cost for CO_2 reduction approximately £8000 U^{-1}, and requiring 625 mg FDH. Such a high cost per unit makes this type of FDH unrealistic for any kind of large-scale use unless there are significant improvements in its activity.

Some practical steps, including pH optimisation and careful enzyme selection, can be taken to maximise activity. For example, CO_2 reduction activity increased 20 times when the pH was decreased from 8.0 to 6.3 in a system containing FDH from *Phaseolus aureus*.[115] In terms of enzyme selection, FDHs from recombinant *Thiobacillus* sp. KNK65MA (*Ts*FDH) have been found to exhibit a six-fold higher formate production rate than the commercially available FDH from *Cb*FDH,[25] up to 0.3 s^{-1} with an activity of 12.2 $mU\, mg^{-1}$, and several structure–function relationships have been proposed to explain these differences. For example, variations to the channels approaching the active site have been identified in the closed conformation, with one channel being blocked by an α-helix in *Ts*FDH, and both channels being larger in *Ts*FDH than in *Cb*FDH. Along with small discrepancies between some active site residues, these channel variations may contribute to the change in rate between the FDHs.[107] Modelling studies have indicated that the activation energy for CO_2 conversion to formate is 10 kcal mol^{-1} lower in *Ts*FDH than in *Cb*FDH, which may relate to the different spatial restraints imposed upon NADH as a result of the differences between the active sites.[107] Enzyme engineering is a field that can improve mechanistic understanding, and has some potential for designed activity and stability improvements; this has been reviewed elsewhere.[116]

2.3.2.6 Metal-dependent Formate Dehydrogenase

In contrast to MI-FDHs, metal-dependent (MD) FDHs tend to be more complex, and many variations exist in terms of metal cluster content and quaternary structure, as reviewed by Maia *et al.*[106] These enzymes are between 80 and 510 kDa, and are naturally found to exist as membrane-bound, cytoplasmic or periplasmic, depending upon their natural function (Figure 2.6).[117]

Those that naturally act as CO_2 reductases have been found in the periplasm[118] or cytoplasm[117] of several examples of prokaryotes, though membrane bound FDHs have also shown CO_2 reduction activity.[7,119] Metal-dependent FDHs typically contain several iron–sulfur clusters, which are essential for electron transport within the enzyme (Figure 2.5b), and are reduced by electron transfer at a donor/acceptor site near the edge of the protein. The active site is largely conserved between variants and consists of Mo or W coordinated by one or two pyranopterin guanine dinucleotide cofactors, a SeCys or Cys and S.[106,117] Two additional His and Arg residues are also highly conserved, and are found to be directly involved in catalysis.[117,120,121] Several different mechanisms for formate oxidation have been proposed and have been summarised in several recent reviews.[106,117] One proposed mechanism (Scheme 2.3) involves formate travelling to the active site through a positively charged channel and then bonding to the oxidised M^{6+} centre oxygen, following a Se–S shift rearrangement at the metal. Further active site rearrangements facilitate oxidative breaking of the formate C–H bond, which has been shown by isotopic substitution studies to be rate limiting.[120] CO_2 release then occurs, and CO_2 travels out of the enzyme, probably *via* a hydrophobic channel. Following CO_2 release, the reduced M^{4+} site is reoxidised.[117,120,122–125] Arg_{333} is thought to play an important role in positioning the substrate and buffering charge, while His_{141} facilitates the formation of the oxidation transition state.[120] Changing the active site SeCys to Cys in FDH-H from *E. coli* reduced k_{cat} 300-fold, which could be related to the different pK_as and subsequent variation in protonation between the two residues.[126] Generally, mechanistic studies have focussed upon CO_2 as the product and it is not certain whether CO_2 reduction

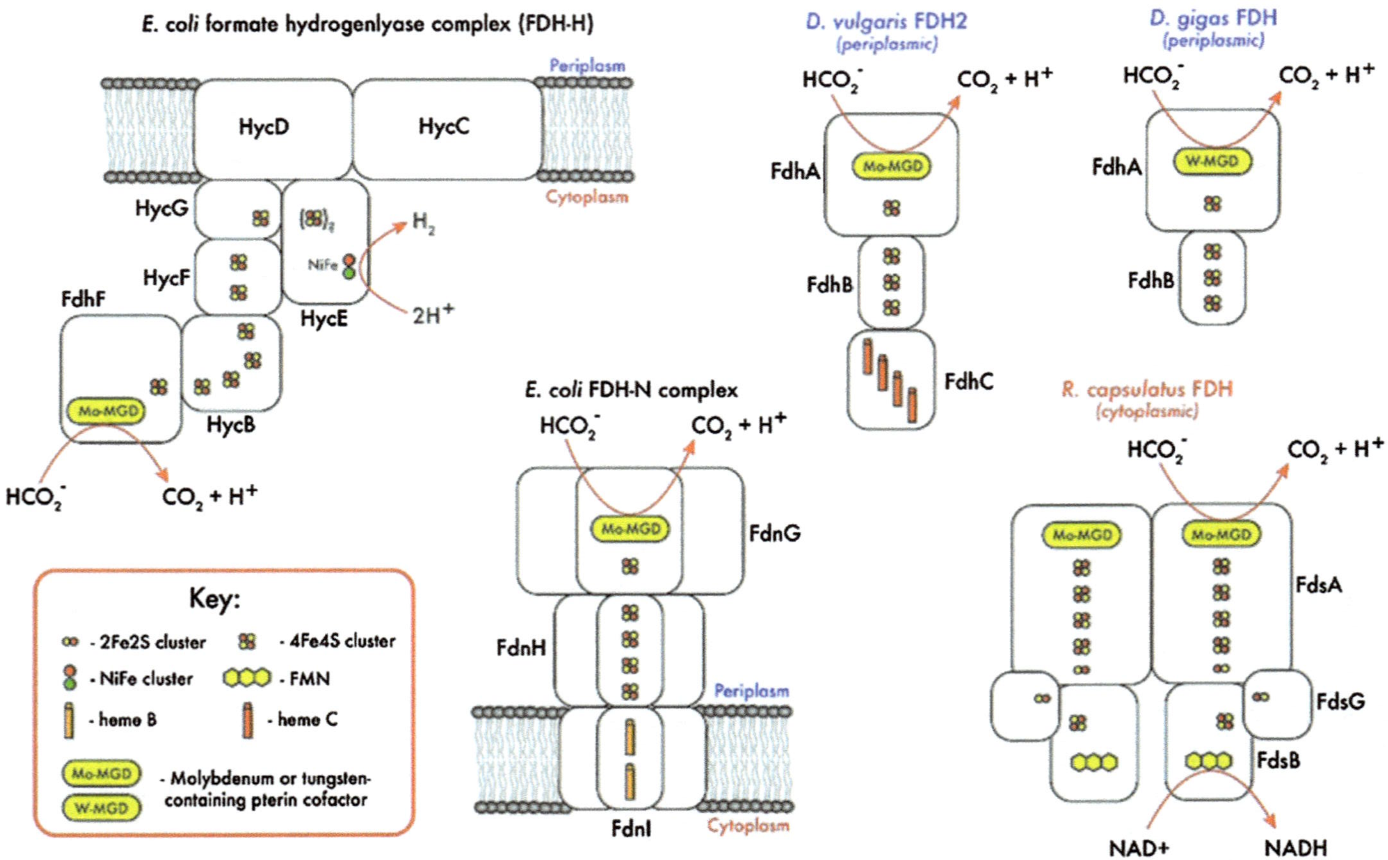

Figure 2.6 Diversity of metal-dependent formate dehydrogenase : carbon dioxide reductases. Representative architectures of various molybdenum and tungsten-dependent multi-subunit complexes from the Gram negative bacteria *E. coli*, *Rhodobacter sphaeroides*, *Desulfovibrio gigas* and *Desulfovibrio vulgaris*. FDH/CO_2Reds domains generally associate with iron–sulfur cluster-containing subunits to enable long distance electron transfer facilitating proton pumping in associated membrane spanning subunits (*e.g.* FDH-H and FDH-N) or the reduction of NAD^+ in the bacterial cytoplasm (*e.g. R. capsulatus* FDH).

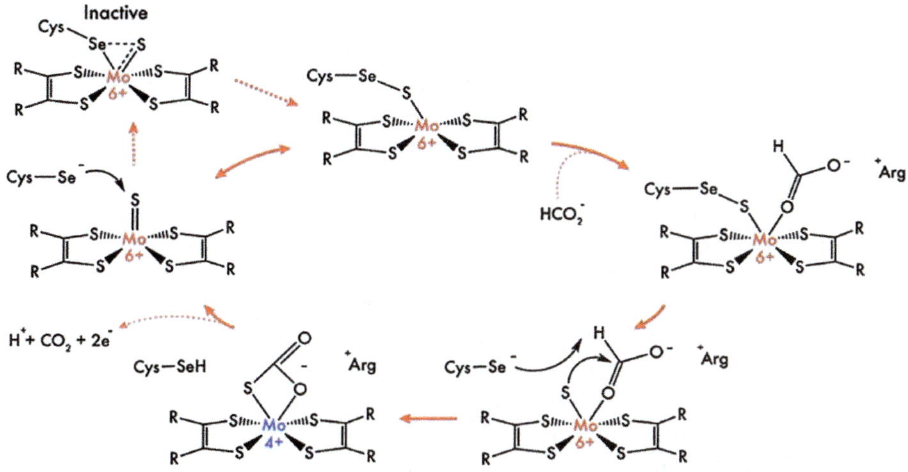

Scheme 2.3 Possible mechanism for formate oxidation at a Mo-FDH active site, involving activating a Se–S shift rearrangement at a metal centre.[106,117,124]

follows the same pathway with a shifted equilibrium position, or an alternative pathway.

2.3.2.7 *Metal-dependent Formate Dehydrogenase: Bio-electrocatalysis*

In MD-FDH, electrons travel between acceptor species near the enzyme surface and the active site *via* iron–sulfur clusters,[127] and artificial electron donor/acceptor molecules such as viologens are able to interact with these terminal acceptor units. Compared to MI-FDHs, these enzymes exhibit much faster CO_2 reduction rates. For example, in the presence of viologens W-FDH1 from *Syntrophobacter fumaroxidans* was shown to catalyse both formate oxidation and CO_2 reduction with activities of 700 and 900 U mg^{-1}, respectively,[118] with turnover rates of 3380 s^{-1} and 282 s^{-1} (pH 7.5).[8]

DET systems incorporating MD-FDHs have been designed, and the Hirst group have examined oxygen sensitive W-FDH[8] and Mo-FDH[7] immobilised onto electrodes. In those works, W-FDH1 incorporating at least nine iron–sulfur clusters was extracted from *Syntrophobacter fumaroxidans*, while Mo-FDH-H incorporating only a single iron–sulfur cluster was produced by overexpression from *Escherichia coli* with both being obtained under anaerobic conditions. When immobilised onto an electrode under anaerobic conditions, both enzymes were reported to reversibly catalyse CO_2 reduction at potentials close to the thermodynamic limit (Figure 2.5c) with high FE (reported values: $102.1 \pm 2.2\%$ at -0.41 V *vs.* SHE and $101.7 \pm 2.0\%$ at -0.5 V *vs.* SHE, respectively[7,8]). Cathodic current, corresponding to turnover rate, increased with cathodic potential up to a maximum of (−)0.08 mA cm^{-2} for W-FDH1, equivalent to 121 s^{-1}, although this was accompanied by a small decrease in FE. At lower overpotentials of 0.25 and 0.15 V, cathodic current densities of (−)5 and (−)80 μA cm^{-2} were achieved for immobilised W-FDH1 and Mo-FDH-H respectively. For Mo-FDH-H, the relative CO_2 reduction activity of the DET *vs.* mediated electron transfer route was much higher than for W-FDH1. This was linked to either a solution inactive form of the Mo species, or the different iron–sulfur cluster. It is possible that, in W-FDH1, the large number of these clusters may be able to act as an electron reservoir, smoothing the supply to the active site, while the single iron–sulfur cluster in Mo-FDH-H may limit the electron transfer process.[7] This raises interesting questions about the use of mediated assays to monitor maximum enzyme activity, and may suggest that in some cases limiting electron transport through enzymes can be partly overcome by immobilisation onto electrodes.

In a slightly different approach, the hydrogen-dependent CO_2 reductase enzyme was purified from the acetogen *Acetobacterium woodii*.[119] This four-subunit protein couples an iron-containing hydrogenase with a Mo-FDH through electron transfer subunits. When tested with excess MV, the FDH activity in the presence of HCO_3^- was 132 U mg^{-1}, but it was possible to use H_2 oxidation at the hydrogenase subunit as the driving force for CO_2 reduction, in which case the (reversible) formate production activity was 10 μmol min^{-1} mg^{-1}. It was also possible to use reduced ferredoxin, which could be generated by CO oxidation at a CODH from *A. woodii*, instead of H_2 to drive CO_2 reduction. However, this enzyme has not yet been incorporated into a system involving electrochemical regeneration of reducing equivalents.

2.3.2.8 NAD-dependent, Metal-dependent Formate Dehydrogenase

A subgroup of metal containing FDHs are the NAD-dependent, metal-dependent FDHs. These tend to have a similar metal containing active site, but usually require NAD to be present as a terminal electron acceptor.[106] Interestingly, the oxygen-tolerant, cytoplasmic NAD-dependent Mo-FDH from *Rhodobacter capsulatus* is capable of both CO_2 reduction and $HCOO^-$ oxidation at rates of 1.5 s^{-1} and 36.5 s^{-1} respectively.[128] Although mechanistic details have not yet been elucidated, its active site is known to contain a cysteine rather than SeCys ligand, indicating that SeCys is not an absolute requirement for catalysis. Very high current densities of up to 15 mA cm^{-2} were achieved at overpotentials of only 40 mV with respect to *E*CO20′ in aerobic, MV mediated systems when using NAD/W-FDH1 from *Methylobacterium extorquens* AM1,[129–131] though product detection for this system has yet to be published.

2.3.2.9 Metal-dependent Formate Dehydrogenase Stability

Generally the metal-dependent FDHs are found to be highly oxygen sensitive, limiting their practical application.[107,118,119,128] This has been linked to the metal-co-ordinating SeCys residue; while some Se-free W-FDHs have been shown to be air-stable,[132] the substitution of SeCys with Cys has also been shown to result in a large decrease in catalytic activity in some cases.[133]

A number of NAD-dependent, metal-dependent FDHs have shown much higher oxygen tolerance. NAD/W-FDH from *Clostridium carboxydivorans* shows stable NADH oxidation activity under aerobic conditions for several days, though formate production was not proved and catalysis was slow ($0.08\ s^{-1}$).[134] Cytoplasmic NAD/Mo-*Rc*FDH incorporates Cys instead of SeCys in the active site, and is found to be oxygen tolerant.[128]

Research into electrochemical CO_2 reduction by MD-FDHs began relatively recently, with only a few studies performed to date.[7,8] While it is apparent that these materials exhibit exceptional properties relating to energy efficiency and specificity, and electrochemical studies of such can provide insight into enzyme properties, problems relating to enzyme production, stability and O_2 sensitivity limit their potential for industrial catalytic applications.

2.3.3 Multiple Enzyme Cascades

Further reduced, high-value products such as methanol can be produced by 'enzyme cascades', incorporating several enzymes that sequentially reduce the initial substrate.[135] By far the most common CO_2-reducing enzymatic cascade in the literature is the three enzyme system comprising FDH, formaldehyde dehydrogenase (F_{ald}DH) and alcohol dehydrogenase (ADH), each of which can be obtained commercially. Such systems generate methanol (a feedstock in several industrial processes, with significant potential as an alternative fuel[105]) by sequential reduction of CO_2, formate and formaldehyde, as shown in Scheme 2.4. The complexity of such schemes means that optimisation requires a compromise between the optimal conditions for each individual enzyme.[46] All of the examples considered here require NADH to drive reactions. This is typically added in excess, however, the incorporation of electrochemical NADH regenerating systems is becoming increasingly feasible, as discussed in Section 2.2.4. The output of such systems has often been reported in terms of MeOH yield as a

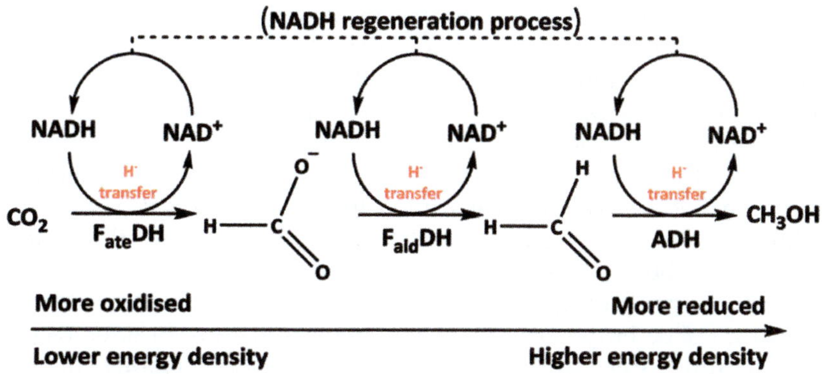

Scheme 2.4 NADH regenerated enzyme cascade incorporating FDH, formaldehyde dehydrogenase and alcohol dehydrogenase for the sequential reduction of CO_2, formate and formaldehyde to methanol.

percentage of initial NADH concentration at a timepoint several hours after the initial NADH addition.[15,136,137]

2.3.3.1 Optimisation

Immobilisation and bio-encapsulation of the three enzymes has been identified as an important technique to stabilise and optimise such systems. This causes partial confinement of the enzymes,[15] which can stabilise and prevent unfolding,[138] whilst still maintaining a degree of enzyme dynamic motion.[136] In multi-enzymatic systems, the close proximity of enzymes also increases the local intermediate substrate concentration,[15,137] though diffusion rates to bulk solution still need to be kept high.[136] For high efficiency it is also important to use enzymes ratios based upon the relative native activity of the three enzymes ($F_{ald}DH \approx FDH < ADH$).[46,136–140]

Several cascade designs have been based on silica gel materials following the initial work by Obert *et al.*, in which MeOH yield was shown to increase from 10 to 20% with free enzyme, to 40–90% upon confinement of enzymes in silica sol–gels.[15] Stability improvements were achieved by using an alginate–SiO_2 hybrid composite material, and cofactor diffusion rates close to bulk solution rates were obtained in this material.[136] Optimisation of the pore size within such silica gel matrices also increased enzyme activity.[139]

Enzyme-containing titania microparticles, synthesised by a biomimetic protamine-mediated solution process, were suggested as an alternative material that could be generated under very mild conditions, and were shown to improve MeOH yield from ~10% to ~50% over the course of an 8 h reaction. The bulk of the improvement occurred in the last four hours of the measurement, suggesting an improvement in the enzyme half-life under the reaction conditions.[138] Co-immobilisation or sequential immobilisation onto three separate flat cellulose/polypropylene membranes by pressured filtration was also shown to be a mild and simple preparation methodology, and gave similar production rates to free enzymes.[140]

In 2014, Wang *et al.* published a system of particular elegance in which ultrathin hybrid micro-capsules containing precisely confined enzymes were fabricated. These capsules incorporated a double-layered shell consisting of a catechol-modified gelatin (GelC) inner layer and silica outer layer of 70 nm with 3.6 nm pores, allowing good substrate/cofactor/product diffusion. Precise control over the relative position and loading of the different enzymes was achieved, with FDH confined to the capsule core void, $F_{ald}DH$ immobilised in the GelC layer and ADH immobilised in the silica layer. In addition, the robust particles were able to be recycled and reused after storage. Such a system improved the methanol yield from 36% to 72% with respect to the free enzymes, which only decreased to 53% after recycling the system nine times.[137] Significantly, the measured change in NADH conversion remained similar in both systems, implying that the encapsulation favours the total reduction of CO_2 through to MeOH. Interestingly, the conversion rate was initially fast in the free enzyme system before decaying, whereas a maximum rate was observed after 4 h in the capsule system. This can be attributed to differences in the movement and local concentrations of substrates and products within the two systems.[137]

It is noted that there is some difficulty in comparing systems between different works in this area as there is a significant variation in MeOH yield between the control 'free enzyme' systems, with values of between 10 and 100% reported.[15,136] A clear trend in factors leading to variations in yield is not observed, although the highest values are generally found for the systems where the temperature is held close to 37 °C and the enzyme concentrations are the lowest.

2.3.3.2 *Coupled NADH Regeneration*

In 2008, El-Zahab *et al.* immobilised FDH, F_{ald}DH and ADH onto polystyrene beads along with NADH.[44] Cofactor regeneration was achieved biochemically *via* glutamate dehydrogenase catalysis coupled with glutamic acid oxidation. Despite the cofactor being immobilised, it was still able to interact with the enzymes (probably through interparticle interactions), although immobilisation reduced the activity by half (from 0.01 U mg^{-1} to 0.005 U mg^{-1}). Cazelles *et al.* compared a number of regeneration systems in three enzyme cascades, and found that at optimised pH conditions for overall CO_2 reduction to MeOH (pH 6.5), phosphite dehydrogenase (coupled with phosphite as a substrate) was the most effective.[46] As shown for previous systems, enzyme immobilisation into silica nanocapsules significantly increased MeOH production activity by up to 55 times that of the free enzyme, with rates up to 1.43 μmol mg^{-1} h^{-1} (mg total enzyme).[46]

An example of a three enzyme cascade coupled to electrochemical NADH regeneration was developed by Addo *et al.*[37] Glassy carbon was modified with poly(neutral red) to reduce the overpotential for NAD^+ reduction (from −1.16 to −0.56 V *vs.* NHE) and coupled with an FDH, F_{ald}DH and ADH free enzyme cascade. Formate production *via* FDH was the limiting process and FDH activity was found to improve when CA was added to the system, possibly because of the increased interconversion of CO_2/bicarbonate. This resulted in an overall increase in the activity for CO_2 reduction to MeOH from 0.05 U mg^{-1} to 0.15 U mg^{-1} (mg total enzyme).

2.3.4 Summary

A number of enzymes present desirable catalytic properties for CO_2 reduction to small molecules. DET measurements give very powerful evidence for the high energetic and faradaic efficiency possible with metal-containing enzymes.[7,65] However, systems have only been shown to remain functional for of the order of hours to days.[13,21,113,134,138] In many mediated systems involving regeneration of reducing equivalents, it is apparent that the regeneration step, and not the enzyme activity, limits the system response (for example, compare rates using excess mediator in ref. 21 with those for regenerated mediator in ref. 13). Surprisingly, DET to some enzymes appears to result in higher activity than that expected from a traditional mediated assay.[7] At present, enzyme selection for this reaction is a trade-off between several desirable properties, and generally the most active enzymes are the least air-stable and most difficult to produce.[7,25]

2.4 Microbial CO_2 Reduction

2.4.1 Microbes in Nature

Microbes are an essential part of the Earth's ecosystem and are important for oxygen production,[141] decomposition of dead matter,[142] as a critical part of the food chain[143] and in symbiotic relationships with other living organisms.[144] As living organisms, several important characteristics including the ability to grow and reproduce make microbes of interest in regards to developing robust self-sustaining systems for controlled CO_2 reduction. In autotrophs such as photosynthetic microbes including cyanobacteria and algae, CO_2 is up-converted into many different carbon-based compounds, generating biomass as described by the Calvin cycle.[145,146] However, some organisms such as methanogens and acetogens utilise CO_2 as an electron acceptor in anaerobic respiration, and generate short chain products such as methane[147] or acetate[148] as by-products of the production of ATP.[149,150]

2.4.1.1 Methanogenesis

Methanogenesis is an example of a microbial CO_2 reduction pathway that generates single carbon by-products, and the catalytic cycle for this process has been well characterised.[149] The process is initiated by the attachment of CO_2 to methanofuran and, subsequently, a two-electron reduction catalysed by formylmethanofuran dehydrogenase to generate formylmethanofuran. The formyl species is then transferred to tetrahydromethanopterin (H_4MPT) to generate formyl-H_4MPT accompanied by regeneration of methanofuran, catalysed by formyl-methanofuran:tetrahydromethanopterin formyltransferase. Formyl-H_4MPT is consecutively converted to methenyl-, methylene- and methyl-H_4MPT by water loss and the action of the reduced F_{420} cofactor. The methyl group is then consecutively transferred by other enzymes to several further cofactors, before being released as methane by the action of methylreductase F_{430} upon CH_3–S–CoM (generating CoB–S–S–CoM). In many cases, the process is driven by H_2 oxidation, in which case four dihydrogen molecules are required to provide the electrons for each CO_2 to CH_4 reduction. This process is described in more detail by Lessner.[149]

2.4.1.2 External Interactions

General microbe–electrode interactions have been well reviewed elsewhere,[151,152] and a number of species, including terminal c-type cytochromes and hydrogenases, have been linked to cathodic electron transfer.[152] In methanogens, which contain no cytochrome species such as *Methanobacterium palustre*,[153] H_2 oxidation has been proposed to power CO_2 reduction through several interactions. It is thought that reduced iron–sulfur containing ferrodoxin (Fd) species used in the initial reductive coupling of CO_2 with methanofuran are produced from H_2 and Fd_{ox} at membrane-bound energy-generating hydrogenase species (*e.g.* Eha). At F_{420} hydrogenase (Frh), H_2 oxidation powers $F_{420,red}$ cofactor production, and these reduced cofactors power the methenyl-, methylene- and methyl-H_4MPT transformations. In addition, H_2 oxidation at a NiFe hydrogenase is also thought to power the reductive regeneration of HS–CoM and HS–CoB.[153] It is not as yet entirely clear how DET might replace H_2 oxidation in such species, particularly as species such as Frh are often cytoplasmic.[154] However, each of the various hydrogenases contain several metal clusters,[153,154] and Fd also contains a metal cluster. As has been shown elsewhere, such species are capable of DET if they can get close enough to an electrode,[155] providing a possible electron transfer route. Interactions *via* mediators such as viologens have also been examined.[152]

2.4.2 Microbial Bio-electrochemical Systems for CO_2 Reduction

Electrochemical methods have been combined with microbial systems for a variety of applications and reactions including fermentation control,[156] understanding fermentation processes,[157] enhancing cell growth,[158] microbial fuel cells,[159,160] and electrobiosynthesis.[10] Typically such systems contain microbes that have been grown in a solution of growth media, and use some form of mediator to shuttle electrons to the microbe.[160–162] However, mediator-less systems have also been produced that utilise some of the many microbes capable of directly transferring electrons to an electrode such as *Geobacter sulfurreducens*[163] and others.[151,164]

Various examples of microbial bio-electrochemical systems (MBEC) for CO_2 reduction exist, many of which utilise methanogens to convert CO_2 to methane (eqn (3.5), $E^{0\prime} = -0.24$ V *vs.* SHE, pH 7.0[63]), powered by H_2 reducing equivalents,[20,165] or mediated[158] or direct electron transfer.[166,167] Microbes follow a distinctive growth pattern when introduced into new conditions, and so are typically grown in a solution of the electrolyte media for a period of several hours to days (in some cases with an applied potential) before being used in such MBEC systems.[20,158,161,162]

Cathodic H_2 generation is one method for driving CO_2 conversion *via* such organisms. However, water splitting for electrochemical H_2 production is usually a low efficiency process due to the kinetically limited four electron transfer anodic oxygen evolution reaction (OER), and so is only feasible with very active catalysts. Several MBEC systems have sought to avoid using H_2 either by utilising DET at an electrode or *via* redox mediators. In a system containing *Actinobacillus succinogenes* in solution, neutral red was used to replace H_2 in mediating reduction and microbial growth. Methane was generated at a total cell potential of 1.5 V, although it was not completely clear whether H_2 was generated at the cathode.[158] Cheng *et al.* found that biofilm formation of (several organisms, including primarily) *Methanobacterium palustre* (*M.P.*) on a carbon cloth cathode (Figure 2.7a) enabled methane production at potentials of −0.5 V *vs.* SHE (Figure 2.7b), whereas no methane was produced even at potentials of −0.8 V in *M.P.* solution in the absence of a biofilm, suggesting that DET to the biofilm was occurring.[166] CH_4 production rates in various published examples of DET MBECs tend to be of the order of

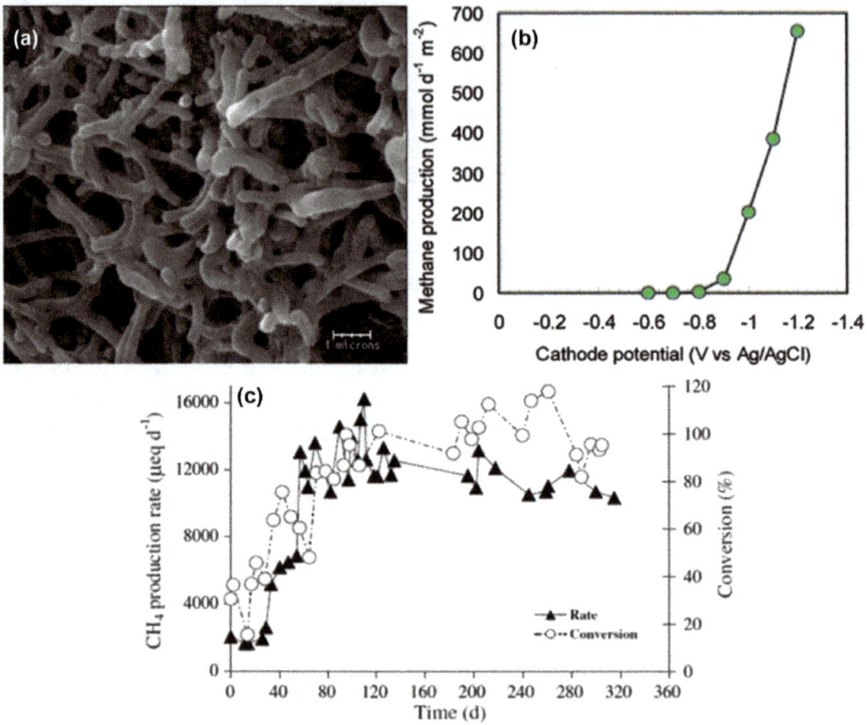

Figure 2.7 (a) SEM image of *Methanobacterium palustre* biofilm on a carbon cloth generated under cathodic conditions in acetate containing growth media, Reprinted (adapted) with permission from S. A. Cheng, D. F. Xing, D. F. Call and B. E. Logan, *Environ. Sci. Technol.*, 2009, **43**, 3953–3958.[166] Copyright 2009 American Chemical Society. (b) Methane production rate as a function of cathode potential at a biofilm cathode similar to that in (a) in a two-compartment microbial electrochemical cell. Reprinted (adapted) with permission from S. A. Cheng, D. F. Xing, D. F. Call and B. E. Logan, *Environ. Sci. Technol.*, 2009, **43**, 3953–3958.[166] Copyright 2009 American Chemical Society. (c) Methane production over time by methanogenic culture supplied with H_2. Reprinted from *Bioresource Technology*, **101**, M. Villano, F. Aulenta, C. Ciucci, T. Ferri, A. Giuliano and M. Majone, Bioelectrochemical reduction of CO_2 to CH_4 *via* direct and indirect extracellular electron transfer by a hydrogenophilic methanogenic culture, 3085–3090,[9] Copyright 2010, with permission from Elsevier.

several hundred $mmol_{CH4}$ day^{-1} m^{-2} (*e.g.* 656 $mmol_{CH4}$ day^{-1} m^{-2} at −1.0 V *vs.* SHE[166] and ∼400 $mmol_{CH4}$ day^{-1} m^{-2} at −0.9 V *vs.* SHE[9]).

Some care must be taken by researchers in claiming H_2 is absent in such systems, as applied potential affects H_2 production at electrodes. For example, on a carbon paper cathode, both DET to microbes and H_2 production increase significantly as the potential is shifted from −0.65 V to −0.9 V *vs.* SHE; the maximum relative contribution of DET has been found to occur at −0.75 V.[9] Such a potential change correlated with a change in methane production rate from 0.062 mmol day^{-1} to 0.32 mmol day^{-1}.[9]

Other CO_2 reduction products can be generated by selection of alternate microorganisms. For example, biofilms of the acetogen *Sporomusa ovata* on graphite electrodes were shown to generate acetate and 2-oxobutyrate with a FE of 85% at a potential of −0.4 V *vs.* SHE.[10] Additionally Hwang *et al.* demonstrated the production of formate by using *Methylobacterium extorquens AM1*.[167] A working electrode potential of −0.55 V *vs.* SHE was required, and at pH 7 formate production remained stable for at least 80 h. The optimised system generated 0.6 mmol formate over 80 h from an initial wet cell mass of 1.9 g, with an average rate of 3.9 μmol h^{-1} g^{-1}. Higher production rates were obtained at decreasing pH, although the system stability decreased.[167]

Successful hybrid systems, which are deliberately designed to utilise electrochemically generated H_2, have recently been reported. Torella *et al.* incorporated a cobalt phosphate based OER catalyst, and (in separate cases) a stainless steel or NiMoZn HER catalyst in a water splitting set-up.[20] Critically, these catalysts require only low overpotentials to perform these reactions under neutral conditions, making high energy efficiency possible. In this system, effective bio-electrochemically driven growth of *Ralstonia eutropha* (*i.e.* CO_2 reduction to biomass) was demonstrated at cell potentials ≥2.7 V. It was calculated that, if connected to an 18% efficient solar cell, this would equate to an overall solar-to-biomass efficiency of 3.2% of the thermodynamic maximum, similar to the efficiency of plants. Furthermore, by using an engineered strain of the bacterium, isopropyl alcohol was generated with a predicted solar-to-fuel yield of 0.7%;[20,168] rational modifications such as this show promise for further enhancing the selectivity of such systems.[169,170] A similar approach using the obligate anaerobe *Methanosarcina barkeri* achieved stable methane production at an average rate of 25 μmol h^{-1} over seven days when supplied with CO_2. Impressive FEs of up to 86% were achieved using a Pt cathode and anode in membrane separated aerobic chambers, and similar performance was observed when using an α-NiS cathode. By incorporating the photocatalysts n-TiO_2 and p-InP, bias-less light-driven methane synthesis was achieved, albeit at much lower rates of 0.6 nmol h^{-1}.[165]

2.4.3 Stability

One of the distinct advantages of microbial systems is that they typically remain active for a long time, provided they are supplied with the correct nutrients. In the work by Nevin *et al.* using *Sporomusa ovata*, healthy biofilms were observed after the cell had been running for three months.[10] In the examples discussed above, MBECs were run for periods of between 8 and 132 h. Extrapolation of the measured product generation trends suggests that the operational stability of such systems could be significantly longer than these experimental periods. Indeed, microbial systems without an electrochemical regeneration system have exhibited stable product generation for over 200 days (Figure 2.7c) after an initial growth period when substrates and reducing equivalents were supplied.[9] In electrochemically biased systems, additional heterogeneous reactions can occur, which decrease efficiency and stability, making the choice of operating potential important. For example, Torella *et al.* demonstrated that, at potentials below those required for water splitting, the generation of reactive oxygen species can be favoured, which can cause cell death.[20]

2.4.4 Summary

Electrochemical systems incorporating microbes for CO_2 reduction have been developed that are capable of generating biomass or short carbon-chain products from CO_2.[10,20,158,165–167] Such living systems require an initial growth phase, but are then able to function effectively for considerable lengths of time (months) without significant loss in activity. However, several hundred millivolts of overpotential are typically required for CO_2 reduction.[167] A variety of electron transfer mechanisms can be exploited, but the development of low overpotential OER catalysts makes H_2 generated by water splitting an attractive reducing equivalent on account of its natural interaction with the methanogenic mechanism.

2.5 Bio-inspired Photo-electrochemical CO_2 Reduction

2.5.1 Photo-electrochemical CO_2 Reduction in Nature

Natural photosynthetic systems are fundamentally important to life on earth. Complex photosynthetic processes that take place in phototrophs (*e.g.* algae) elegantly convert energy, in the form of radiation, into high-energy chemical forms utilising only very simple, low energy compounds including CO_2 and H_2O as substrates. Research into artificial photosynthesis has sought to incorporate key features of this biochemical process into synthetic designs.[11,171,172] Many photosynthetic organisms achieve light-to-biomass conversion efficiency of around 1.5% under natural conditions,[20,173] and various strategies and system architectures have been explored in attempts to artificially achieve similar or greater efficiencies.

2.5.1.1 Natural Photosynthesis

The transfer of energy from photons to reducing equivalents, and its subsequent use in powering reduction processes, is described for plants by the

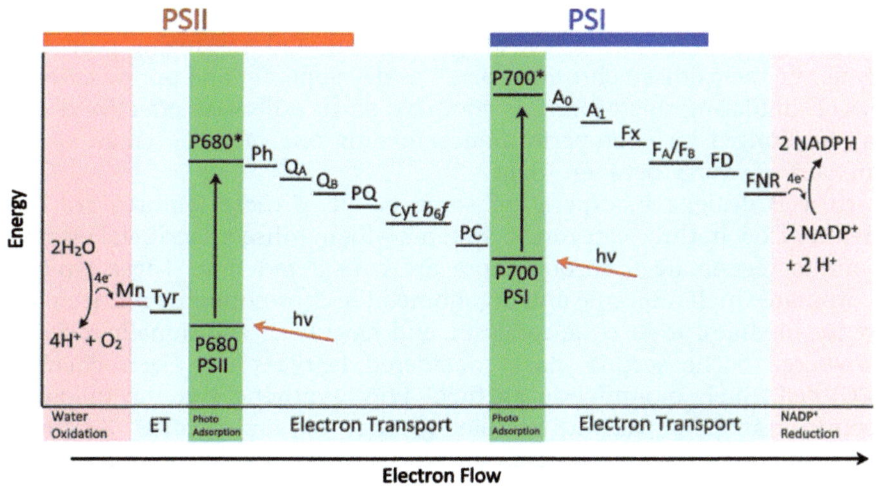

Figure 2.8 Electron transfer and energy flow in photosynthesis for the generation of reduced NADPH. Light adsorption occurs at the membrane bound PSI and PSII photocentres and powers electron flow down the potential gradient. Water acts as a terminal electron donor, $NADP^+$ as a terminal electron acceptor. A proton gradient across the membrane is generated when electron transfer occurs at PQ and enables ATP synthesis at membrane bound ATP synthase (not shown).

Z-scheme of photosynthesis shown in Figure 2.8. Photons are first absorbed at the thylakoid membrane-bound protein complexes known as photosystems I and II (PSI and PSII), generating excited states. Excited state PSII transfers an electron to the acceptor pheophytin, and this electron is quickly transported away from the PSII reaction centre through a chain of species with increasing positive redox potentials that ultimately connect to PSI. This transport chain facilitates charge separation and minimises electron/hole recombination at PSII. Oxidised PSII is regenerated by electrons generated by water oxidation catalysed at the PSII manganese centre. Excited PSI transfers an electron, firstly to chlorophyll A_0, and subsequently *via* adjacent redox species to ferredoxin-NADP oxidoreductase, which reduces NAD^+ to NADH. This reducing equivalent can then diffuse elsewhere and interact with enzymes such as FDH to reduce CO_2. Oxidised PSI is regenerated by the electron transported from PSII.[12,171] Additionally, the movement of an electron through cytochrome b_6f in the transport chain between PSII and PSI causes proton translocation across the thylakoid membrane, contributing to proton gradients that power the production of ATP.[60]

2.5.1.2 *Key Features*

The natural process described has several key features. Broadly, these are the chromophore(s), which harvest photons; the electron transport chain(s), which moves electrons away from the excited chromophore after light adsorption; the catalytic centre, which converts electronic energy into reduced species (and ultimately CO_2 reduction); and a reducing process to regenerate the oxidised chromophore. The development and optimisation of devices emulating this process is complex, as an otherwise effective design may be limited by poor performance in only one of these areas, or the transport of energy between them.

Artificial designs incorporating some or all of these features are considered below in three categories. Systems which utilise a sacrificial electron donor to regenerate a chromophore are considered first. These typically incorporate small molecule and organometallic chromophores connected by a redox mediator to enzyme catalysts, and have been additionally reviewed elsewhere.[172] The second class considered is the photo-electrochemical (PEC) cell. Such bio-mimetic artificial photosynthetic systems follow an electron transport pathway similar to that of a simplified Z-scheme, employing one or two photocatalysts. No sacrificial species are required, as these are regenerated through anodic water oxidation. Finally, systems where the light absorber is indirectly connected to the catalytic cell are considered. These typically consist of a semiconductor photovoltaic cell connected to an electrochemical cell (PV–EC).[174–176] These three classes of device are represented diagrammatically in Figures 2.9–2.11.

It is noted that, in most artificial systems, electron transport between the excited chromophore and a catalytic site differs from that of natural photosynthesis in that it occurs in a single material. Efficient charge separation is driven by the energetic difference between a molecular excited state (or conduction band in a semiconductor) and the formal potential of the catalyst. Poor charge separation will limit the device if this potential is too small,[23,177] and optimisation of this energy gap is important for maximising the activity and efficiency of these systems.[177]

2.5.2 Class 1. Sacrificial Chromophore Regeneration

2.5.2.1 Formate and Methanol Production

Molecular species such as triethanolamine (TEOA)[178,179] and the buffer 2-(*N*-morpholino)ethanesulfonic acid (MES)[100,101] have been used as sacrificial electron donors in these systems (Figure 2.9a), as these minimise charge recombination close to the chromophore by quickly regenerating the reduced ground state after electron injection. One example of a system in this class was developed by Yadav *et al.*[180] In this example, a highly substituted porphyrin chromophore coupled to reduced graphene oxide (RGO) was added to buffer solution containing a rhodium complex, NADH and FDH. CO_2 was reduced at FDH, powered by NADH. The rhodium complex regenerated NADH, and was itself regenerated at the graphene oxide, which accepted electrons from the excited chromophore after photon adsorption. Interestingly, the attachment of RGO to the chromophore more than doubled the NADH production compared to the reduced graphene oxide free

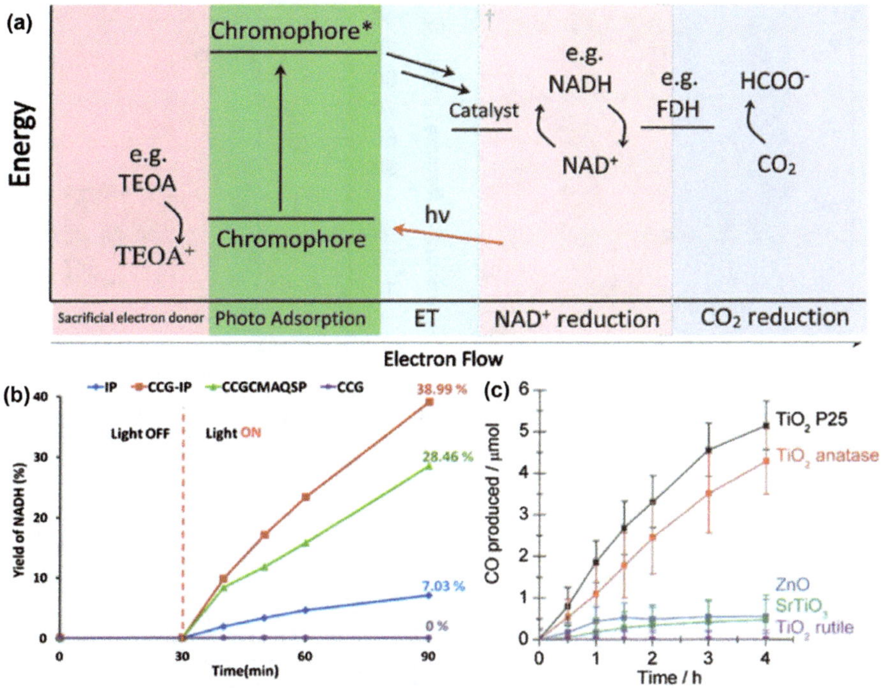

Figure 2.9 (a) Sacrificial chromophore regeneration system energy diagram. The nanoparticle $CO_2 \rightarrow CO$ systems discussed exclude a mediator step. (b) NADH regeneration activity of isatin-substituted porphyrin photo-absorber (IP, blue) anthraquinone-substituted (green, CMAQSP) and IP-substituted (red, IP) porphyrin photo-absorber grafted onto RGO and bare RGO (purple) upon illumination in a pH 7.0 phosphate buffer containing rhodium catalyst and a TEOA sacrificial electron donor. Reprinted (adapted) with permission from R. K. Yadav, G. H. Oh, N. J. Park, A. Kumar, K. J. Kong and J. O. Baeg, *J. Am. Chem. Soc.*, 2014, **136**, 16728–16731.[16] Copyright 2014 American Chemical Society. (c) CO production from several ChCODH I modified metal oxide nanoparticles under visible illumination (pH 6 MES solution under a 98% CO_2/2% CH_4 atmosphere). Limitations in the $SrTiO_3$ system may be caused by the high $SrTiO_3$ CB position resulting in a low driving force for electron injection from the dye. Reproduced from ref. 23 with permission from The Royal Society of Chemistry.

system (Figure 2.9b), and resulted in a similar increase in formate production up to 111 μmol $mg_{photocat}^{-1}$ h^{-1}, or 18.4 μmol U_{FDH}^{-1} h^{-1}.[180] Further optimisation of the chromophore improved NADH regeneration yields by 30% and this system retained 94% of its regeneration activity after 270 min of illumination (60 min) and dark (30 min) operating cycles.[16] In this optimised system using an isatin-substituted porphyrin (IP), methanol was produced at 0.07 μmol $mg_{photocat}^{-1}$ h^{-1} (0.0039 μmol $U_{FDH,FaldDH,ADH}^{-1}$ h^{-1}) in a three enzyme cascade system.

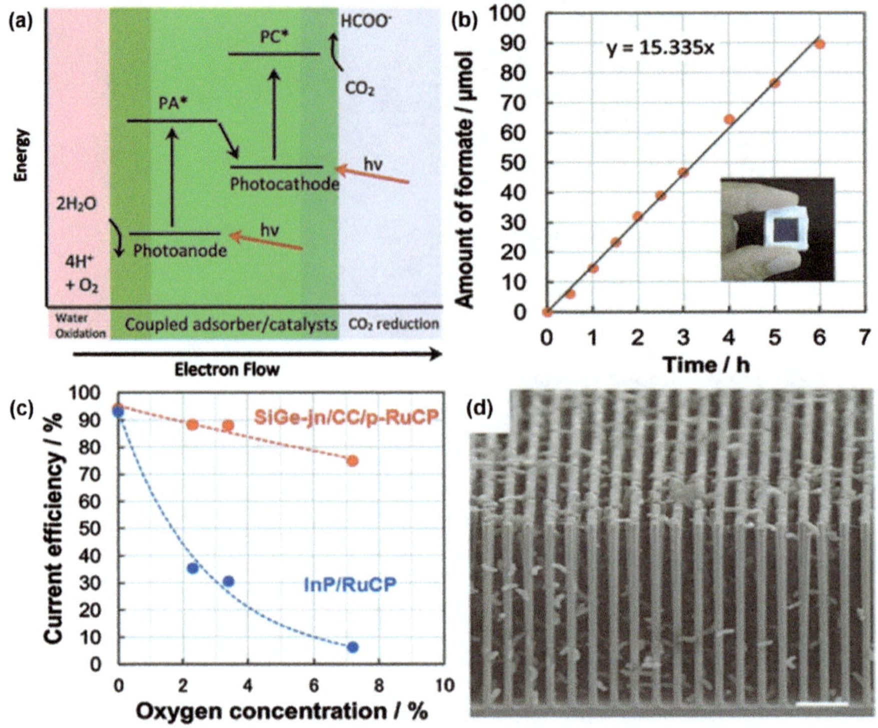

Figure 2.10 (a) Photo-electrochemical cell energy diagram. (b) Formate generation by monolithic IrO_x/SiGe-jn/CC/p-RuCpx device (shown in insert) under 1 sun illumination in a CO_2 saturated pH 6.4 buffer. Reproduced from ref. 185 with permission from The Royal Society of Chemistry. (c) Influence of oxygen concentration upon the faradic efficiency of formate generation under illumination for SiGe-jn/CC/p-RuCpx and InP/RuCpx photocathodes. Reproduced from ref. 185 with permission from The Royal Society of Chemistry. (d) SEM image of a TiO_2-coated Si nanowire–bacteria hybrid array (scale bar 5 μm). Reprinted (adapted) with permission from C. Liu, J. J. Gallagher, K. K. Sakimoto, E. M. Nichols, C. J. Chang, M. C. Y. Chang, *et al.*, *Nano. Lett.*, 2015, **15**, 3634–3639.[186] Copyright (2015) American Chemical Society.

Optimisation of such systems requires identification of the limiting step in the process. In the system incorporating RGO,[16,180] a large potential was present to drive charge separation ($E_{\text{Fermi,RGO}} - E_{\text{LUMO-chromophore}} = 1$ eV), so efficient charge transfer might be expected. However, the decay lifetime of free IP was 3.7 ns, while when bound to RGO the (bifunctional) decay had similar lifetimes of 1.1 and 9.0 ns, suggesting that the IP–RGO charge separation and IP recombination processes compete.[16] This indicates that charge separation and transport is still a possible area for improvement, although RGO did increase the system activity. Chromophore adsorption and energy level optimisation is not considered here, although significant progress in this area has been made in fields such as the dye-sensitised solar

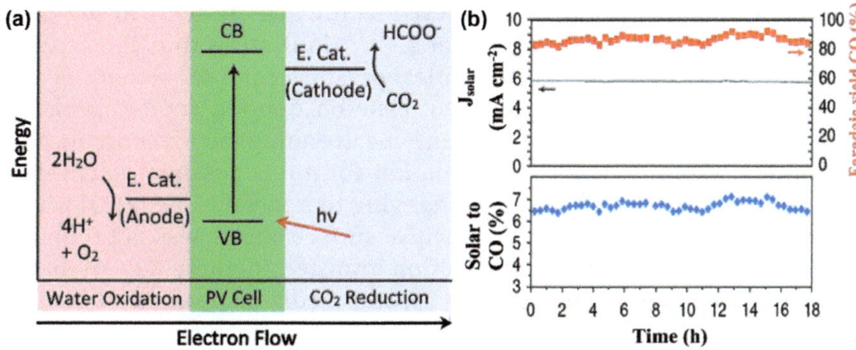

Figure 2.11 (a) Photovoltaic coupled electrocatalysis cell energy diagram. (b) Characterisation of a single-compartment electrochemical cell incorporating an IrO_2 anode and a Au cathode powered by three perovskite solar cells, showing PV current output, faradaic CO yield and total solar-to-CO conversion efficiency under AM 1.5G 1 sun illumination. The EC cell contained CO_2 saturated bicarbonate buffer at pH 7.2.
Reproduced (adapted) from M. Schreier, L. Curvat, F. Giordano, L. Steier, A. Abate, S. M. Zakeeruddin, J. Luo, M. T. Mayer and M. Gratzel, *Nat. Commun.*, 2015, **6**, 7326, http://dx.doi.org/10.1038/ncomms8326.[174]

cell.[177] In terms of the enzyme, commercial *Cb*FDH has been shown to oxidise NADH in the presence of CO_2 at a rate of 54 h^{-1} and have an activity of 1.6 mU mg^{-1} (1 U = 1 $\mu mol_{NADH}\,min^{-1}$, 12 200 times slower than its ability to reduce NAD^+).[25] For 1 $U_{(NAD+reduction)}$ FDH, a limiting rate of 5 $nmol_{CO_2}\,h^{-1}$ is expected, which is clearly far surpassed in the systems examined above. The reason for this is not clear, and to determine whether the enzyme activity is limiting in these RGO systems, a control activity test using excess NAD^+ is required under the same conditions.

2.5.2.2 Carbon Monoxide Production

A series of photo-enzymatic CO_2 to CO reduction systems have been developed by workers from the Armstrong group, which directly couple *Ch*CODH I to dye-sensitised metal oxide nanoparticles[23,100] or CdS nanoparticles.[101] A comparison of the activity of two phases of TiO_2 (rutile and anatase) with differing conduction band positions (Figure 2.9c) demonstrated that the semiconductor nanoparticle did not just act as a support material, but was electronically involved in the electron transport pathway between chromophore and catalyst. Turnovers of 0.15 s^{-1} were achieved on TiO_2 P25 particles (80% anatase) sensitised by a Ru complex in the presence of the buffer and sacrificial mediator MES under visible illumination, and by replacing TiO_2 with non-sensitised CdS quantum dots (QDs) a turnover of 1.23 s^{-1} per CODH was achieved.[101]

System activity was below that achieved with CODH systems in the presence of an excess NADH cofactor (45 s^{-1}),[76] indicating that the enzyme turnover rate was not limiting. One critical parameter in DET systems is that of enzyme orientation, as the terminal electron donor/acceptor species is usually only found on one side of the enzyme, meaning incorrect orientation can inhibit electron transfer.[181] Orientation control is possible *via* covalent or electrostatic methods.[24] However changing to a metal oxide (ZnO) with a higher isoelectric point to increase positive surface charge was not found to improve enzyme orientation for electron transfer in these CODH based systems (Figure 2.9c).[23] Addition of a second mediator resulted in a slight improvement in activity in some cases, indicating that chromophore regeneration may also partially limit the turnover.[23]

2.5.3 Class 2. Bio-inspired: Photo-electrochemical Cells

By utilising water as the terminal electron donor, the problem of using sacrificial electron donor species is avoided, and the mechanism is brought closer to that of photosynthesis. Water oxidation to form O_2 and H^+ (the oxygen evolution reaction, OER) is a complex, multi-electron reaction.[182,183] Many examples of separate photocathodes and photo-anodes for CO_2 reduction and OER exist, however coupling these is not a simple task, and several key challenges have to be overcome. Examples that have successfully linked these processes have taken inspiration from the Z-scheme of photosynthesis, and have often been described as 'artificial photosynthesis'.[171] These mimic the key points of the photosynthetic system in that they feature two photocatalysts, each consisting of a light absorbing material directly coupled to a catalytic site, attached in such a way that there is fast charge transfer between the two. CO_2 reduction occurs at the photocathode and water oxidation occurs at the photo-anode.[184]

2.5.3.1 Challenges

The coupled reduction of CO_2 to formic acid (eqn (2.2)) and oxidation of water to O_2 (eqn (2.6))[28] requires a minimum total potential of 1.42 V. In practice, significantly more than this is needed due to overpotential losses, resistance losses and the energy required to separate charges.[187] In the case of semiconductor photo-absorbers, which are used in many of these systems, the band gaps must not only be large enough to supply this energy, but the band edges should be in a suitable position relative to the donor/acceptor species to facilitate fast charge transfer and minimise electron–hole recombination. Additionally the materials must absorb at practical wavelengths, and in systems with multiple absorbers losses due to overlapping adsorption spectra should be minimised. Compromises between all these requirements must often be made. Such problems are not exclusive to this type of system, and have been the subject of extensive research in the design of photovoltaics for decades.[177,188,189]

Many CO_2 photocatalysts have been developed for use in non-aqueous conditions where the solubility of CO_2 is high.[190–192] In aqueous conditions the potential for H^+ reduction (eqn (2.7)[28]) is more positive than that for two electron CO_2 reduction to formate,[187] meaning the design of a photocathode with a high specificity for CO_2 reduction and high overpotential for H^+ reduction is a research priority.[174] Indeed, many artificial photosynthesis systems have been designed that generate H_2 from H_2O reduction,[12,193] but fewer examples currently exist for CO_2 reduction.

$$H_2O \rightarrow \frac{1}{2}O_2 + 2H^+ + 2e^- \qquad E^{0\prime} = +0.82\ \text{V}\ (\textit{vs.}\ \text{NHE, pH 7, 25 °C}) \tag{2.6}$$

$$2H^+ + 2e^- \rightarrow H_2 \qquad E^{0\prime} = -0.42\ \text{V}\ (\textit{vs.}\ \text{NHE, pH 7, 25 °C}) \tag{2.7}$$

2.5.3.2 Formate Production

One of the most effective CO_2 reducing PEC systems has been developed by Arai and Sato. In 2010 they demonstrated that under aqueous conditions a polymeric InP photo-absorber coupled to a ruthenium complex (p-InP|Ru(cpx)) reduced CO_2 at a decreased bias when under illumination.[194] Alternatively, this could be coupled to the (sustainable) absorber material, copper zinc tin sulphide.[195] Importantly, this photocathode showed 80% selectivity for CO_2 reduction to formate in aqueous conditions,[196] with the best example generating 4.7 μmol cm^{-2} formate when biased at −0.2 V *vs.* SHE in CO_2-saturated solution.[197] The optimised photocatalyst was incorporated into a fully light-powered, unbiased Z-scheme-like system with a platinum-loaded TiO_2 photo-anode. This membrane-separated, two-compartment device exhibited stable performance under irradiation over 24 h, with a CO_2-to-formate conversion efficiency of 70% at an average rate of 0.2 μmol cm^{-2} h^{-1}. However, the total solar-to-chemical energy conversion efficiency was only 0.03%.[197] Using a reduced $SrTiO_3$ photo-anode capable of selective H_2O oxidation in the presence of formate enabled a one-compartment cell to be fabricated.[184] This set-up yielded an improved solar-to-chemical energy conversion efficiency of 0.08%, which was further increased to 0.14% in a two-compartment set-up.[184] Isotope analysis proved that the formate was being generated from the CO_2 supplied, and was not a contaminant.[184] Incorporation of FDH into a PEC has been achieved using highly stable electrodes consisting of electrodeposited compact films containing MI-FDH and NADH within a polydopamine matrix, and a 0.04% solar-to-formate efficiency has been achieved when coupling such biocathodes with a $BiVO_4$/CoP photo-anode.[55]

Recently, a step-change improvement was realised, and reproducible solar-to-chemical energy conversion efficiencies for formate of up to 4.6% (average 4.3%, $n = 3$) were reported.[185] This was obtained using a monolithic system incorporating colloidal IrO_x, triple junction amorphous-SiGe and porous carbon cloth coated with polymeric Ru(cpx) (Figure 2.10b). The photo-anode was formed of IrO_x on a-SiGe while the photocathode was p-Ru(cpx)|(pCC)|a-SiGe. Interestingly, this no longer has two light absorber species as in the Z-scheme, but is powered by the single bandgap of a-SiGe, which has an open circuit voltage of 2.1 V (illumination: 1 sun, AM1.5) avoiding some adsorption losses. Importantly, the photocathode showed improved selectivity for CO_2 reduction in the presence of O_2 compared to the previous systems (Figure 2.10c). Some loss of efficiency was still observed as O_2 concentration increased, and hence this is an area where further improvements can be made. This system continuously generated formate at 7 μmol h^{-1} cm^{-2} over a six hour period, with no apparent loss of activity.

2.5.3.3 *Other Reduction Products*

PECs have also been used to generate other CO_2 reduction products. Park *et al.* utilised sodium trititanate ($Na_xH_2 - xTi_3O_7$) nanotubes coated with elemental Cu deposits and CdS QDs to generate C_xH_y products.[198] Water oxidation was thought to occur at the CdS QDs, while CO_2 reduction occurred at Cu(0). After irradiation, the $CH_3^{\bullet}$ intermediate species was detected *via* EPR, and products were found to include a number of different C_xH_y species (and no H_2), which originated from CO_2 reduction (based on isotopic analysis). In other work, Liu *et al.* incorporated bacteria with Si nanowire arrays to make photocathodes that reduced CO_2 under illumination when connected to TiO_2 photo-anodes in a two compartment cell.[186] The Si nanowires were coated with a 30 nm TiO_2 layer and supported networks of the anaerobic acetogen *Sporomusa ovata* (Figure 2.10d) after incubation of the electrodes in nutrient-containing brackish water. Illumination of the cell under anaerobic conditions resulted in stable operation over 120 h, and acetate was produced from CO_2 reduction with a selectivity of 86% and energy conversion efficiency of 0.38%. Interestingly the cell remained functional under aerobic conditions if Pt was incorporated in the nanowires, which was attributed to the creation of an anaerobic environment within the nanowires.

2.5.4 Class 3. Bio-inspired: Photovoltaic Coupled to Electrochemical Cell (PV-EC)

The third class of system couples a photovoltaic cell with a separate electrochemical cell incorporating electrocatalysts (PV-EC), and is also often discussed with reference to artificial photosynthesis.[175,176] This architecture relaxes some of the design constraints present when light adsorption and electrocatalysis are more intimately coupled, and allows the use of commercially available PVs.

2.5.4.1 *Formate Production*

An example of the incorporation of commercial PV is a formate generating PV-EC device designed by White *et al.*[175] In this case, a 200 W commercial polycrystalline Si cell was connected to three 109 cm^2 series-connected two-compartment EC cells containing IrO_2 anodes and In-based cathodes, separated by a Nafion membrane.[175] When using commercial components, care must be taken to impedance match the PV output and EC input to ensure high efficiency, and in this device a DC–DC converter was used to step down the PV voltage to 12 V. Under this configuration, the set-up ran at high currents of 9 A over 90 min in direct sunlight, achieving a reasonable FE of 67% for CO_2-to-formate, and a solar-to-formate conversion efficiency of 1.8%. A small decrease in the device voltage over this time indicated the build-up of internal resistance related to the formation of a pH gradient across the membrane between the two cell compartments. However this device is a good example of what can be achieved when scaling up the active catalyst area to several hundred cm^2.

2.5.4.2 Carbon Monoxide Production

On a smaller scale, higher FE and solar-to-product efficiencies have been achieved for CO generation. Jeon *et al.* attached an air-exposed CIGS photo-absorber *via* wires (ohmic contact) to a two compartment cell containing Co_3O_4 water oxidation catalyst deposited onto a Ti/Pt anode, and a nano-structured gold cathode, separated by a cation exchange membrane.[176] Under illumination, CO_2 reduction to CO occurred with an FE of 91.2% under the operating conditions of 7.12 mA at 2.34 V, and a solar-to-CO conversion efficiency of 4.2% was achieved, although this dropped to 2.7% after five hours.

The most efficient PV-EC device for CO_2 conversion to CO to date was developed by Schreier *et al.*, and achieved 6.5% solar-to-CO efficiency, as well as 80–90% FE for CO.[174] A series connection of three perovskite cells having a V_{OC} of 3.1 V and J_{SC} of 6.15 mA cm^{-2} provided the power source, which was connected to an IrO_2 anode and porous Au cathode within a single compartment cell. The device demonstrated stable performance over 18 h (Figure 2.11b).

2.5.5 Summary

In solar-to-hydrogen conversion systems, an efficiency of 10% is considered the minimum required for commercial implementation, along with millimole production of product per hour per gram of catalyst.[199,200] On the basis of a stable system activity of 150 μmol h^{-1} cm^{-2}, as achieved in the highest performing PV-EC systems discussed here, a formate production rate of 10 mmol h^{-1} requires a catalyst area of 67 cm^2, or 15 m^2 for 1 kg h^{-1}. Most of the CO_2 photo-reduction systems discussed above are currently a long way off such performance targets, although this highest performing system is closing in on commercially practical output.

In enzyme-containing systems incorporating sacrificial hole acceptor species, enzyme-limited rates have not yet been reached using CODH, while further studies are required to determine this for FDHs. The efficiency of charge transfer between the excited chromophore and the catalytic centre is one area where further optimisations could be made in some systems, and by changing the materials used for various components, incremental optimisations have been achieved.[16,180] If high activity, metal-containing enzymes could be harnessed in these systems close to their maximal activity (*e.g.* 50 mmol h^{-1} mg^{-1} from W-*Sf*FDH[118]), then 10 $mmol_{formate}$ h^{-1} could be generated with approximately 0.2 mg catalyst (1 $kg_{formate}$ h^{-1} with less than 500 mg catalyst). However, producing even milligram quantities of W-*Sf*FDH is extremely challenging,[7] and O_2 sensitivity remains a problem.[8]

Bio-inspired synthetic systems incorporating materials such as transition metal photocatalysts and metal oxides generally exhibit greater stability than enzymatic systems. The PV-EC system in particular was also significantly more active, although product selectivity in these synthetic systems tends to be below the 100% possible with enzymes.[8,174] Compared to enzymes, these artificial materials are simpler to scale up and present a more feasible route to large scale CO_2 photo-reduction to a variety of products.

2.6 Outlook and Future Perspective

BEC studies utilising DET have clearly demonstrated the energetic efficiency of various CO_2 reducing enzymes, and continue to provide insight into enzymatic behaviour and inform catalyst design. However significant challenges remain in finding aerobically stable, high activity enzymes that can be simply obtained and purified at low cost. These problems are currently prohibitive to utilising enzyme-based systems in practical large-scale CO_2-reducing applications. Optimisation of the electron transport chain between electrode and enzyme is a viable strategy for improving activity in DET devices, whilst the development of energetically efficient, high-activity NADH regeneration electrocatalysts is key in improving system activity with NADH-dependent enzymes. Improvements to O_2 tolerance and degradation stability may be achieved in the future through the use of bio-compatible support matrices.[55,138]

Microbial BEC systems are significantly more robust than those based purely on enzymes as a consequence of the protecting microbial environment, within which enzymes are generated and replaced or repaired as necessary. As shown in Table 2.1, microbial CO_2 reduction systems can operate stably for months, and when coupled into BEC systems regularly demonstrate stable activity for days.[20,165,167] On a mass basis microbes are significantly less active than CO_2 reducing enzymes, but are easier to produce on a larger scale.[9,60] Genetic engineering has been demonstrated as a successful technique for controlling the reduction product output, and rational genetic modification to select for desired reaction pathways within the organism is an increasingly practical approach to achieving designed activity improvements in bacteria.[20,169,170]

Comparison between different photo-electrochemical approaches to CO_2 reduction indicates that bio-inspired artificial photosynthetic systems based

Table 2.1 Comparative summary of CO_2 reducing systems examined in the text. Activity and efficiencies are maximum values based on the literature covered in this chapter. Variation between system composition (*e.g.* enzymes *vs.* metallic electrode) and catalyst availability complicates comparisons, and typical lab scale availability is included to allow an approximate comparison of practically achievable performance.

System class	Activity[a]	Typical lab scale catalyst availability	Typical products	Stability	Efficiency
Enzymatic	$121\ s^{-1}$[8] Several $mmol\ h^{-1}\ g^{-1}$[b]	μg–mg	CO Formate MeOH	Hours–days	FE: 100%[8]
Microbial	$3\ \mu mol\ h^{-1}\ cm^{-2}$[166] $4\ \mu mol\ h^{-1}\ g^{-1}$[167]	Biofilm: ∼10 cm^2 Cell mass: g	Methane Acetate Formate IPA Biomass	Months	Biomass:[c] 17.8%[20] Product:[c] 3.9%[20]
Photo-enzymatic	$1.23\ s^{-1}$[101] $20\ \mu mol\ h^{-1}\ U_{enzyme}^{-1}$	mg	CO Formate MeOH	Hours	FE: 100%[55]
PEC	$7\ \mu mol\ h^{-1}\ cm^{-2}$[185]	<5 cm^2	Formate C_xH_y	Hours +[d]	S-P η[e]: 4.6%
PV-EC	$150\ \mu mol\ h^{-1}\ cm^{-2}$[f]	∼5 cm^2[g]	CO Formate	Hours +[h]	S-P η[e]: 6.5% FE: 90%

[a]Values displayed give a comparison of what has been achieved in practical lab scale systems.
[b]For electrochemically regenerated systems—significantly higher activities can be achieved based on excess mediator based initial rate assays.
[c]Percentage of maximum thermodynamic yield.[20]
[d]Optimal system showed no decrease in activity over six hours.[185]
[e]Solar to chemical product thermodynamic efficiency.
[f]Calculated from raw data; FE: 90%, *J*: 8 $mA\ cm^{-2}$ over 18 h.[174]
[g]Size used in experiment. Straightforward scale up possible.
[h]Optimal system showed no decrease in activity over 18 h.[174]

on inorganic materials hold the greatest promise in terms of stability and scalability, and high selectivity has also been achieved.[174,176] The best performing PV-EC systems use catalysts made from readily available metals and metal oxides, and are capable of stable CO generation over 18 h with activity over 100 μmol h^{-1} cm^{-2}, around 90% FE and a solar-to-product efficiency of 6.5%.[174] Although not currently at the levels required for commercialisation, improvement trends suggest that 10% efficiency with mmol h^{-1} cm^{-2} activity may be achievable in the not too distant future. Enzyme coupled systems currently achieve the highest selectivity, and in some cases have not yet reached enzyme-limited activities, suggesting that research focussed upon optimising electron transfer between light absorber and enzyme should result in improved activity.

In summary, electrochemical CO_2 reduction by BEC and bio-inspired methods is clearly achievable as shown by the variety of successful approaches in this chapter. Integrated BEC CO_2 reduction devices incorporating simple organisms or extracted enzymes are capable of generating a range of carbon-based products with high faradaic efficiency. Bio-inspired systems such as artificial photosynthesis can also be tailored to a range of products, and solar-to-product efficiencies of over 6% have been achieved.[174] However, several key challenges remain for each approach if commercially useful stability, rate and efficiency is to be achieved, and particularly in the case of enzymatic systems these may ultimately prove prohibitive to such a goal.

References

1. M. D. Archer and J. Barber, *Molecular to Global Photosynthesis*, Imperial College Press, London, 2004.
2. G. F. Swiegers, *Bioinspiration and Biomimicry in Chemistry: Reverse-Engineering Nature*, Wiley, Hoboken, NJ, 2012.
3. C. K. Mathews, K. E. Van Holde and K. G. Ahern, *Biochemistry*, Benjamin Cummings, San Francisco, CA; Harlow, 3rd edn, 2000.
4. D. C. Ducat and P. A. Silver, *Curr. Opin. Chem. Biol.*, 2012, **16**, 337–344.
5. G. Fuchs, *Annu. Rev. Microbiol.*, 2011, **65**, 631–658.
6. A. Alissandratos and C. J. Easton, *Beilstein J. Org. Chem.*, 2015, **11**, 2370–2387.
7. A. Bassegoda, C. Madden, D. W. Wakerley, E. Reisner and J. Hirst, *J. Am. Chem. Soc.*, 2014, **136**, 15473–15476.
8. T. Reda, C. M. Plugge, N. J. Abram and J. Hirst, *Proc. Natl. Acad. Sci. U. S. A.*, 2008, **105**, 10654–10658.
9. M. Villano, F. Aulenta, C. Ciucci, T. Ferri, A. Giuliano and M. Majone, *Bioresour. Technol.*, 2010, **101**, 3085–3090.
10. K. P. Nevin, T. L. Woodard, A. E. Franks, Z. M. Summers and D. R. Lovley, *mBio*, 2010, **1**, 1–4.
11. D. Kim, K. K. Sakimoto, D. C. Hong and P. D. Yang, *Angew. Chem., Int. Ed.*, 2015, **54**, 3259–3266.

12. Y. Tachibana, L. Vayssieres and J. R. Durrant, *Nat. Photonics*, 2012, **6**, 511–518.
13. W. Shin, S. Lee, J. Shin, S. Lee and Y. Kim, *J. Am. Chem. Soc.*, 2003, **125**, 14688–14689.
14. S. Srikanth, M. Maesen, X. Dominguez-Benetton, K. Vanbroekhoven and D. Pant, *Bioresour. Technol.*, 2014, **165**, 350–354.
15. R. Obert and B. C. Dave, *J. Am. Chem. Soc.*, 1999, **121**, 12192–12193.
16. R. K. Yadav, G. H. Oh, N. J. Park, A. Kumar, K. J. Kong and J. O. Baeg, *J. Am. Chem. Soc.*, 2014, **136**, 16728–16731.
17. F. Harnisch, U. Schroder and F. Scholz, *Environ. Sci. Technol.*, 2008, **42**, 1740–1746.
18. W. W. Li, G. P. Sheng, X. W. Liu and H. Q. Yu, *Bioresour. Technol.*, 2011, **102**, 244–252.
19. E. A. Hernandez-Pagan, N. M. Vargas-Barbosa, T. Wang, Y. Zhao, E. S. Smotkin and T. E. Mallouk, *Energy Environ. Sci.*, 2012, **5**, 7582–7589.
20. J. P. Torella, C. J. Gagliardi, J. S. Chen, D. K. Bediako, B. Colón, J. C. Way, P. A. Silver and D. G. Nocera, *Proc. Natl. Acad. Sci. U. S. A.*, 2015, **112**, 2337–2342.
21. S. A. Ensign, *Biochemistry*, 1995, **34**, 5372–5381.
22. V. C. C. Wang, M. Can, E. Pierce, S. W. Ragsdae and F. A. Armstrong, *J. Am. Chem. Soc.*, 2013, **135**, 2198–2206.
23. T. W. Woolerton, S. Sheard, E. Pierce, S. W. Ragsdale and F. A. Armstrong, *Energy Environ. Sci.*, 2011, **4**, 2393–2399.
24. M. Kato, T. Cardona, A. W. Rutherford and E. Reisner, *J. Am. Chem. Soc.*, 2013, **135**, 10610–10613.
25. H. Choe, J. C. Joo, D. H. Cho, M. H. Kim, S. H. Lee and K. D. Jung, *et al.*, *PLoS One*, 2014, **9**, e103111.
26. B. Kumar, M. Llorente, J. Froehlich, T. Dang, A. Sathrum and C. P. Kubiak, *Annu. Rev. Phys. Chem.*, 2012, **63**, 541–569.
27. For example, in cases where enzyme activity is determined by spectroscopic assay based upon reduced cofactor concentration changes.
28. W. M. Haynes and D. R. Lide, in *CRC Handbook of Chemistry and Physics: A Ready-reference Book of Chemical and Physical Data*, ed. W. M. Haynes and D. R. Lide, CRC, Boca Raton, FL; London, 92nd edn, 2011.
29. R. C. Ropp, *Encyclopedia of the Alkaline Earth Compounds*, Elsevier, Amsterdam, 2013.
30. T. J. Erb, *Appl. Environ. Microb.*, 2011, 77, 8466–8477.
31. M. H. Oleary and J. D. Hermes, *Anal. Biochem.*, 1987, **162**, 358–362.
32. M. Bott and R. K. Thauer, *Z. Naturforsch., C*, 1989, **44**, 392–396.
33. R. K. Thauer, B. KÄUfer and G. Fuchs, *Eur. J. Biochem.*, 1975, **55**, 111–117.
34. S. Lindskog, *Inorg. Chim. A-Bioinorg.*, 1973, **79**, 36–37.
35. M. R. Badger and G. D. Price, *Annu. Rev. Plant. Physiol.*, 1994, **45**, 369–392.
36. Y. Z. Wang, M. F. Li, Z. P. Zhao and W. F. Liu, *J. Mol. Catal. B: Enzym.*, 2015, **116**, 89–94.

37. P. K. Addo, R. L. Arechederra, A. Waheed, J. D. Shoemaker, W. S. Sly and S. D. Minteer, *Electrochem. Solid-State Lett.*, 2011, **14**, E9–E13.
38. S. W. Ragsdale and E. Pierce, *Biochim. Biophys. Acta, Proteins Proteomics*, 2008, **1784**, 1873–1898.
39. S. Kim, M. K. Kim, S. H. Lee, S. Yoon and K. D. Jung, *J. Mol. Catal. B: Enzym.*, 2014, **102**, 9–15.
40. T. Quinto, V. Kohler and T. R. Ward, *Top. Catal.*, 2014, **57**, 321–331.
41. C. Kohlmann, W. Markle and S. Lutz, *J. Mol. Catal. B: Enzym.*, 2008, **51**, 57–72.
42. V. Uppada, S. Bhaduri and S. B. Noronha, *Curr. Sci. India*, 2014, **106**, 946–957.
43. X. Wang and H. H. P. Yiu, *ACS Catal.*, 2016, **6**, 1880–1886.
44. B. El-Zahab, D. Donnelly and P. Wang, *Biotechnol. Bioeng.*, 2008, **99**, 508–514.
45. J. M. Vrtis, A. K. White, W. W. Metcalf and W. A. van der Donk, *Angew. Chem., Int. Ed.*, 2002, **41**, 3257–3259.
46. R. Cazelles, J. Drone, F. Fajula, O. Ersen, S. Moldovan and A. Galarneau, *New J. Chem.*, 2013, **37**, 3721–3730.
47. S. Shipovskov, A. Bonamore, A. Boffi and E. E. Ferapontova, *Chem. Commun.*, 2015, **51**, 16096–16098.
48. P. J. Elving, W. T. Bresnahan, J. Moiroux and Z. Samec, *Bioelectrochem. Bioenerg.*, 1982, **9**, 365–378.
49. I. Ali, M. McArthur, N. Hordy, S. Coulombe and S. Omanovic, *Int. J. Electrochem. Sci.*, 2012, 7, 7675–7683.
50. I. Ali, B. Soomro and S. Omanovic, *Electrochem. Commun.*, 2011, **13**, 562–565.
51. C. O. Schmakel, K. S. V. Santhanam and P. J. Elving, *J. Am. Chem. Soc.*, 1975, **97**, 5083–5092.
52. I. Ali, A. Gill and S. Omanovic, *Chem. Eng. J.*, 2012, **188**, 173–180.
53. A. Azem, F. Man and S. Omanovic, *J. Mol. Catal. A: Chem.*, 2004, **219**, 283–299.
54. I. Ali, T. Khan and S. Omanovic, *J. Mol. Catal. A: Chem.*, 2014, **387**, 86–91.
55. S. Y. Lee, S. Y. Lim, D. Seo, J.-Y. Lee and T. D. Chung, *Advanced Energy Materials*, 2016.
56. G. Hilt, B. Lewall, G. Montero, J. H. P. Utley and E. Steckhan, *Liebigs Ann./Recl.*, 1997, 2289–2296.
57. H. J. Lee, S. H. Lee, C. B. Park and K. Won, *Chem. Commun.*, 2011, **47**, 12538–12540.
58. P. L. Hentall, N. Flowers and T. D. H. Bugg, *Chem. Commun.*, 2001, 2098–2099.
59. S. Graslund, P. Nordlund, J. Weigelt, J. Bray, B. M. Hallberg and O. Gileadi, *et al.*, *Nat. Methods*, 2008, **5**, 135–146.
60. D. Voet and J. G. Voet, *Biochemistry*, Wiley, Hoboken, NJ, 4th edn, 2011.
61. S. Fradette, J. Gingras, N. Voyer, J. Carley, G. Kelly and O. Ceperkovic, *Process for CO2 capture using micro-particles comprising biocatalysts, US Pat.*, 8846377, 2014.

62. T. W. Woolerton, S. Sheard, Y. S. Chaudhary and F. A. Armstrong, *Energy Environ. Sci.*, 2012, **5**, 7470–7490.
63. E. E. Benson, C. P. Kubiak, A. J. Sathrum and J. M. Smieja, *Chem. Soc. Rev.*, 2009, **38**, 89–99.
64. A. Bachmeier, V. C. C. Wang, T. W. Woolerton, S. Bell, J. C. Fontecilla-Camps and M. Can, *et al.*, *J. Am. Chem. Soc.*, 2013, **135**, 15026–15032.
65. A. Parkin, J. Seravalli, K. A. Vincent, S. W. Ragsdale and F. A. Armstrong, *J. Am. Chem. Soc.*, 2007, **129**, 10328–10329.
66. F. A. Armstrong and J. Hirst, *Proc. Natl. Acad. Sci. U. S. A.*, 2011, **108**, 14049–14054.
67. A. M. Appel, J. E. Bercaw, A. B. Bocarsly, H. Dobbek, D. L. DuBois and M. Dupuis, *et al.*, *Chem. Rev.*, 2013, **113**, 6621–6658.
68. W. B. Tolman, *Activation of Small Molecules: Organometallic and Bioinorganic Perspectives*, Wiley-VCH, Weinheim, 2006.
69. C. Higman and M. V. d. Burgt, *Gasification*, Gulf Professional, Oxford, 2nd edn, 2008.
70. D. Kim, J. Resasco, Y. Yu, A. M. Asiri and P. Yang, *Nat. Commun.*, 2014, **5**, 4948.
71. R. Hille, S. Dingwall and J. Wilcoxen, *J. Biol. Inorg. Chem.*, 2015, **20**, 243–251.
72. W. K. Russell and P. A. Lindahl, *Biochemistry*, 1998, **37**, 10016–10026.
73. J. Heo, C. R. Staples and P. W. Ludden, *Biochemistry*, 2001, **40**, 7604–7611.
74. M. S. M. Jetten, A. J. M. Stams and A. J. B. Zehnder, *Eur. J. Biochem.*, 1989, **181**, 437–441.
75. V. Svetlitchnyi, C. Peschel, G. Acker and O. Meyer, *J. Bacteriol.*, 2001, **183**, 5134–5144.
76. M. Kumar, W. P. Lu and S. W. Ragsdale, *Biochemistry*, 1994, **33**, 9769–9777.
77. P. A. Lindahl, E. Munck and S. W. Ragsdale, *J. Biol. Chem.*, 1990, **265**, 3873–3879.
78. M. Can, F. A. Armstrong and S. W. Ragsdale, *Chem. Rev.*, 2014, **114**, 4149–4174.
79. M. Kumar, W. P. Lu, L. F. Liu and S. W. Ragsdale, *J. Am. Chem. Soc.*, 1993, **115**, 11646–11647.
80. E. Brignole, E. Wittenborn, S. Thompson, S. Ragsdale and C. Drennan, *FASEB J.*, 2015, **29**(1 supplement), 573.37.
81. T. I. Doukov, T. M. Iverson, J. Seravalli, S. W. Ragsdale and C. L. Drennan, *Science*, 2002, **298**, 567–572.
82. C. Darnault, A. Volbeda, E. J. Kim, P. Legrand, X. Vernede and P. A. Lindahl, *et al.*, *Nat. Struct. Biol.*, 2003, **10**, 271–279.
83. J. Fesseler, J. H. Jeoung and H. Dobbek, *Angew. Chem., Int. Ed.*, 2015, **54**, 8560–8564.
84. C. L. Drennan, J. Y. Heo, M. D. Sintchak, E. Schreiter and P. W. Ludden, *Proc. Natl. Acad. Sci. U. S. A.*, 2001, **98**, 11973–11978.
85. Y. Kung, T. I. Doukov, J. Seravalli, S. W. Ragsdale and C. L. Drennan, *Biochemistry*, 2009, **48**, 7432–7440.

86. Z. G. Hu, N. J. Spangler, M. E. Anderson, J. Q. Xia, P. W. Ludden, P. A. Lindahl and E. Munck, *J. Am. Chem. Soc.*, 1996, **118**, 830–845.
87. H. Dobbek, V. Svetlitchnyi, L. Gremer, R. Huber and O. Meyer, *Science*, 2001, **293**, 1281–1285.
88. E. J. Kim, J. Feng, M. R. Bramlett and P. A. Lindahl, *Biochemistry*, 2004, **43**, 5728–5734.
89. A. Ciaccafava, D. Tombolelli, L. Domnik, J. Fesseler, J.-H. Jeoung, H. Dobbek, M. A. Mroginski, I. Zebger and P. Hildebrandt, *Chem. Sci.*, 2016, 7, 3162–3171.
90. J. H. Jeoung and H. Dobbek, *Science*, 2007, **318**, 1461–1464.
91. Schrodinger, LLC, The PyMOL Molecular Graphics System, Version 1.8, 2015.
92. W. Gong, B. Hao, Z. Wei, D. J. Ferguson, T. Tallant, J. A. Krzycki and K. K. Chan, *Proc. Natl. Acad. Sci. U. S. A.*, 2008, **105**, 9558–9563.
93. J. L. Boer, S. B. Mulrooney and R. P. Hausinger, *Arch. Biochem. Biophys.*, 2014, **544**, 142–152.
94. E. L. Maynard and P. A. Lindahl, *J. Am. Chem. Soc.*, 1999, **121**, 9221–9222.
95. A. Majumdar, *Dalton Trans.*, 2014, **43**, 12135–12145.
96. D. M. Fraser and P. A. Lindahl, *Biochemistry*, 1999, **38**, 15706–15711.
97. M. E. Anderson and P. A. Lindahl, Biochemistry, 1996, **35**, 8371–8380.
98. J. L. Craft, P. W. Ludden and T. C. Brunold, *Biochemistry*, 2002, **41**, 1681–1688.
99. S. V. Hexter, F. Grey, T. Happe, V. Climent and F. A. Armstrong, *Proc. Natl. Acad. Sci. U. S. A.*, 2012, **109**, 11516–11521.
100. T. W. Woolerton, S. Sheard, E. Reisner, E. Pierce, S. W. Ragsdale and F. A. Armstrong, *J. Am. Chem. Soc.*, 2010, **132**, 2132–2133.
101. Y. S. Chaudhary, T. W. Woolerton, C. S. Allen, J. H. Warner, E. Pierce, S. W. Ragsdale and F. A. Armstrong, *Chem. Commun.*, 2012, **48**, 58–60.
102. MarketsandMarkets, Formic Acid Market by Types (Grades of 85%, 94%, 99%, and others) by Application (Agriculture, Leather & Textile, Rubber, Chemical & Pharmaceuticals, & others) & by Geography – Global Trends, Forecasts to 2019, 2014.
103. F. Joo, *Chemsuschem*, 2008, **1**, 805–808.
104. A. S. Agarwal, Y. M. Zhai, D. Hill and N. Sridhar, *Chemsuschem*, 2011, **4**, 1301–1310.
105. G. A. Olah, *Angew. Chem., Int. Ed.*, 2005, **44**, 2636–2639.
106. L. B. Maia, J. J. G. Moura and I. Moura, *J. Biol. Inorg. Chem.*, 2015, **20**, 287–309.
107. H. Choe, J. M. Ha, J. C. Joo, H. Kim, H. J. Yoon, S. Kim, S. H. Son, R. M. Gengan, S. T. Jeon, R. Chang, K. D. Jung, Y. H. Kim and H. H. Lee, *Acta Crystallogr., Sect. D*, 2015, **71**, 313–323.
108. V. O. Popov and V. S. Lamzin, *Biochem. J.*, 1994, **301**, 625–643.
109. V. I. Tishkov and V. O. Popov, *Biochemistry*, 2004, **69**, 1252–1267.
110. V. S. Lamzin, Z. Dauter, V. O. Popov, E. H. Harutyunyan and K. S. Wilson, *J. Mol. Biol.*, 1994, **236**, 759–785.

111. I. G. Shabalin, K. M. Polyakov, V. I. Tishkov and V. O. Popov, *Acta Nat.*, 2009, **1**, 89–93.
112. J. S. Blanchard and W. W. Cleland, *Biochemistry*, 1980, **19**, 3543–3550.
113. Y. Amao, S. Takahara and Y. Sakai, *Int. J. Hydrogen Energy*, 2014, **39**, 20771–20776.
114. Sigma-Aldrich, Product F8649-500UN, 02.12.15.
115. D. Peacock and D. Boulter, *Biochem. J.*, 1970, **120**, 763–769.
116. V. I. Tishkov and V. O. Popov, *Biomol. Eng.*, 2006, **23**, 89–110.
117. T. Hartmann, N. Schwanhold and S. Leimkuhler, *Biochim. Biophys. Acta, Proteins Proteomics*, 2015, **1854**, 1090–1100.
118. F. A. M. de Bok, P. L. Hagedoorn, P. J. Silva, W. R. Hagen, E. Schiltz, K. Fritsche and A. J. M. Stams, *Eur. J. Biochem.*, 2003, **270**, 2476–2485.
119. K. Schuchmann and V. Muller, *Science*, 2013, **342**, 1382–1385.
120. C. S. Mota, M. G. Rivas, C. D. Brondino, I. Moura, J. J. G. Moura, P. J. Gonzalez and N. M. F. S. A. Cerqueira, *J. Biol. Inorg. Chem.*, 2011, **16**, 1255–1268.
121. J. C. Boyington, V. N. Gladyshev, S. V. Khangulov, T. C. Stadtman and P. D. Sun, *Science*, 1997, **275**, 1305–1308.
122. H. Raaijmakers, S. Macieira, J. M. Dias, S. Teixeira, S. Bursakov, R. Huber, J. J. G. Moura, I. Moura and M. J. Romao, *Structure*, 2002, **10**, 1261–1272.
123. J. F. Shi, Y. J. Jiang, Z. Y. Jiang, X. Y. Wang, X. L. Wang, S. H. Zhang, P. Han and C. Yang, *Chem. Soc. Rev.*, 2015, **44**, 5981–6000.
124. N. M. F. S. A. Cerqueira, P. J. Gonzalez, P. A. Fernandes, J. J. G. Moura and M. J. Ramos, *Acc. Chem. Res.*, 2015, **48**, 2875–2884.
125. N. M. F. S. A. Cerqueira, P. A. Fernandes, P. J. Gonzalez, J. J. G. Moura and M. J. Ramos, *Inorg. Chem.*, 2013, **52**, 10766–10772.
126. M. J. Axley and D. A. Grahame, *J. Biol. Chem.*, 1991, **266**, 13731–13736.
127. M. Jormakka, S. Tornroth, B. Byrne and S. Iwata, *Science*, 2002, **295**, 1863–1868.
128. T. Hartmann and S. Leimkuhler, *FEBS J.*, 2013, **280**, 6083–6096.
129. K. Sakai, B.-C. Hsieh, A. Maruyama, Y. Kitazumi, O. Shirai and K. Kano, *Sens. Bio-Sens. Res.*, 2015, **5**, 90–96.
130. K. Sakai, Y. Kitazumi, O. Shirai and K. Kano, *Electrochem. Commun.*, 2016, **65**, 31–34.
131. M. Laukel, L. Chistoserdova, M. E. Lidstrom and J. A. Vorholt, *Eur. J. Biochem.*, 2003, **270**, 325–333.
132. M. J. Almendra, C. D. Brondino, O. Gavel, A. S. Pereira, P. Tavares, S. Bursakov, R. Duarte, J. Caldeira, J. J. Moura and I. Moura, *Biochemistry*, 1999, **38**, 16366–16372.
133. V. N. Gladyshev, S. V. Khangulov, M. J. Axley and T. C. Stadtman, *Proc. Natl. Acad. Sci. U. S. A.*, 1994, **91**, 7708–7711.
134. A. Alissandratos, H. K. Kim, H. Matthews, J. E. Hennessy, A. Philbrook and C. J. Easton, *Appl. Environ. Microbiol.*, 2013, **79**, 741–744.
135. E. Ricca, B. Brucher and J. H. Schrittwieser, *Adv. Synth. Catal.*, 2011, **353**, 2239–2262.

136. S. W. Xu, Y. Lu, J. Li, Z. Y. Jiang and H. Wu, *Ind. Eng. Chem. Res.*, 2006, **45**, 4567–4573.
137. X. L. Wang, Z. Li, J. F. Shi, H. Wu, Z. Y. Jiang, W. Y. Zhang, X. Song and Q. Ai, *ACS Catal.*, 2014, **4**, 962–972.
138. Q. Y. Sun, Y. J. Jiang, Z. Y. Jiang, L. Zhang, X. H. Sun and J. Li, *Ind. Eng. Chem. Res.*, 2009, **48**, 4210–4215.
139. H. Wu, S. Huang and Z. Jiang, *Catal. Today*, 2004, **98**, 545–552.
140. J. Q. Luo, A. S. Meyer, R. V. Mateiu and M. Pinelo, *New Biotechnol.*, 2015, **32**, 319–327.
141. G. C. Dismukes, V. V. Klimov, S. V. Baranov, Y. N. Kozlov, J. DasGupta and A. Tyryshkin, *Proc. Natl. Acad. Sci. U. S. A.*, 2001, **98**, 2170–2175.
142. R. G. Burns, J. L. DeForest, J. Marxsen, R. L. Sinsabaugh, M. E. Stromberger, M. D. Wallenstein, M. N. Weintraub and A. Zoppini, *Soil Biol. Biochem.*, 2013, **58**, 216–234.
143. E. B. Sherr and B. F. Sherr, *Trends Ecol. Evol.*, 1991, **6**, 50–54.
144. C. Huttenhower, D. Gevers, R. Knight, S. Abubucker, J. H. Badger and A. T. Chinwalla, *et al.*, *Nature*, 2012, **486**, 207–214.
145. H. Li and J. C. Liao, *Energy Environ. Sci.*, 2013, **6**, 2892–2899.
146. X. H. Zeng, M. K. Danquah, X. D. Chen and Y. H. Lu, *Renewable Sustainable Energy Rev.*, 2011, **15**, 3252–3260.
147. J. C. Lee, J. H. Kim, W. S. Chang and D. Pak, *J. Chem. Technol. Biotechnol.*, 2012, **87**, 844–847.
148. J. A. Breznak and J. M. Switzer, *Appl. Environ. Microbiol.*, 1986, **52**, 623–630.
149. D. J. Lessner, in *eLS*, John Wiley & Sons, Ltd, 2001.
150. M. Dworkin and S. Falkow, in *The Prokaryotes: A Handbook on the Biology of Bacteria*, ed. M. Dworkin and S. Falkow, *et al.*, Springer, New York, London, 3rd edn, 2006.
151. F. Kracke, I. Vassilev and J. O. Kromer, *Front. Microbiol.*, 2015, **6**, 575.
152. M. Rosenbaum, F. Aulenta, M. Villano and L. T. Angenent, *Bioresour. Technol.*, 2011, **102**, 324–333.
153. R. K. Thauer, A. K. Kaster, H. Seedorf, W. Buckel and R. Hedderich, *Nat. Rev. Microbiol.*, 2008, **6**, 579–591.
154. D. J. Mills, S. Vitt, M. Strauss, S. Shima and J. Vonck, *eLife*, 2013, **2**, e00218.
155. I. J. McPherson and K. A. Vincent, *J. Braz. Chem. Soc.*, 2014, **25**, 427–441.
156. A. Ishizaki, H. Shibai and Y. Hirose, *Agric. Biol. Chem.*, 1974, **38**, 2399–2406.
157. M. Hongo, M. Uyeda and A. Ishizaki, *Agric. Biol. Chem.*, 1972, **36**, 141–145.
158. D. H. Park, M. Laivenieks, M. V. Guettler, M. K. Jain and J. G. Zeikus, *Appl. Environ. Microbiol.*, 1999, **65**, 2912–2917.
159. K. Rabaey, G. Lissens, S. D. Siciliano and W. Verstraete, *Biotechnol. Lett.*, 2003, **25**, 1531–1535.
160. H. Liu and B. Logan, *Abstr. Pap. Am. Chem. Soc.*, 2004, **228**, U622.
161. X. X. Cao, X. Huang, N. Boon, P. Liang and M. Z. Fan, *Electrochem. Commun.*, 2008, **10**, 1392–1395.

162. E. E. Powell, M. L. Mapiour, R. W. Evitts and G. A. Hill, *Bioresour. Technol.*, 2009, **100**, 269–274.
163. D. R. Bond and D. R. Lovley, *Appl. Environ. Microbiol.*, 2003, **69**, 1548–1555.
164. H. J. Kim, H. S. Park, M. S. Hyun, I. S. Chang, M. Kim and B. H. Kim, *Enzyme Microb. Technol.*, 2002, **30**, 145–152.
165. E. M. Nichols, J. J. Gallagher, C. Liu, Y. D. Su, J. Resasco, Y. Yu, Y. Sun, P. Yang, M. C. Y. Chang and C. J. Chang, *Proc. Natl. Acad. Sci. U. S. A.*, 2015, **112**, 11461–11466.
166. S. A. Cheng, D. F. Xing, D. F. Call and B. E. Logan, *Environ. Sci. Technol.*, 2009, **43**, 3953–3958.
167. H. Hwang, Y. J. Yeon, S. Lee, H. Choe, M. G. Jang, D. H. Cho, S. Park and Y. H. Kim, *Bioresour. Technol.*, 2015, **185**, 35–39.
168. E. Grousseau, J. N. Lu, N. Gorret, S. E. Guillouet and A. J. Sinskey, *Appl. Microbiol. Biotechnol.*, 2014, **98**, 4277–4290.
169. J.-W. Chwa, W. J. Kim, S. J. Sim, Y. Um and H. M. Woo, *Plant Biotechnol. J.*, 2016, **14**, 1768–1776.
170. X. Gao, F. Gao, D. Liu, H. Zhang, X. Nie and C. Yang, *Energy Environ. Sci.*, 2016, **9**, 1400–1411.
171. S. Berardi, S. Drouet, L. Francas, C. Gimbert-Surinach, M. Guttentag, C. Richmond, T. Stoll and A. Llobet, *Chem. Soc. Rev.*, 2014, **43**, 7501–7519.
172. S. H. Lee, J. H. Kim and C. B. Park, *Chem. – Eur. J.*, 2013, **19**, 4392–4406.
173. X. G. Zhu, S. P. Long and D. R. Ort, *Annu. Rev. Plant Biol.*, 2010, **61**, 235–261.
174. M. Schreier, L. Curvat, F. Giordano, L. Steier, A. Abate, S. M. Zakeeruddin, J. Luo, M. T. Mayer and M. Gratzel, *Nat. Commun.*, 2015, **6**, 7326.
175. J. L. White, J. T. Herb, J. J. Kaczur, P. W. Majsztrik and A. B. Bocarsly, *J. Co2 Util.*, 2014, **7**, 1–5.
176. H. S. Jeon, J. H. Koh, S. J. Park, M. S. Jee, D. H. Ko, Y. J. Hwang and B. K. Min, *J. Mater. Chem. A*, 2015, **3**, 5835–5842.
177. A. Hagfeldt, G. Boschloo, L. C. Sun, L. Kloo and H. Pettersson, *Chem. Rev.*, 2010, **110**, 6595–6663.
178. R. Miyatani and Y. Amao, *J. Mol. Catal. B: Enzym.*, 2004, **27**, 121–125.
179. Y. Amao, R. Abe and S. Shiotani, *J. Photochem. Photobiol., A*, 2015, **313**, 149–153.
180. R. K. Yadav, J. O. Baeg, G. H. Oh, N. J. Park, K. J. Kong, J. Kim, D. W. Hwang and S. K. Biswas, *J. Am. Chem. Soc.*, 2012, **134**, 11455–11461.
181. M. Kato, T. Cardona, A. W. Rutherford and E. Reisner, *J. Am. Chem. Soc.*, 2012, **134**, 8332–8335.
182. X. Rong, J. Parolin and A. M. Kolpak, *ACS Catal.*, 2016, **6**, 1153–1158.
183. M. W. Kanan and D. G. Nocera, *Science*, 2008, **321**, 1072–1075.
184. T. Arai, S. Sato, T. Kajino and T. Morikawa, *Energy Environ. Sci.*, 2013, **6**, 1274–1282.
185. T. Arai, S. Sato and T. Morikawa, *Energy Environ. Sci.*, 2015, **8**, 1998–2002.

186. C. Liu, J. J. Gallagher, K. K. Sakimoto, E. M. Nichols, C. J. Chang, M. C. Y. Chang and P. Yang, *Nano Lett.*, 2015, **15**, 3634–3639.
187. J. Zhao, X. Wang, Z. C. Xu and J. S. C. Loo, *J. Mater. Chem. A*, 2014, **2**, 15228–15233.
188. L. T. Dou, J. B. You, J. Yang, C. C. Chen, Y. J. He, S. Murase, T. Moriarty, K. Emery, G. Li and Y. Yan, *Nat. Photonics*, 2012, **6**, 180–185.
189. P. D. Würfel, *Physics of Solar Cells: From Principles to New Concepts*, Wiley-VCH, Weinheim, 2005.
190. T. Ogata, S. Yanagida, B. S. Brunschwig and E. Fujita, *Energy Convers. Manage.*, 1995, **36**, 669–672.
191. G. T. Zeng, J. Qiu, B. Y. Hou, H. T. Shi, Y. J. Lin, M. Hettick, A. Javey and S. B. Cronin, *Chem. – Eur. J.*, 2015, **21**, 13502–13507.
192. L. J. Chen, Z. G. Guo, X. G. Wei, C. Gallenkamp, J. Bonin, E. Anxolabehere-Mallart, K. C. Lau, T. C. Lau and M. Robert, *J. Am. Chem. Soc.*, 2015, **137**, 10918–10921.
193. K. Maeda, K. Teramura, D. L. Lu, T. Takata, N. Saito, Y. Inoue and K. Domen, *Nature*, 2006, **440**, 295.
194. T. Arai, S. Sato, K. Uemura, T. Morikawa, T. Kajino and T. Motohiro, *Chem. Commun.*, 2010, **46**, 6944–6946.
195. T. Arai, S. Tajima, S. Sato, K. Uemura, T. Morikawa and T. Kajino, *Chem. Commun.*, 2011, **47**, 12664–12666.
196. T. Arai, S. Tajima, S. Sato, K. Uemura, T. Morikawa and T. Kajino, *Chem. Commun.*, 2011, **47**, 12664–12666.
197. S. Sato, T. Arai, T. Morikawa, K. Uemura, T. M. Suzuki, H. Tanaka and T. Kajino, *J. Am. Chem. Soc.*, 2011, **133**, 15240–15243.
198. H. Park, H. H. Ou, A. J. Colussi and M. R. Hoffmann, *J. Phys. Chem. A*, 2015, **119**, 4658–4666.
199. J. Highfield, *Molecules*, 2015, **20**, 6739–6793.
200. B. A. Pinaud, J. D. Benck, L. C. Seitz, A. J. Forman, Z. B. Chen, T. G. Deutsch, B. D. James, K. N. Baum, G. N. Baum, S. Ardo, H. Wang, E. Millere and T. F. Jaramillo, *Energy Environ. Sci.*, 2013, **6**, 1983–2002.

CHAPTER 3

Copper Catalysts for the Electrochemical Reduction of Carbon Dioxide

HYUNG MO JEONG,[a] BOON SIANG YEO[b] AND YOUNGKOOK KWON*[c,d]

[a] Department of Nano Applied Engineering, Kangwon National University, Chuncheon 24341, Republic of Korea; [b] Department of Chemistry, Faculty of Science, National University of Singapore, 3 Science Drive 3, Singapore 117543; [c] Carbon Resources Institute, Korea Research Institute of Chemical Technology, Daejeon 34114, Republic of Korea; [d] Advanced Materials and Chemical Engineering, University of Science & Technology, Daejeon 34113, Republic of Korea
*Email: ykwon@krict.re.kr

3.1 Introduction

For several decades, electrocatalysts have been studied and developed for the efficient conversion of carbon dioxide (CO_2) to value-added chemicals such as formate, CO and hydrocarbons.[1,2] Electrocatalysts for CO_2 reduction are mostly metal-based and the reduction reactions are performed in CO_2-saturated aqueous solutions. Based on the reaction products, CO_2 reduction electrocatalysts can be divided into three groups:[2–4] (i) Hg, Cd, Pb, Tl, In and Sn generate formic acid (or formate) with a high overpotential for the hydrogen evolution reaction; (ii) Au, Ag and Zn favour the formation of CO; (iii) Cu is the only metal capable of generating CO, and further reducing it to

Energy and Environment Series No. 21
Electrochemical Reduction of Carbon Dioxide: Overcoming the Limitations of Photosynthesis
Edited by Frank Marken and David Fermin

Published by the Royal Society of Chemistry, www.rsc.org

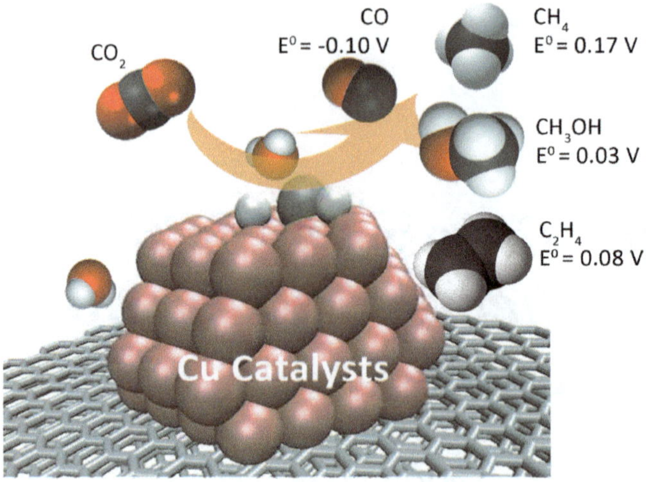

Figure 3.1 Schematic illustration of the CO_2 reduction reaction on Cu catalysts.

hydrocarbons and alcohols (Figure 3.1). This unique property of Cu can be explained by the Sabatier principle, *i.e.* key reaction intermediates such as CO binds optimally on Cu surfaces. In contrast, excessively strong binding of the intermediates will lead to catalyst poisoning (*i.e.* Pt, Rh, Ni), while weak binding of the intermediates prohibits the reduction reaction to continue to give hydrocarbon and alcohol products (*i.e.* Au, Ag).[5]

This chapter introduces state-of-the-art Cu catalysts such as Cu nanoparticles, oxide-derived Cu and Cu composites for the electrocatalytic reduction of CO_2. Experimental conditions for the electrolysis will be evaluated in a critical manner, aiming at a better understanding of results. For better comparison, experimental parameters such as cell geometry, volume, operating conditions, analysis methods, *etc.* are summarized in Table 3.1.

3.2 Reactivity of Copper

Since the landmark discovery of Hori *et al.* in 1985 that copper has the unique ability to electrochemically reduce CO_2 to hydrocarbons with good Faradaic efficiency (FE),[1] substantial efforts have been invested in elucidating the crucial design criteria from theoretical and experimental approaches.[5,6] Peterson and Nørskov proposed plausible reaction intermediates and reaction pathways on a Cu(211) surface using first-principles density functional theory.[7,8] Among the metallic electrodes, Cu sites near the top of the volcano-type plot with an optimal binding energy of CO. This facilitates the protonation of the key intermediate *CO to *CHO or *COH (the symbol "*" refers to the surface without adsorbates), which can then further react to give hydrocarbons and alcohols (Figure 3.2).

Experimentally, it was proven by Hori[9,10] and Koper[5,11] that CO_2 and CO share the same reaction pathway to hydrocarbons on Cu single crystals.

Table 3.1 Measurement conditions of electrochemical CO_2 reduction reaction on various type of Cu electrodes.

Electrode	Electrode geometry[a]	Cell volume[b] (ml)	Potential (V *vs.* RHE)	Product analysis	Electrolyte	pH[c]	Temperature (°C)	IR[d]	Ref.
Cu sheet (purity 99.999%)	20×20 mm	60	−21.36	GC/LC	CO in 0.1 M $KHCO_3$	9	19	Y	9
Cu single crystal (10 nm)	20×20 mm	60	−1.37 to −1.52	GC/LC	0.1 M $KHCO_3$	6.8	18	Y	10, 13
Cu nanoparticle	3 cm^2	25	0 to −1.1	GC	0.1 M $KHCO_3$	6.8	RT[f]	Y	16
n-Cu/C	5.2 cm^2	5	−0.95 to −1.45	GC	0.1 M $NaHCO_3$	6.8	RT[f]	NA[g]	17
Cu NP monolayer	—	40	−0.7 to −1.1	GC/NMR	0.1 M $KHCO_3$	6.8	RT[f]	NA[g]	18
Cu nanocube	—	—	−1.1	GC/LC	0.1 M $KHCO_3$	6.8	RT[f]	NA[g]	20
Annealed Cu	—	20	−0.2 to −1	GC/NMR	0.5 M $NaHCO_3$	7.2	RT[f]	NA[g]	22
Cu nanowire on Cu mesh	—	11	−0.2 to −1	GC	0.1 M $KHCO_3$	6.8	RT[f]	Y	25
Cu nanowire array	—	—	−0.7 to −1.1	GC/NMR	0.1 M $KHCO_3$	6.83 (5.9–6.8)	RT[f]	NA[g]	26
Plasma treated Cu_2O	—	120	−0.45 to −1	GC/LC	0.1 M $KHCO_3$	6.8	RT[f]	Y	27
Cu_2O film	—	10	−0.59 to −1.19	GC/NMR	0.1 M $KHCO_3$	6.8	RT[f]	Y	31
Electroreduced CuO_x cube with halide ion	1.13 cm^2	1.3	−0.6 to −1.1	GC/LC	0.1 M $KHCO_3$	6.8	RT[f]	Y	32
3D porous Cu fibre[e]	Inner: 0.8 mm, outer: 2 mm, length: 4±0.5 cm	100	−0.15 to −0.55	GC/LC	0.3 M $KHCO_3$	6.8	RT[f]	NA[g]	33
Biphasic Cu_2O–Cu	—	300	−0.6 to −1.8	GC/LC/NMR	0.1 M KCl		RT[f]	NA[g]	35

Table 3.1 *(Continued)*

Electrode	Electrode geometry[a]	Cell volume[b] (ml)	Potential (V *vs.* RHE)	Product analysis	Electrolyte	pH[c]	Temperature (°C)	IR[d]	Ref.
Electrodeposited Cu NP	—	85	−0.2 to −1.1	GC/LC	0.1 M $KHCO_3$	6.8	RT[f]	Y	40
Electroreduced Cu nanocrystal	0.865 cm^2	10	−0.65 to −1.25	GC/NMR	0.1 M $KHCO_3$	6.8	RT[f]	Y	41
Electroreduced Cu mesocrystal	0.385 cm^2	10	−0.24 to −1.74	GC/NMR	0.1 M $KHCO_3$	6.8	RT[f]	Y	42
Electroreduced Cu nanocube	0.5 cm^2	—	0.9 to −1.15	OLEMS	0.1 M $KHCO_3$	6.8	RT[f]	Y	43
Cu NP on N-graphene	—	—	−0.8 to −1.1	GC/NMR	0.5 M $KHCO_3$	7.3	RT[f]	NA[g]	44
Trace level of Cu in carbon materials	1 cm^2	0.5	−1.3	GC/LC	0.1 M $NaHCO_3$	6.8	RT[f]	Y	45
Molecular Cu catalyst	1 cm^2 (carbon fibre)	15	−0.7 to −1	NMR	0.5 M $KHCO_3$	—	RT[f]	Y	46

[a]Exposed surface to the electrolyte.
[b]Electrolyte volume except head space.
[c]Electrolyte pH after CO_2 saturation.
[d]IR compensation.
[e]Gas phase reaction.
[f]RT: room temperature.
[g]NA: not applicable.

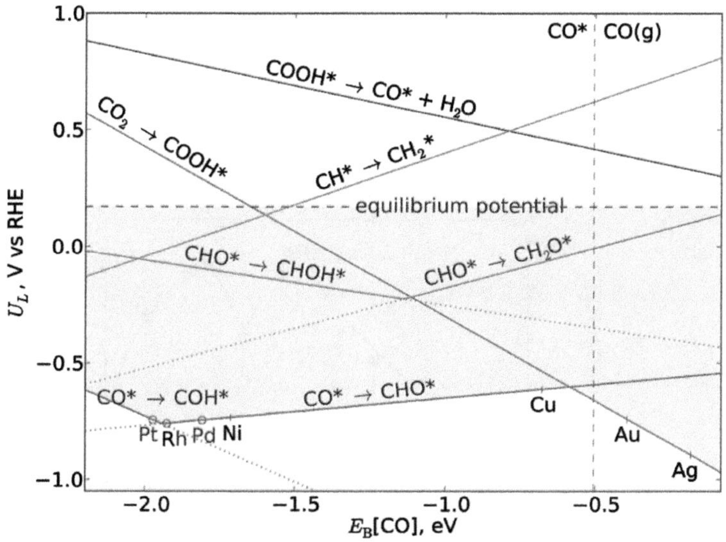

Figure 3.2 Limiting potentials (U_L) allowing for competitive reaction mechanisms. Competitive pathways are shown as lines of the same colour, and the more favourable route is shown as a solid line, while the less favourable route is shown as a dotted line. Also shown is the binding energy at which adsorbed CO is in equilibrium with gaseous CO (at a partial pressure of 1%); the CO would be predicted to desorb at binding energies weaker than this line. The E_B[CO] values are shown for each metal. The open circles correspond to surfaces that have an experimental exchange current density in the hydrogen evolution reaction greater than 10^{-4} A cm^{-2}.
Reprinted with permission from A. A. Peterson and J. K. Nørskov, *J. Phys. Chem. Lett.*, 2012, **3**, 251.[7] Copyright 2012 American Chemical Society.

The formation of methane and ethylene was found to be preferred on Cu(111) and Cu(100) surfaces, respectively. In a separate gas-phase study, Vollmer *et al.* studied the binding energies of CO using temperature programmed desorption on (i) low index surfaces, *i.e.* Cu(111), Cu(100) and Cu(110), (ii) stepped and kinked surfaces, *i.e.* Cu(211), Cu(221) and Cu(532), and (iii) polycrystalline Cu films, and concluded that CO binds more strongly on Cu step edges and kinks than on terrace sites.[12] Hori *et al.* also performed electrochemical CO_2 reduction on Cu single crystals with diverse step densities.[13] These experiments demonstrated that CO_2 catalytic activity and product selectivity depends strongly on the atomic configuration of the Cu surface (Figure 3.3).

On the basis of these theoretical predictions and experimental evidence, tuning the binding energy of the key reaction intermediates by nanostructuring the surface structure seems to be essential for promoting the formation of valuable CO_2 reduction products with high selectivity. However, there are several issues for Cu catalysts to overcome arising from thermodynamic and kinetic limitations. Firstly, the minor differences of

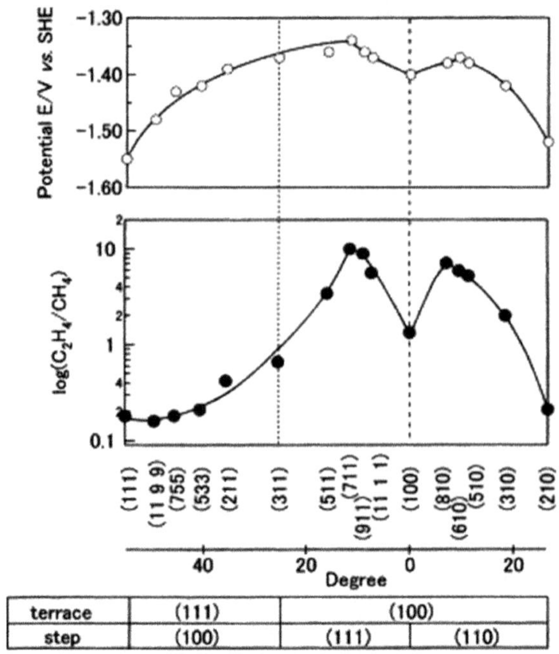

Figure 3.3 Variation of $\log(C_2H_4/CH_4)$ in terms of the FE and the electrode potential with the angle of the crystal orientation with the reference Cu(100). Reprinted from *Journal of Molecular Catalysis A: Chemical*, **199**, Y. Hori, I. Takahashi, O. Koga and N. Hoshi, Electrochemical reduction of carbon dioxide at various series of copper single crystal electrodes, 39–41,[13] Copyright 2003, with permission from Elsevier.

thermodynamic potentials for the formation of the products, as well as the myriad of possible reaction pathways available to the CO intermediates, make it difficult to selectively reduce CO_2 to a desirable product. Secondly, CO_2 reduction always competes with the hydrogen evolution reaction in aqueous solution. Jaramillo and co-workers had reported that 16 different products were generated on a Cu foil electrode, and major products with FE higher than 10% were only methane, ethylene, carbon monoxide, formate and hydrogen.[14] In addition to the poor product selectivity, the durability of Cu catalysts[15] must be improved for practical industrial applications. Therefore, it is essential to be able to design durable Cu catalysts that require low overpotentials for the conversion of CO_2 to desired products with high efficiency and selectivity. There have been recent promising achievements using nanostructured Cu catalysts, which we shall introduce in this chapter.

3.3 Types of Copper Catalysts

The electrochemical CO_2 reduction pathway is greatly affected by the structure of the Cu surface. "Nanostructuring" is thus envisioned to play an important role in controlling catalyst activity and product selectivity. Herein, we introduce recent advances in the development of nanostructured Cu catalysts such as Cu nanoparticles, oxide-derived Cu and Cu composites.

3.3.1 Cu Nanoparticles

Fabricating Cu nanoparticles for electrocatalytic CO_2 reduction has attracted great interest because of their potential scalability and enhanced catalytic activity, as well as to elucidate new nanoscale catalytic properties. The surface chemisorption and resulting catalytic activities and selectivities can be tuned by varying the size and/or shapes of the Cu nanoparticles.

Strasser and Cuenya demonstrated the "catalytic particle size effects" in CO_2 electroreduction on Cu nanoparticles.[16] This work also enabled the catalytic activity of Cu nanoparticles with various surface populations of under-coordinated atoms and crystal facets to be evaluated (Figures 3.4a

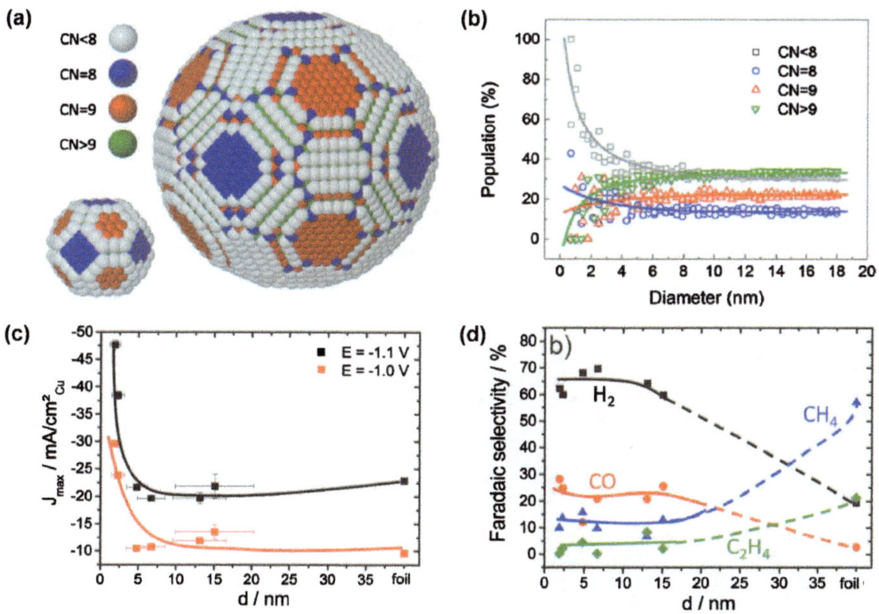

Figure 3.4 (a) Models of spherical Cu nanoparticles with diameters of 2.2 and 6.9 nm. Surface atoms are colour-coded according to their first neighbour coordination number (CN), CN < 8 (grey), CN = 8 (blue), CN = 9 (red), CN > 9 (green). (b) Population (relative ratio) of surface atoms with a specific CN as a function of particle diameter. (c) Particle size dependent current densities at −1.1 and −1.0 V *vs.* RHE. (d) Faradic selectivities of reaction products during the CO_2 electroreduction on Cu nanoparticles. Conditions: 0.1 M $KHCO_3$, $E = -1.1$ V *vs.* RHE, 25 °C. Reprinted with permission from R. Reske, H. Mistry, F. Behafarid, B. R. Cuenya and P. Strasser, *J. Am. Chem. Soc.*, 2014, **136**, 6978.[16]

and b). Nanometre-sized Cu nanoparticles were found to show an increase in overall catalytic activity with an increased selectivity in the formation of CO and H_2. It was further argued that the FE of hydrocarbon formation on 2–15 nm sized nanoparticles decreased compared to Cu foil, because of an excessive number of strong binding sites (surface atoms with coordination number less than 8), which reduced the mobility of the H and CO for further reaction to give hydrocarbons (Figures 3.4c and d). While it is noteworthy that there is no clear trend for product selectivity for particles below 15 nm in size, this study shows that size effects of Cu nanoparticles need to be considered when designing advanced Cu catalysts.

Alivisatos and co-workers[17] presented the enhanced electrochemical formation of methane using dispersed Cu nanoparticles. Isolated Cu nanoparticles supported on glassy carbon substrates (*n*-Cu/C) exhibited four times greater current density for methane compared to high purity copper foil and achieved the highest methane FE (80%) at −1.35 V *vs.* RHE. Furthermore, it was observed that thin films composed of isolated nanoparticles showed enhanced methane selectivity (Figures 3.5a and b). In contrast, a thicker film composed of highly interconnected and aggregated Cu nanoparticles exhibited a low methane FE (Figures 3.5c and d), similar to a polycrystalline Cu electrode (Figure 3.5e). This work showed that isolated nanoparticles have many active sites that lead to selective methane production with a high selectivity. However we note that the catalytic activities of Cu nanoparticles and foil including onset potential are biased against the representative Cu foil[14] and Cu nanoparticles,[18] and the potential region where methane selectivity is maximized suffers polarization losses,[19] which also affects the reaction mechanism.

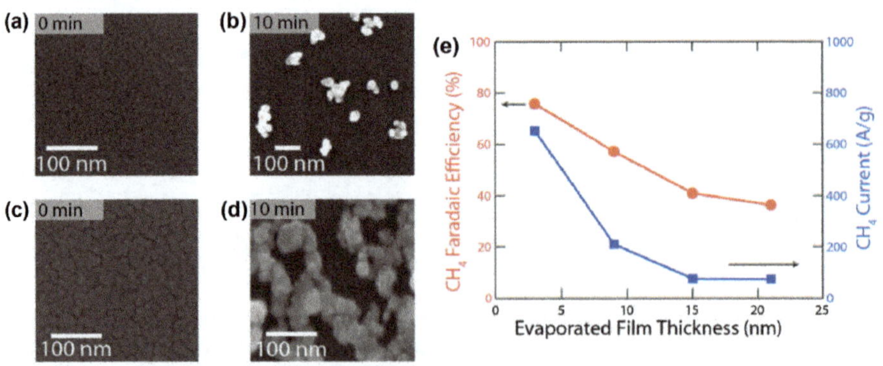

Figure 3.5 Continuum of catalytic behavior between nanoparticle-like and foil-like electrodes: 3 nm evaporated Cu film (a) prior to and (b) following polarization at −1.25 V for 10 min. 15 nm evaporated Cu film (c) prior to and (d) following polarization at −1.25 V for 10 min. (e) FE and gravimetric CH_4 current as a function of evaporated Cu film thickness. Reprinted with permission from K. Manthiram, B. J. Beberwyck and A. P. Alivisatos, *J. Am. Chem. Soc.*, 2014, **136**, 13319.[17] Copyright 2014 American Chemical Society.

The rational synthesis of spherical or cubical nanoparticles with various sizes was achieved by colloidal synthesis.[20] 44 nm size cubical structures exhibited the highest activity and a high rate of C–C coupling reactions to give ethylene (FE = 41%). This was attributed to the high density of edge sites present on the cubical nanoparticles, which is believed to be more important than Cu(100) sites. This result is well matched with single-crystal works[13,21] and theoretical prediction.[11] However, the production of the C_1 product, methane, is not negligible, indicating that further modification is still necessary to increase the yield of ethylene.

3.3.2 Oxide-derived Copper

Oxide-derived Cu (OD-Cu) has shown promising progress for CO_2 reduction by reducing the onset potentials and increasing the selectivity for the formation of formic acid, CO and ethylene. The unique properties of OD-Cu such as its morphology, presence of metastable active sites and (possibly) sub-layer oxide might have increased the catalytic activity and stability. There are several synthetic methodologies to prepare OD-Cu by oxidation and subsequent reduction processes using thermal treatment, electrodeposition and electrochemical cycling in the presence of halide ions.

3.3.2.1 Thermal Treatment

Thermal oxidation and reduction of Cu electrodes is a widely used process for inducing a nanostructured morphology. Kanan *et al.* reported that Cu electrodes with nanowire morphologies can be fabricated by thermal annealing of the disk electrode at 500 °C for 12 h (Figure 3.6a) and subsequent electrochemical reduction.[22] The resultant Cu catalyst exhibited an enhanced catalytic activity at low overpotentials (>1 mA cm^{-2} at overpotentials of <0.4 V), high selectivity of CO (>45% FE at −0.35 V *vs.* RHE) and stability (stable current and CO efficiency for 7 h) compared to a polycrystalline Cu catalyst (Figures 3.6b–d). In a separate work, the electrochemical reduction of CO was performed on OD-Cu in alkaline solution and multi-carbon oxygenates (ethanol, acetate and *n*-propanol) were generated with 57% FE at modest potential ranges of −0.25 to −0.5 V *vs.* RHE.[23] The improved activity of OD-Cu may arise from the metastable sites on the grain boundaries, which increases the binding energy of CO, thus promoting further C–C coupling.[24] However, the rational control of these active sites has not yet been achieved because of the limited control resolution. The currents drawn (corresponding to the yield of products) were also rather modest at 1 mA cm^{-2}.

Cu nanowires with controllable densities and lengths have been designed and fabricated by Wang[25] and Smith.[26] Highly dense Cu nanowires were directly grown on Cu mesh by thermal treatment and these exhibited high CO_2 reduction activities.[25] The direct growth of the nanowires on the Cu

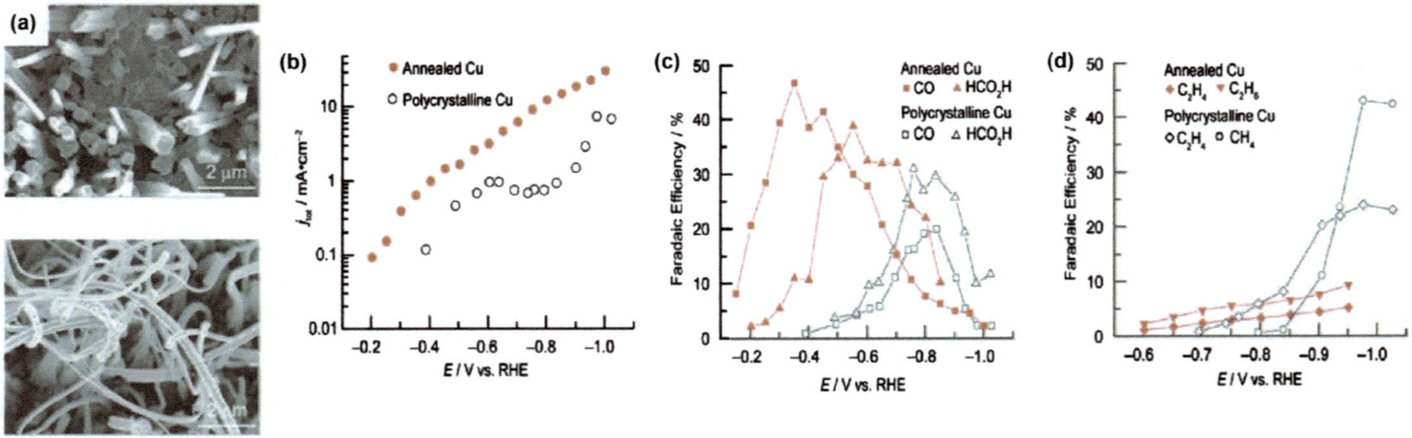

Figure 3.6 (a) Top: SEM image of a Cu electrode after annealing at 500 °C for 12 h. Bottom: analogous data for the same electrode after CO_2 reduction electrocatalysis at −0.5 V *vs.* RHE. Comparison of the electrocatalytic activities of polycrystalline Cu and Cu annealed at 500 °C for 12 h. (b) Total current density, (c) FEs for CO and HCO_2H, and (d) FEs for CH_4, C_2H_4 and C_2H_6 as a function of applied potentials.
Reprinted with permission from C. W. Li and M. W. Kanan, *J. Am. Chem. Soc.*, 2012, **134**, 7231.[22] Copyright 2012 American Chemical Society.

mesh reduced the contact resistance and increased the density of the reactive surface area, achieving a low overpotential of ~300 mV to reach 1 mA cm^{-2} geometric current densities for CO formation with ~60% FE (Figure 3.7). It was also proposed that the density and length of the nanowire may alter the product selectivity induced by local pH.[26] This work offers the methodology of thermal nanostructuring and the relationship between structural properties and CO_2 reduction reactivity. However, the selectivity of hydrocarbon products needs to be improved and the mechanism of CO_2 reduction on the nanowire needs to be updated.

Cuenya and co-workers introduced plasma treatment for the oxidation of Cu foil. The wire type surface structure was constructed and controlled by plasma treatment durations and reactive gases (Figure 3.8).[27] These electrodes displayed high surface roughness. The resultant catalyst exhibited ethylene selectivity up to 60% at −0.9 V *vs.* RHE. It was suggested that the main driving force for the exceptionally high ethylene selectivity is the presence of oxidized Cu species. However, from many thermodynamic studies of transition metal oxides,[28–30] Cu oxides are hardly stable under CO_2 reduction conditions. Moreover, recent studies using *in situ* Raman spectroscopy[31,32] proved that Cu_2O film reduced rapidly and remained in metallic Cu state during the CO_2 reduction. It is necessary to further clarify the oxidation state of Cu during the CO_2 reduction reaction. *In situ* ambient pressure X-ray photoelectron spectroscopy could be very useful towards this end.

It is noteworthy that the above three works generated similar wire-type structures through oxidation of Cu based metal, followed by a reduction process, while each of them demonstrates different results, most probably due to the different active sites. However, we are unable to exclude the fact that differences in experimental conditions, *i.e.* purity of the catalyst, electrolyte type and volume, cell geometry, *etc.*, can significantly influence the experimental results. Therefore, it is highly recommended that the electrochemical CO_2 reduction reaction be performed in a standardized fashion for better comparison.

Low solubility of CO_2 in the aqueous electrolyte (33 mM in water) limits mass transport and thereby the maximum current density that can be driven. Direct reduction of gaseous CO_2 is a solution to overcome this limitation. Mul *et al.* prepared a three-dimensional (3D) porous hollow Cu fibre by assembling copper particles (1–2 μm diameter), subsequent annealing at 600 °C and reduction under H_2 gas.[33] The performance of the hollow Cu fibre electrode was significantly enhanced. At overpotentials between 200 and 400 mV, the FE of CO reached up to 75% (Figure 3.9). The high porosity of the electrodes can offer the gaseous CO_2 better accessibility to the catalytic active sites. The large number of micro-sized pore structures also offers the effects of small size and metastable Cu active sites, which favor the formation of the *COOH intermediate. This gas-diffusion electrode inspired Cu electrode is an excellent solution to overcome the limitation of poor solubility of CO_2 in aqueous media.

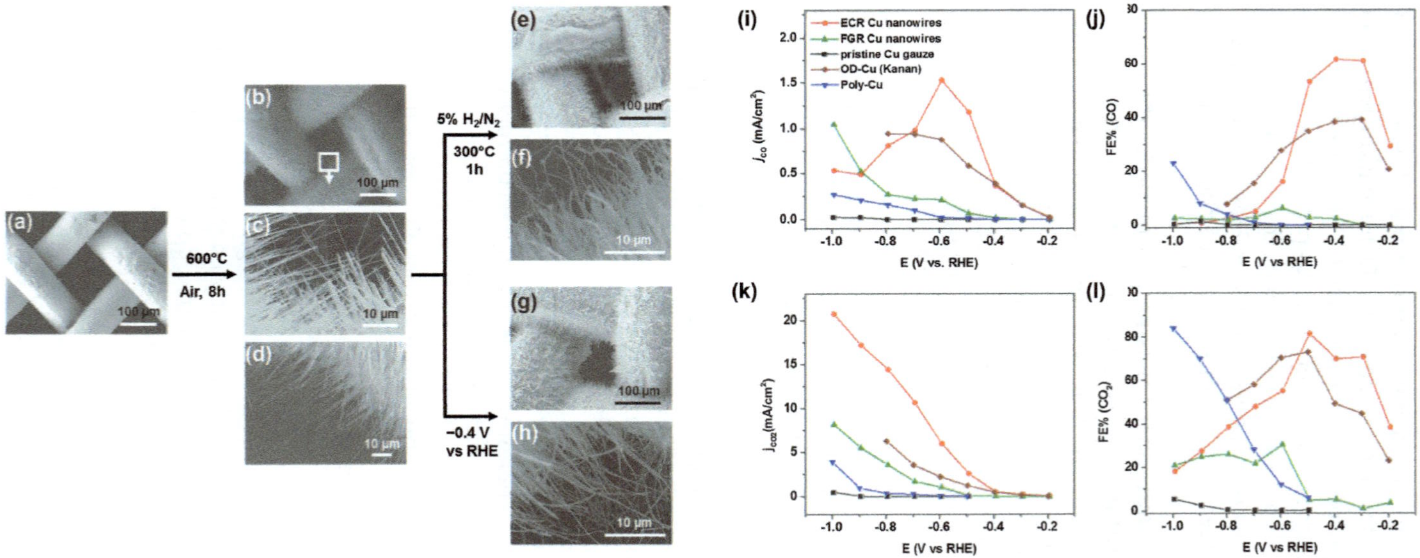

Figure 3.7 Reaction scheme showing the process for fabricating Cu nanowires: Cu mesh (a) was oxidized in air at 600 °C to grow CuO nanowires (b)–(d), which were then reduced to Cu by either annealing in the presence of hydrogen at 300 °C (e), (f) or electrochemical reduction at −0.4 V *vs.* RHE (g), (h). The catalytic performance of the Cu nanowires and pristine Cu gauze reported in this study in comparison to oxide derived Cu and polycrystalline Cu: (i) Partial current densities (j_{CO}), (j) FEs for CO production, (k) total current densities (j_{CO_2}) and (l) FEs of CO_2 reduction at various electrochemical potentials.
Reprinted with permission from D. Raciti, K. J. Livi and C. Wang, *Nano Lett.*, 2015, **15**, 6829.[25] Copyright 2015 American Chemical Society.

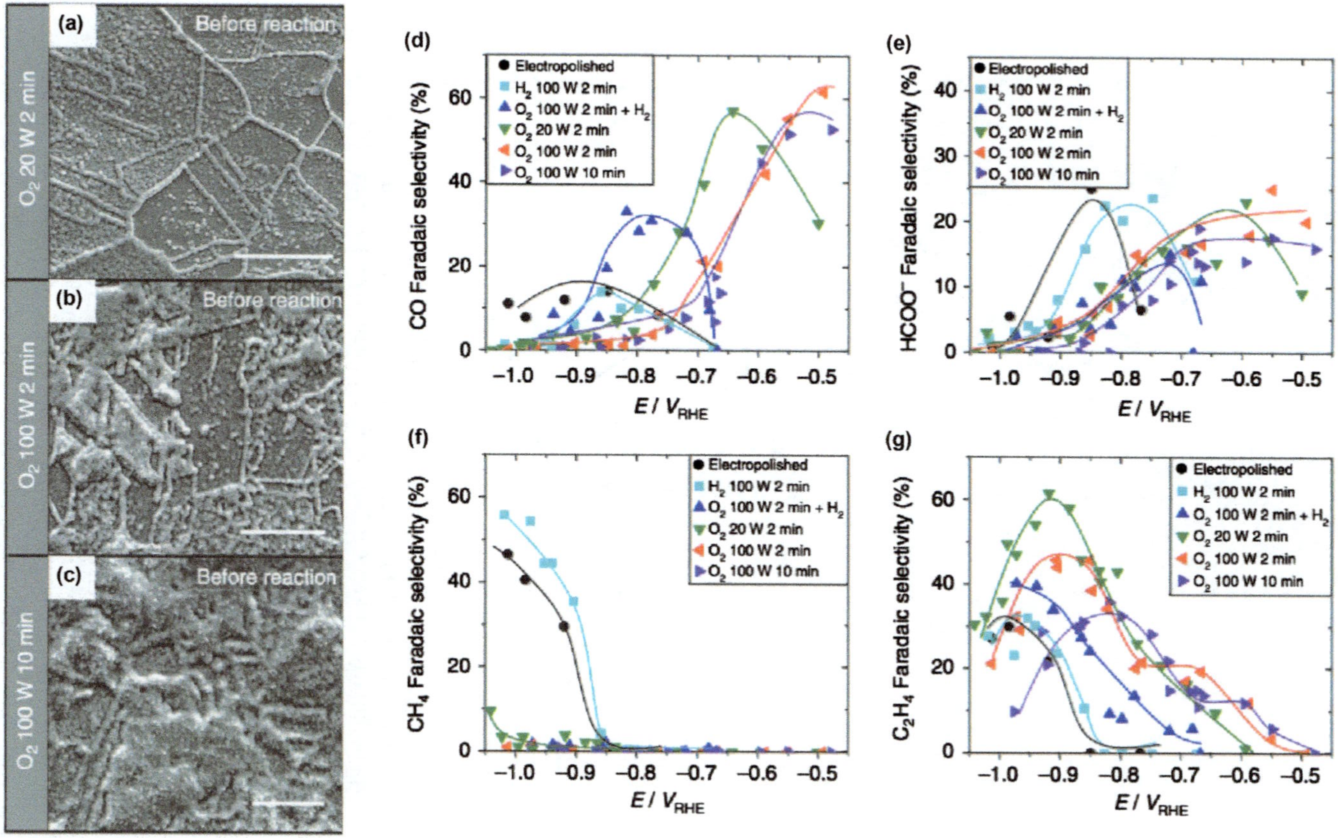

Figure 3.8 SEM images of Cu foils treated with O_2 plasma for (a) 20 W 2 min, (b) 100 W 2 min and (c) 100 W 10 min. FEs of the main CO_2RR products; data were acquired after 60 min of CO_2 electrolysis at a constant potential in CO_2 saturated 0.1 M $KHCO_3$: (d) CO, (e) formate, (f) CH_4, and (g) C_2H_4.

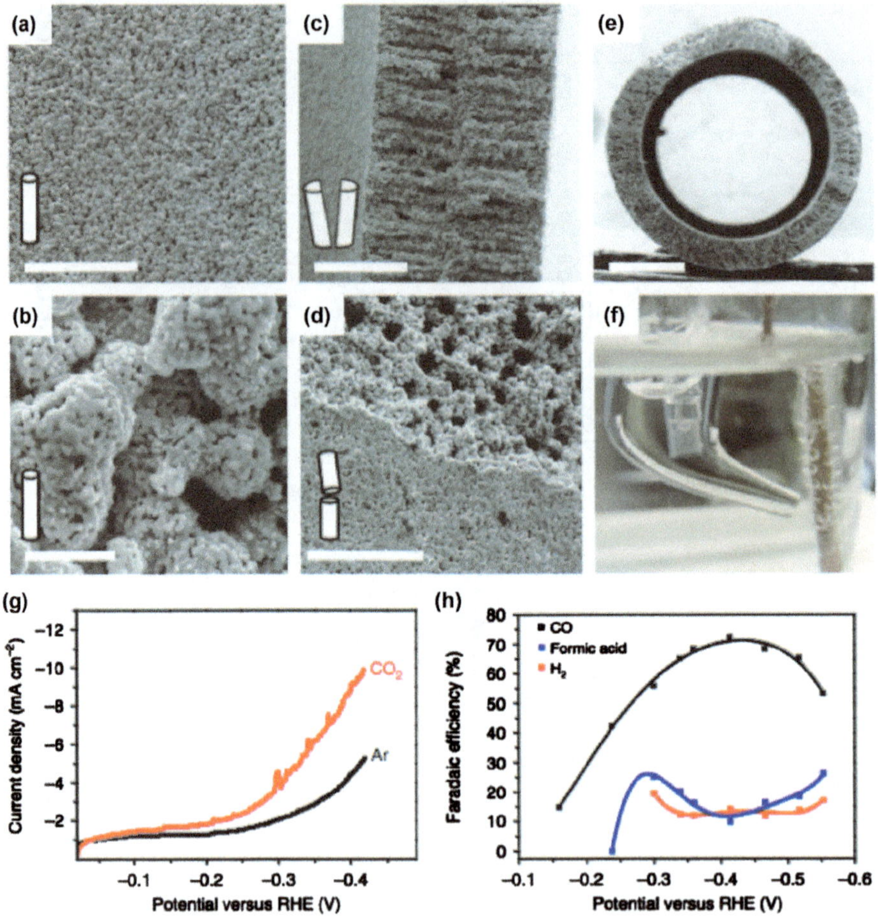

Figure 3.9 Physical characterization of Cu hollow fibres. (a) SEM images of low and (b) high magnification of the outer surface of the Cu hollow fibre. Scale bars: 50 and 2 mm, respectively. (c) Cross-sectional image of a perpendicularly broken Cu hollow fibre. Scale bar: 100 mm. (d) Outer surface and cross-section of a Cu hollow fibre in the parallel direction to the length of the hollow fibre. Scale bar: 50 mm. (e) Cross-sectional image of the Cu hollow fibre taken at low magnification. Scale bar: 500 mm. (f) Cu hollow fibre employed as an electrode at 20 ml min^{-1} gas flow. Electrocatalytic performance of the Cu hollow fibres. (g) Linear polarization curves obtained for Cu hollow fibres when CO_2 or Ar was purged in 0.3 M $KHCO_3$ electrolyte (scan rate: 50 mV s^{-1}). (h) FE of CO, formic acid and H_2 as a function of applied potential, using a CO_2 purge of 20 ml min^{-1}.

3.3.2.2 Electrodeposition

Electrodeposition, a process that uses electric current to reduce metal cations to a metal coating on an electrode, has been widely used to synthesize nanostructured Cu_2O or Cu electrodes because of the following advantages; (i) many parameters for the nanostructure can be controlled precisely; (ii) unique (three-dimensional) morphology can be synthesized directly on the template or substrate; (iii) high production efficiency can be obtained by rapid growth rate, and (iv) a relatively low-cost process can be established as expensive equipment such as vacuum chambers or high-temperature controllers are not needed.[34]

Lee *et al.* prepared Cu_2O nanoparticles *via* electrodeposition for the efficient conversion of CO_2 to hydrocarbons.[35] The as-prepared Cu_2O was reduced in KCl solution so that Cl ion was incorporated into the Cu catalyst. The resultant Cu nanostructures were found to highly enhance the formation of multi-carbon products, which includes C_3 and C_4 products. It was demonstrated that the FE of *n*-propanol (C_3H_7OH) reached up to 8.7% at an overpotential of −1.6 V *vs.* RHE. The residence time of the CO intermediate on the surface was reported to be prolonged by the presence of Cl adatoms.[36] Similar to Cuenya,[27] Lee proposed that Cu_2O remains during CO_2 reduction. This was supported by *in situ* X-ray absorption spectroscopy (XAS) data obtained from their previous work in which the Cu_2O–Cu interface was proposed to be the active site for enhanced C–C coupling. In this earlier work, Cu^+ species were reported to persist during CO_2 reduction, as evidenced by depth profiling using *ex situ* X-ray photoelectron spectroscopy (XPS).[37] We note that the electrochemical CO_2 reduction occurs at the top atomic layers of the catalyst so that it is necessary to employ surface sensitive techniques to prove the presence of Cu^+ species or subsurface oxide. For instance, the depth of the incident beam is dependent on the kinetic energy of photoelectrons of each analysis technique such as XPS (~10 nm)[38] and XAS (bulk, 50–200 nm at surface-specific modes).[39] In addition, the oxidation rate of nanostructured copper is fast enough to change the oxidation state of Cu during sample preparation/transfer for *ex situ* analysis. Therefore, for the oxidation state of an active site to be further asserted, it is desirable to make a more concrete conclusion based on surface sensitive *in situ* analysis techniques.

Baltrusaitis and Mul prepared highly roughened Cu_2O films by electrodeposition as shown in Figure 3.10.[40] Variable thickness of Cu_2O films with (100), (110) and (111) orientations were electrodeposited on Cu plates and resulted in the selective formation of ethylene with high ethylene-to-methane ratios (~8 to 12). Importantly, the product was selectivity found to be largely dependent on the parent Cu_2O film thickness, rather than on the initial crystal orientation. Moreover, a nanoparticulate Cu morphology was generated after reduction of the Cu_2O films, and X-ray diffraction measurement indicated the complete reduction of oxidized Cu species (Figure 3.10c). Yeo and co-workers also reported that faradic yields of

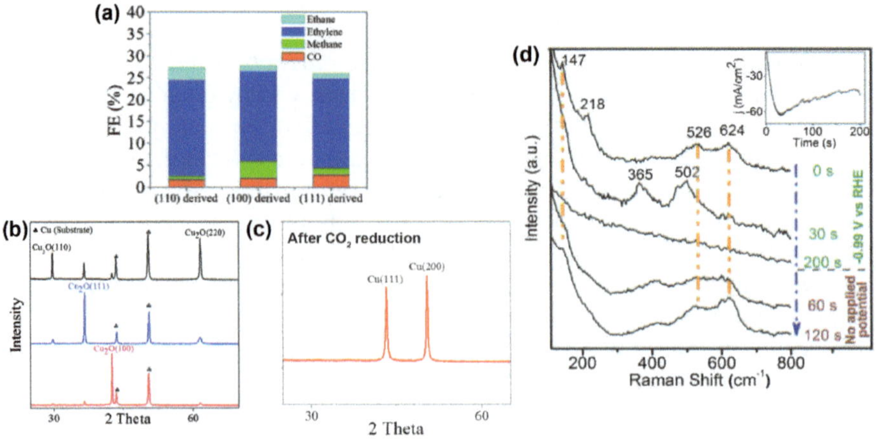

Figure 3.10 (a) FEs of gaseous products from CO_2 reduction at −1.1 V *vs.* RHE in 0.1 M $KHCO_3$ solution for oxides having different crystal orientation. X-ray diffraction patterns of (b) Cu_2O films electrochemically deposited on a Cu substrate and (c) coatings after electrochemical reduction at −0.6 V *vs.* RHE. Reproduced from ref. 40 with permission from the PCCP Owner Societies. (d) *In situ* Raman spectra and corresponding chrono-amperogram (inset) of 1.7 μm film at −0.99 V in 0.1 M $KHCO_3$. Reprinted with permission from D. Ren, Y. Deng, A. D. Handoko, C. S. Chen, S. Malkhandi and B. S. Yeo, *ACS Catal.*, 2015, **5**, 2814.[31] Copyright 2015 American Chemical Society.

ethylene and ethanol can be systematically varied by changing the thickness of the electrodeposited Cu_2O films.[31] 1.7–3.6 μm thickness of film exhibited the best selectivity for these C_2 compounds at −0.99 V *vs.* RHE, with FEs of 34–39% for ethylene and 9–16% for ethanol. Less than 1% methane was formed and a high C_2H_4/CH_4 ratio (up to 100) was achieved. *In situ* Raman spectroscopy revealed that the Cu_2O film was reduced rapidly and remained as metallic Cu particles during the CO_2 reduction (Figure 3.10d).

It was further reported by Yeo that agglomerates of 15 nm sized Cu nanocrystals, synthesized by the electrochemical reduction of $Cu_2O/Cu(OH)_2$ nanoparticles, were efficacious for reducing CO_2 to *n*-propanol (Figure 3.11).[41] At −0.95 V *vs.* RHE, *n*-propanol was formed on the Cu nanocrystals with a partial current density of $-1.74\ \mathrm{mA\,cm^{-2}}$ (FE = 9%), which was ~25× larger than that found on Cu^0 nanoparticles at the same applied potential. The Cu nanocrystals were also catalytically stable for at least six hours. It was shown that the *n*-propanol was formed through the C–C coupling of carbon monoxide and ethylene precursors. The enhanced activity of the Cu nanocrystals towards *n*-propanol formation could be correlated linearly to their surface population of defect sites, which was estimated using cyclic voltammetry.

As part of the electrodeposition method, a highly roughened nanostructured electrode could be prepared by electrochemical cycling of Cu foil in the presence of halide ions. By repeated electrochemical cycling, Cu or

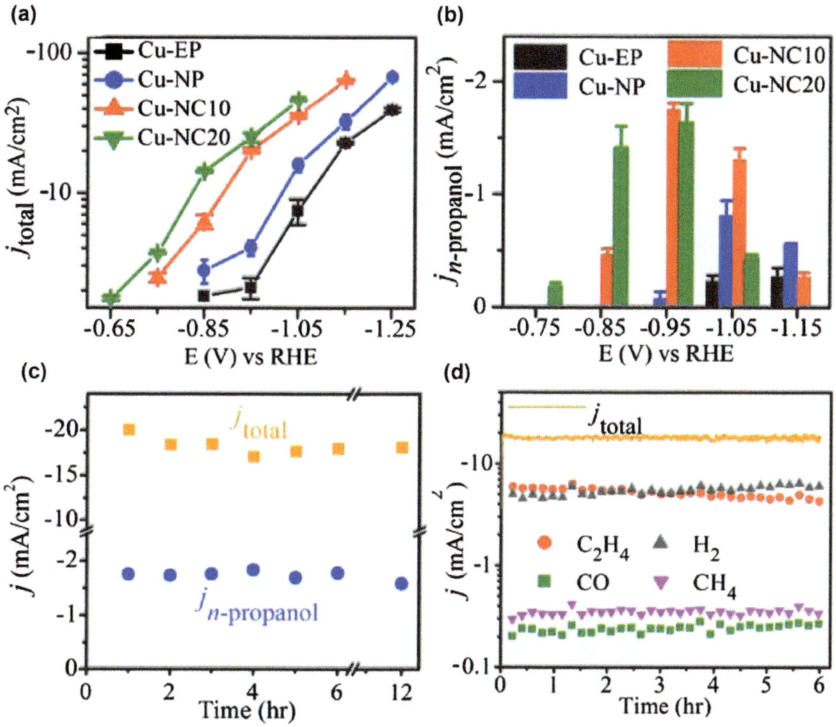

Figure 3.11 (a) Total current density measured during CO_2 reduction and (b) partial current density of *n*-propanol formation on electropolished Cu and Cu nanoparticles, and Cu nanocrystals (Cu-NC10 and Cu-NC20; 10 and 20 refer to the duration for which the Cu nanoparticles were oxidized to Cu oxide in minutes), respectively, at different potentials. (c) Total current density during CO_2 reduction and partial current density of *n*-propanol formation on Cu-NC10 at −0.95 V as a function of electrolysis time. (d) Chrono-amperogram and partial current densities of major gaseous products during 6 h of electrolysis on Cu-NC10 at −0.95 V. Electrolyte used: 0.1 M $KHCO_3$.
Reprinted with permission from D. Ren, N. T. Wong, A. D. Handoko, Y. Huang and B. S. Yeo, *J. Phys. Chem. Lett.*, 2016, 7, 20.[41] Copyright 2016 American Chemical Society.

Cu_2O starts to grow from the metal ion seed to nanoparticles or nanocubes. Yeo *et al.* prepared Cu mesocrystals by the *in situ* reduction of a thin CuCl film during CO_2 electroreduction.[42] Cu mesocrystals with active sites such as Cu(100), steps and edges were generated and these are believed to reduce CO_2 efficiently to C_2H_4, which formed ~81% of the total carbonaceous product (FE ≈ 27%) (Figure 3.12). The accumulation of CO on the active sites is believed to aid in their C–C coupling to form C_2 intermediates. Nilsson *et al.* also prepared a uniquely shaped Cu nanocube in the presence of a Cl ion in bicarbonate solution and the onset potentials for ethylene on the Cu nanocube (−0.6 V) was highly improved compared to the several Cu single

Figure 3.12 Time resolved *ex situ* SEM micrographs of Cu mesocrystals after (a) 10 s and (b) 4200 s of CO_2 electroreduction. (c) FEs of the CO_2 electroreduction products of Cu mesocrystals as a function of applied electrochemical potential. Reproduced from ref. 42 with permission from The Royal Society of Chemistry.

crystals ((100): −0.73 V, (211): −0.79 V and (111): −0.96 V) and polycrystalline Cu (−0.74 V).[43]

The different sizes of halide ions also affect the electrochemical cycling process. Bell *et al.* reported that electrochemical cycling of the Cu foil in the presence of fluoride (F), chloride (Cl), bromide (Br), or iodide (I) anions formed similar cubical structures and the density of the CuO_x cubes follows the order $Cl^- > F^- > Br^- > I^-$.[32] The nanostructured cubical structure increased the ethylene FE by a factor of 1.5 higher than that for polycrystalline Cu (at −1.0 V *vs.* RHE). It was noted that the CuO_x cubical structure was completely nanoparticulated upon applying a negative potential and, more importantly, CuO_x was completely reduced to metallic Cu monitored by *in situ* Raman spectroscopy, even before CO_2 reduction reaction starts.

3.3.3 Copper Composites

Composite catalysts show highly enhanced performance in many fields of energy conversion due to the synergetic effects of combined materials. Cu composites for electrochemical CO_2 reduction are usually synthesized on carbon materials because of their synergetic effects such as conductivity, defects sites of the substrate, dispersion and metal–carbon bridging.

Monodisperse Cu nanoparticles anchored on a pyridinic-N rich graphene (p-NG) substrate exhibited an improved electrochemical CO_2 reduction activity toward hydrocarbon formation.[44] Cu nanoparticles were synthesized by the reduction of Cu precursors (copper(I) acetate) with surfactant, and assembled on N-rich graphene layers to build a composite. The pyridinic-N is expected to serve as CO_2 and proton absorber, and to induce hydrogenation and C–C coupling reactions for the formation of ethylene. This work demonstrated a new strategy to anchor Cu nanoparticles on the functionalized conductive substrate, to improve the rate of C–C coupling (19% FE of C_2H_4 at −0.9 V *vs.* RHE) (Figure 3.13). However, hydrogen and formate generation comprised the majority of the electrochemical reactions.

Extremely small amounts of Cu on a carbon based substrate can also exhibit significant electrochemical CO_2 reduction activity.[45] In the work by Bell and Ager, graphene oxide (GO) containing ~120 ppm Cu exhibited an activity for methane formation, whereas the GO washed by ultrapure nitric acid lost the activity of hydrocarbon (Figure 3.14). Furthermore, 0.6 μg of Cu loaded on the cleaned graphene substrate demonstrated high current density and selectivity to CH_4 with respect to that of the Cu nanoparticle. From this study, it is noteworthy that trace amounts of impurities in the catalyst support or the electrolyte might lead to an over- or under-estimation of electrochemical CO_2 reduction activity of catalysts.

Cu based molecular catalysts that combine the Cu ion with organic ligands such as polypyridine, phosphine, cyclam, porphyrin, phthalocyanine and other macrocyclic structures having a negative site for the Cu atom can be used as Cu composites for electrochemical CO_2 reduction. Wang *et al.*

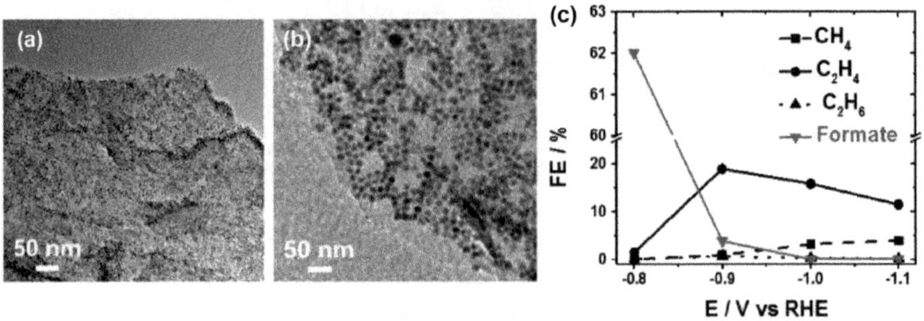

Figure 3.13 TEM images of (a) p-NG–Cu-7 and (b) p-NG–Cu-13 catalysts after Cu nanoparticle assembly on the p-NG and *n*-butylamine/ethanol/water washing to remove capping agents for electrochemical reduction of CO_2. (c) Reduction potential dependent FEs of the p-NG–Cu-7 catalysed electrochemical reduction of CO_2 to various hydrocarbons.
Reprinted from *Nano Energy*, **24**, Q. Li, W. Zhu, J. Fu, H. Zhang, G. Wu, S. Sun, Controlled assembly of Cu nanoparticles on pyridinic-N rich graphene for electrochemical reduction of CO_2 to ethylene, 1–9,[44] Copyright 2016, with permission from Elsevier.

reported that a Cu–porphyrin complex (Cu-(II)-5,10,15,20-tetrakis(2,6-dihydroxyphenyl)porphyrin) displayed promising catalytic properties for electrochemical CO_2 reduction in aqueous media.[46] The Cu–porphyrin complex exhibited a highly enhanced reaction activity (varied by measurement time) and around 50% of FE for hydrocarbons (CH_4 and C_2H_4) at −0.976 V *vs.* RHE (Figure 3.15). They suggested that the oxidation state of the Cu centre and the built-in hydroxyl groups in the porphyrin ligand are critical factors contributing to the electrocatalytic performance. However, hydrogen generation occupied over 50% in total electrochemical reactions and the catalyst stability originating from the organic electrode is still unclear. In addition, elucidating the detailed reaction mechanism may provide a crucial factor to design a superior catalyst for electrochemical CO_2 reduction.

3.4 Summary and Future Work

The electrochemical conversion of CO_2 to valuable chemicals and fuels is a promising way to ensure a sustainable future for humanity. The unique properties of the Cu catalyst in binding CO and related intermediates has enabled it to reduce CO_2 to hydrocarbons or alcohols. From previous experimental and theoretical works, it is suggested that a nanostructuring Cu surface would enhance CO_2 reduction activity and product selectivity. This chapter has introduced the overall features of CO_2 reduction on various types of Cu electrodes as summarized below:

1. Cu nanoparticles with diverse shapes and sizes exhibit variable catalytic activities because of the high density of exposed surface active

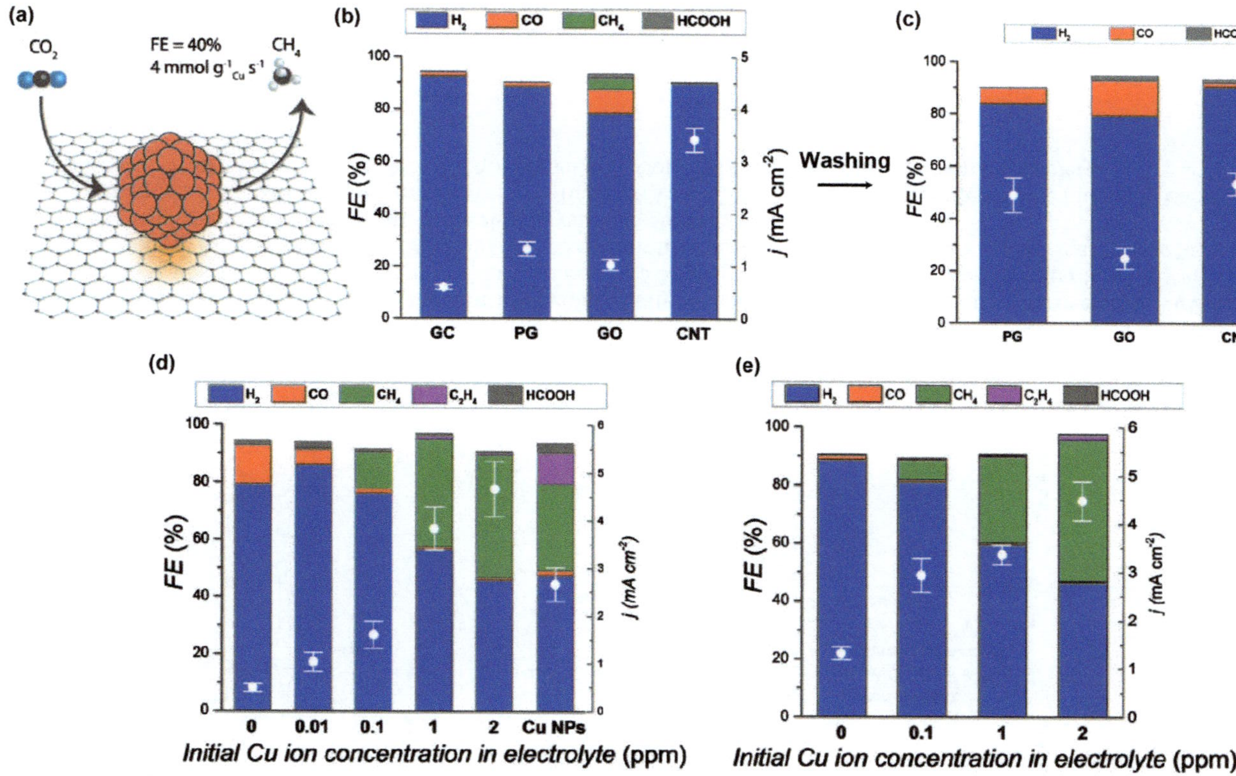

Figure 3.14 (a) Illustration of an extremely small number of particles on graphene showing high selectivity of hydrocarbons. (b) FE and current densities of a glassy-carbon electrode (GC) and as-received carbon supports dispersed on glassy-carbon electrodes: high purity graphite (PG), graphene oxide (GO), and carbon nanotubes (CNT). (c) FE and current densities of high-purity nitric acid washed carbon supports dispersed on GC electrodes. FE and current densities of (d) cleaned GO with the indicated concentrations of Cu ion initially "spiked" in the electrolyte and Cu nanoparticles (Cu nanoparticles, 9.69 μg loading) loaded on GC and (e) as-received PG with various concentrations of Cu ion initially "spiked" in the electrolyte. Electrolysis was carried out at −1.3 V *vs.* RHE in 0.1 M $NaHCO_3$ solution for 2 h.

Reprinted with permission from Y. Lum, Y. Kwon, P. Lobaccaro, L. Chen, E. L. Clark, A. T. Bell and J. W. Ager, *ACS Catal.*, 2016, **6**, 202.[45]

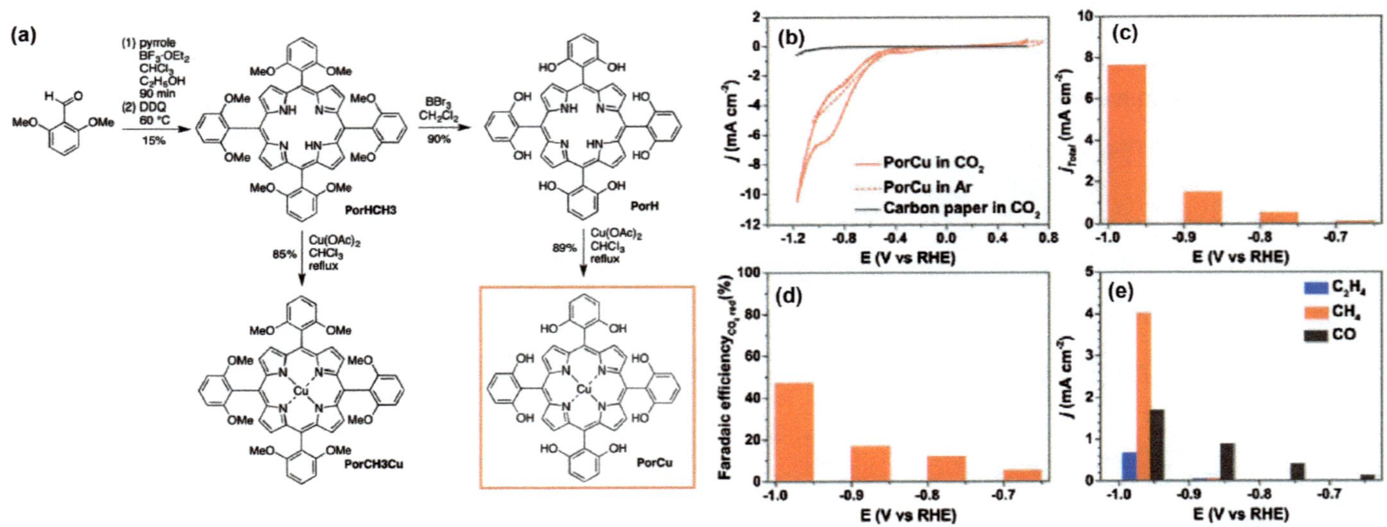

Figure 3.15 (a) Synthetic routes for Cu–porphyrin molecular catalysts. Electrochemical CO_2 reduction over the PorCu electrode with a catalyst mass loading of 0.25 mg cm^{-2} in aqueous 0.5 M $KHCO_3$. (b) CV curves at a scan rate of 100 mV s^{-1}. (c) Total current densities, (d) CO_2 reduction FEs and (e) distribution of CO_2 reduction products in the gas phase at various electrode potentials; all measured 8 min after electrolysis was started.
Reprinted with permission from Z. Weng, J. Jiang, Y. Wu, Z. Wu, X. Guo, K. L. Materna, W. Liu, V. S. Batista, G. W. Brudvig and H. Wang, *J. Am. Chem. Soc.*, 2016, **138**, 8076.[46] Copyright 2016 American Chemical Society.

sites for tuning the binding energies of the important reaction intermediates. Further lowering the overpotential requirement and enhancing the selectivity of the reaction to targeted products need to be further achieved.
2. Oxide-derived Cu has played a major part in recent advances in Cu electrodes prepared by thermal treatment and electrodeposition. The Cu nanostructure with active sites exhibited a dramatically enhanced activity and controlled surface reactions for hydrocarbons. There are several proposed activity descriptors such as oxidation state and grain boundary, and it is desirable to make a conclusion based on surface sensitive *in situ* analysis.
3. Cu composite catalysts demonstrate synergetic effects between the Cu nanoparticle/atom and substrate/supports, controlling CO_2 reduction activity and a reaction pathway toward a specific hydrocarbon.

Nanostructuring Cu catalysts has meant remarkable progress for enhanced catalytic activity and product selectivity. However, several crucial issues such as the reliability of performance, stability, mass-scale problems and cost-effectiveness still remain for future work.

Acknowledgements

This work has been performed as project SI1701-05 supported by the Korea Research Institute of Chemical Technology (KRICT) and "Next Generation Carbon Upcycling Project" (Project No. 2017M1A2A2043122) through the National Research Foundation (NRF) funded by the Ministry of Science and ICT, Republic of Korea.

References

1. Y. Hori, K. Kikuchi and S. Suzuki, *Chem. Lett.*, 1985, **14**, 1695.
2. Y. Hori, H. Wakebe, T. Tsukamoto and O. Koga, *Electrochim. Acta*, 1994, **39**, 1833.
3. Y. Hori, in *Modern Aspects of Electrochemistry*, ed. C. G. Vayenas, R. E. White and M. E. Gamboa-Aldeco, Springer, New York, NY, 2008, p. 89.
4. J.-P. Jones, G. K. S. Prakash and G. A. Olah, *Isr. J. Chem.*, 2014, **54**, 1451.
5. R. Kortlever, J. Shen, K. J. P. Schouten, F. Calle-Vallejo and M. T. M. Koper, *J. Phys. Chem. Lett.*, 2015, **6**, 4073.
6. M. Gattrell, N. Gupta and A. Co, *J. Electroanal. Chem.*, 2006, **594**, 1.
7. A. A. Peterson and J. K. Nørskov, *J. Phys. Chem. Lett.*, 2012, **3**, 251.
8. A. A. Peterson, F. Abild-Pedersen, F. Studt, J. Rossmeisl and J. K. Nørskov, *Energy Environ. Sci.*, 2010, **3**, 1311.
9. Y. Hori, R. Takahashi, Y. Yoshinami and A. Murata, *J. Phys. Chem. B*, 1997, **101**, 7075.
10. Y. Hori, I. Takahashi, O. Koga and N. Hoshi, *J. Phys. Chem. B*, 2002, **106**, 15.

11. F. Calle-Vallejo and M. T. M. Koper, *Angew. Chem., Int. Ed.*, 2013, **52**, 7282.
12. S. Vollmer, G. Witte and C. Wöll, *Catal. Lett.*, 2001, **77**, 97.
13. Y. Hori, I. Takahashi, O. Koga and N. Hoshi, *J. Mol. Catal. A: Chem.*, 2003, **199**, 39.
14. K. P. Kuhl, E. R. Cave, D. N. Abram and T. F. Jaramillo, *Energy Environ. Sci.*, 2012, **5**, 7050.
15. Y.-G. Kim, J. H. Baricuatro, A. Javier, J. M. Gregoire and M. P. Soriaga, *Langmuir*, 2014, **30**, 15053.
16. R. Reske, H. Mistry, F. Behafarid, B. R. Cuenya and P. Strasser, *J. Am. Chem. Soc.*, 2014, **136**, 6978.
17. K. Manthiram, B. J. Beberwyck and A. P. Alivisatos, *J. Am. Chem. Soc.*, 2014, **136**, 13319.
18. D. Kim, J. Resasco, Y. Yu, A. M. Asiri and P. Yang, *Nat. Commun.*, 2014, **5**, 4948.
19. M. R. Singh, E. L. Clark and A. T. Bell, *Phys. Chem. Chem. Phys.*, 2015, **17**, 18924.
20. A. Loiudice, P. Lobaccaro, E. A. Kamali, T. Thao, B. H. Huang, J. W. Ager and R. Buonsanti, *Angew. Chem., Int. Ed.*, 2016, **55**, 5789.
21. K. J. P. Schouten, Z. Qin, E. P. Gallent and M. T. M. Koper, *J. Am. Chem. Soc.*, 2012, **134**, 9864.
22. C. W. Li and M. W. Kanan, *J. Am. Chem. Soc.*, 2012, **134**, 7231.
23. C. W. Li, J. Ciston and M. W. Kanan, *Nature*, 2014, **508**, 504.
24. A. Verdaguer-Casadevall, C. W. Li, T. P. Johansson, S. B. Scott, J. T. McKeown, M. Kumar, I. E. L. Stephens, M. W. Kanan and I. Chorkendorff, *J. Am. Chem. Soc.*, 2015, **137**, 9808.
25. D. Raciti, K. J. Livi and C. Wang, *Nano Lett.*, 2015, **15**, 6829.
26. M. Ma, K. Djanashvili and W. A. Smith, *Angew. Chem., Int. Ed.*, 2016, **55**, 6680.
27. H. Mistry, A. S. Varela, C. S. Bonifacio, I. Zegkinoglou, I. Sinev, Y.-W. Choi, K. Kisslinger, E. A. Stach, J. C. Yang, P. Strasser and B. R. Cuenya, *Nat. Commun.*, 2016, **7**, 12123.
28. S. G. Bratsch, *J. Phys. Chem. Ref. Data*, 1989, **18**, 1.
29. A. J. Bard, R. Parsons and J. Jordan, *Standard Potentials in Aqueous* Solutions, Marcel Dekker, New York, 1985.
30. G. Milazzo, S. Caroli and V. K. Sharma, *Tables of Standard Electrode Potentials*, Wiley, Chichester, 1978.
31. D. Ren, Y. Deng, A. D. Handoko, C. S. Chen, S. Malkhandi and B. S. Yeo, *ACS Catal.*, 2015, **5**, 2814.
32. Y. Kwon, Y. Lum, E. L. Clark, J. W. Ager and A. T. Bell, *ChemElectroChem*, 2016, **3**, 1012.
33. R. Kas, K. K. Hummadi, R. Kortlever, P. de Wit, A. Milbrat, M. W. J. Luiten-Olieman, N. E. Benes, M. T. M. Koper and G. Mul, *Nat. Commun.*, 2016, **7**, 10748.
34. L. Gal-Or, I. Silberman and R. Chaim, *J. Electrochem. Soc.*, 1991, **138**, 1939.

35. S. Lee, D. Kim and J. Lee, *Angew. Chem., Int. Ed.*, 2015, **54**, 14701.
36. L.-Y. Gan and Y.-J. Zhao, *J. Chem. Phys.*, 2010, **133**, 094703.
37. D. Kim, S. Lee, J. D. Ocon, B. Jeong, J. K. Lee and J. Lee, *Phys. Chem. Chem. Phys.*, 2015, **17**, 824.
38. V. Young and G. Hoflund, *Handbook of Surface and Interface Analysis*, CRC Press, 2009, p. 19.
39. A. Gerson, D. Cookson and K. Prince, *Handbook of Surface and Interface Analysis*, CRC Press, 2009, p. 193.
40. R. Kas, R. Kortlever, A. Milbrat, M. T. M. Koper, G. Mul and J. Baltrusaitis, *Phys. Chem. Chem. Phys.*, 2014, **16**, 12194.
41. D. Ren, N. T. Wong, A. D. Handoko, Y. Huang and B. S. Yeo, *J. Phys. Chem. Lett.*, 2016, 7, 20.
42. C. S. Chen, A. D. Handoko, J. H. Wan, L. Ma, D. Ren and B. S. Yeo, *Catal. Sci. Technol.*, 2015, **5**, 161.
43. F. S. Roberts, K. P. Kuhl and A. Nilsson, *Angew. Chem., Int. Ed.*, 2015, **127**, 5268.
44. Q. Li, W. Zhu, J. Fu, H. Zhang, G. Wu and S. Sun, *Nano Energy*, 2016, **24**, 1.
45. Y. Lum, Y. Kwon, P. Lobaccaro, L. Chen, E. L. Clark, A. T. Bell and J. W. Ager, *ACS Catal.*, 2016, **6**, 202.
46. Z. Weng, J. Jiang, Y. Wu, Z. Wu, X. Guo, K. L. Materna, W. Liu, V. S. Batista, G. W. Brudvig and H. Wang, *J. Am. Chem. Soc.*, 2016, **138**, 8076.

CHAPTER 4

Single-crystal Surfaces as Model Electrocatalysts for CO_2 Reduction

ADAM KOLODZIEJ,[a] PARAMACONI RODRIGUEZ*[a] AND ANGEL CUESTA*[b]

[a] School of Chemistry, University of Birmingham, Edgbaston B15 2TT, UK; [b] Department of Chemistry, School of Natural and Computing Sciences, University of Aberdeen, AB24 3UE, Aberdeen, Scotland, UK
*Email: p.b.rodriguez@bham.ac.uk; angel.cuestaciscar@abdn.ac.uk

4.1 Introduction and General Principles

Single-crystal electrode surfaces have been the subject of intense study ever since the pioneering work by Clavilier *et al.*[1] that opened the field to electrocatalyst optimisation based on a deep knowledge of the relationship between surface structure and surface reactivity. Single-crystal surfaces not only provide key mechanistic information about reaction pathways and reaction intermediates, they also provide the only means to perform studies on surfaces with only one kind of active site (high-quality basal planes single-crystal electrodes), or with a controlled combination of sites with different activities (*e.g.*, by using stepped single-crystal electrodes). Single-crystal surfaces are also the only experimental systems comparable to the ideal surfaces used in computational calculations. Single-crystalline systems for electrocatalysis studies can be fabricated by orienting and then cutting a macroscopic single crystal along the desired crystallographic plane, or by

Energy and Environment Series No. 21
Electrochemical Reduction of Carbon Dioxide: Overcoming the Limitations of Photosynthesis
Edited by Frank Marken and David Fermin

Published by the Royal Society of Chemistry, www.rsc.org

controlled growth of single-crystalline nanoparticles. In the latter case, the type and relative number of surface sites will depend on the orientation of the crystal faces enclosing the crystal, which can often be controlled by the synthesis method.[2,3]

The effect of the chemical nature of the catalyst on the activity and product selectivity of the CO_2 reduction reaction (CO_2RR) was studied by Hori *et al.* in his seminal work.[4] Pure-metal electrocatalysts can be grouped according to their product selectivity. Three types of metals can be identified:[5] (i) those producing high yields of hydrogen (Pt-group metals, Ni, Fe and Ti), (ii) those producing high yields of CO (Au, Ag, Zn and Pd, Ga) and (iii) those producing high yields of formic acid (Pb, Hg, In, Sn, Cd and Tl). Cu appears as the maverick, being the only metal producing hydrocarbons such as methane, ethylene and short-chain alcohols as main products at low temperatures and high overpotentials (HCOOH, CO and H_2 are the products at low overpotentials).[6,7] These differences in activity and selectivity have to be due to differences in the stability of the several possible reaction intermediates, as well as to differences in the activation barriers that have to be overcome when transiting from one intermediate to the next.

It is well accepted that the mechanism of the CO_2RR in aqueous solution, and the nature of the intermediates involved, is independent of the metal catalyst employed. Figure 4.1 shows proposed mechanistic routes and key intermediates for four possible products of the CO_2RR (methane, methanol, formic acid and ethylene), as recently proposed.[8] As can be seen in Figure 4.1, adsorbed CO is expected to play a critical role in the formation of hydrogenated C1 and C2 products. Consequently, the CO reduction reaction (CORR) is embedded within the CO_2RR, and understanding the latter requires understanding the former.

As in the case of any other electrocatalytic reaction, it is well known that not only the chemical nature of the surface, but also its atomic structure, plays a relevant role in determining these factors. Here we review relevant work carried out over the last few years on the role of atomic sites of metal single-crystal surfaces in the electrochemical conversion of CO_2.

4.2 Role of the Surface Structure on the Reduction of CO_2 on Pt-group Single-crystal Electrodes

A variety of single-crystal electrodes of Pt, Pd, Rh and Ir have been evaluated for the electrocatalytic conversion of CO_2 to hydrocarbons and CO. Pt-group metals are very active catalysts for the hydrogen evolution/oxidation reactions (HERs/HORs) and, in aqueous media, H_2 production dominates over any CO_2 reduction pathway. In spite of the resulting low faradaic efficiency, CO (the main product of the CO_2RR on these materials) accumulates on the surface of these metals when polarised at potentials within the hydrogen evolution region in the presence of CO_2. This is due to the strong bond between adsorbed CO and surface atoms of the Pt-group metals, and results

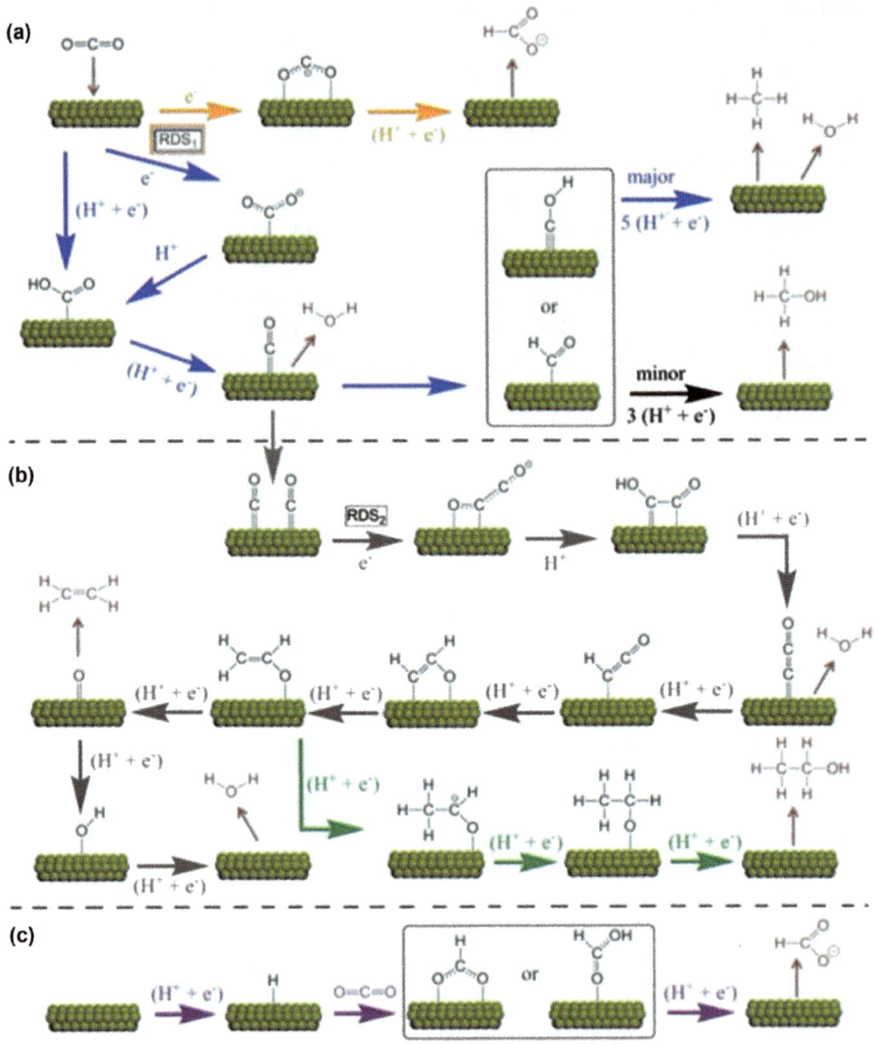

Figure 4.1 Possible reaction pathways for the electrocatalytic reduction of CO_2 on transition metals and molecular catalysts: (a) pathways from CO_2 to CO, CH_4 (blue arrows), CH_3OH (black arrows) and $HCOO^-$ (orange arrows); (b) pathways from CO_2 to ethylene (grey arrows) and ethanol (green arrows); (c) pathway of CO_2 insertion into a metal–H bond yielding formate (purple arrows). Species in black are adsorbates, while those in red are reactants or products in solution. Potentials are reported *versus* RHE, while RDS indicates rate-determining steps and $(H^+ + e^-)$ indicates steps in which either concerted or separated proton–electron transfer takes place.
Adapted with permission from R. Kortlever, J. Shen, K. J. P. Schouten, F. Calle-Vallejo and M. T. M. Koper, *J. Phys. Chem. Lett.*, 2015, **6**, 4073.[8]

in a poisoning of the catalyst for the HER. CO adsorbs irreversibly on these materials, and can only be removed from their surface by oxidation at very positive overpotentials. This, however, provides the means to estimate the catalytic activity of the surface investigated for the reduction of CO_2 to CO, by simply integrating the CO-stripping peak after polarising the electrode at a negative enough potential for a given time. This was the approach employed by Hoshi *et al.* in order to explore structure–activity relationships in the CO_2RR on Pt-group metal single-crystal electrodes.[9–15] Structure effects are expected on these metals, not only due to the effect of the surface structure on the binding energy of CO_2 or the CO_2^- radical (believed to be the first reaction intermediate), but also to the well-known dependence of the CO binding energy and adsorption geometry. Hoshi *et al.* demonstrated that Pt flat surfaces (Pt(100) and Pt(111)) have the lowest activity for the CO_2RR, and that the activity increases when steps or kinks are introduced in the surface.[9,12–14] Kinks are more active than steps, and Pt(S)-[n(100)×(110)] surfaces, which contain densely packed kink atoms along step lines, were found to be the most active.[9,12] For stepped surfaces, the combination of (111) terraces with (111) steps was found to be more active than the combination of (111) terraces with (100) steps, or that of (100) terraces with (111) steps (Figure 4.2). The order of activity for the flat and stepped surfaces was found to be Pt(111) < Pt(100) < Pt(S)-[n(111)×(100)] = Pt(S)-[n(100)×(111)] < Pt(S)-[n(111)×(111)] < Pt(110).[9,13,14] It is not surprising that Pt(110) is the most active surface within this series, since it actually corresponds to the Pt(S)-[n(111)×(111)] surface with the highest possible density of (111) steps. Hori *et al.* suggested that the mechanism of the CO_2RR on Pt involves adsorbed hydrogen,[9,12,14] and that the higher activity of (111) steps and, in particular, of kinks, is due to the higher reactivity of hydrogen adsorbed at the bottom of the step or the kink, as compared to hydrogen adsorbed on a terrace site.[9,12,14] Hori's group also studied CO_2RR on Pd,[11] Rh[10] and Ir[15] single-crystal electrodes. Among the three basal planes: (111), (100) and (110), Pd (110) showed the highest activity,[11] as was also the case for Pt,[9] Rh[10] and Ir.[15] However, interestingly, Pd(111) was shown to be more active than Pd(100), although in all the other three cases the (100) surface proved to be more active.[9,10,15] The activity of Pd(110) was found to be two orders of magnitude higher than that of Pt(110),[11] which was found to be more active than Ir(110).[15] In the case of Rh, they reported an effect of the electrolyte anion on the catalytic activity of the single-crystal electrodes. The activity order in $HClO_4$ was found to be Rh(110) > Rh(100) > Rh(111), while in H_2SO_4 it follows the trend Rh(100) > Rh(110) > Rh(111).[10] The catalytic activity of Rh(110) is higher than that of Pt(110) and Ir(110) at $E < 0.1$ V *vs.* RHE in 0.1 M $HClO_4$.[10] However, it decreases continuously with increasing potential in the case of Rh(110), while in the case of Pt(110) it peaks at 0.2 V *vs.* RHE, and, as a consequence, between 0.1 and 0.3 V *vs.* RHE the catalytic activity of Pt(110) is higher than that of Rh(110).[10]

Rodes *et al.* studied the CO_2RR on Pt single-crystal electrodes using a combination of cyclic voltammetry and *in situ* infrared reflection–absorption

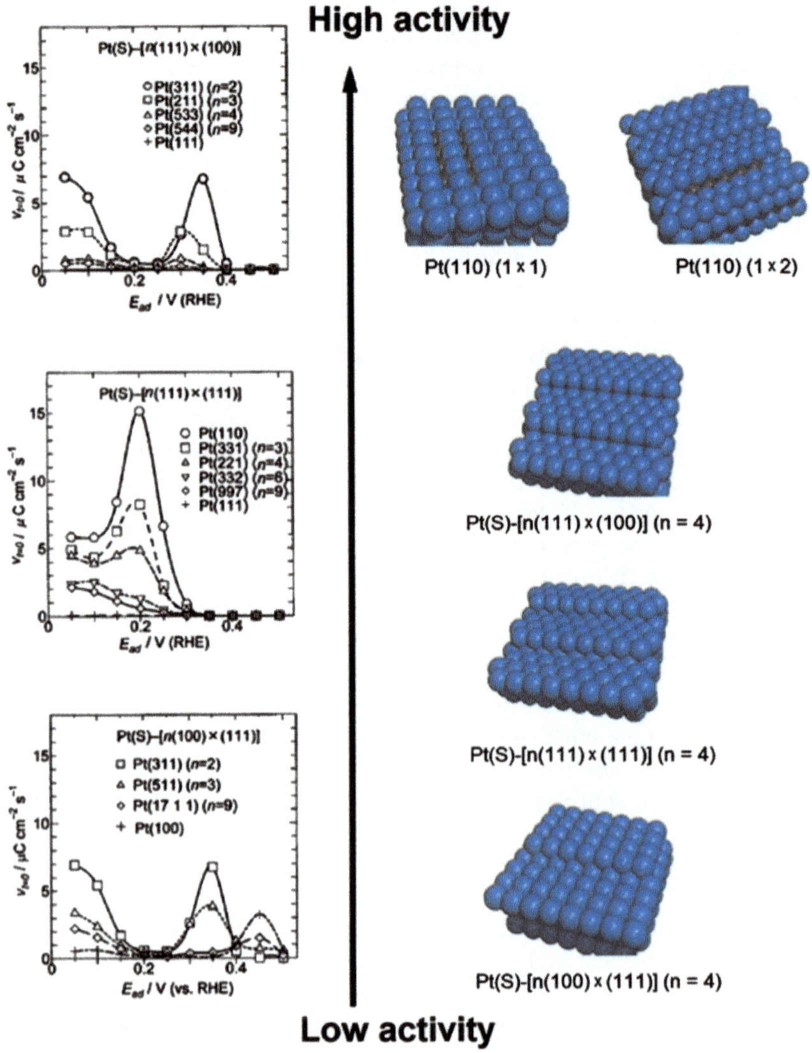

Figure 4.2 Initial reaction rates of CO formation on Pt(111), Pt(100), Pt(110) and various stepped surfaces in 0.1 M $HClO_4$ as a function of potential. Adapted from *Electrochima Acta*, **45**, N. Hoshi and Y. Hori, Electrochemical reduction of carbon dioxide at a series of platinum single crystal electrodes, 4263–4270,[9] Copyright 2000, with permission from Elsevier.

spectroscopy, providing spectroscopic evidence of the adsorption site of the reaction product (adsorbed CO).[16–18] On Pt(110), essentially linearly adsorbed CO and traces of multibonded CO were found,[16] in good agreement with previous work by Nikolic *et al.*[19] The nature of the adsorbed reduction products was found to be independent of the reaction potential, the reaction time, and the electrode preparation, although higher CO_2 reduction rates with air-cooled Pt(110) electrodes were attributed to the expected higher density of surface defects. By contrast, well-ordered Pt(111) electrodes were found to be inactive for the CO_2RR, but inducing surface disorder increased their activity.[17] It was also found that increasing the density of (100) steps by using stepped single-crystal electrodes did not lead to a similar activity increase as by introducing random defects. This points to a higher activity of (111) steps, as also found by Hoshi *et al.*[9,13,14] On Pt(100), multibonded CO was the only adsorbate detected after CO_2 reduction at $E > 0.20$ V *vs.* RHE, with a small amount of linearly bonded CO being formed after CO_2 reduction at 0.08 V.[18] The latter was related to the presence of adsorbed hydrogen on residual surface defects (*i.e.*, step or kink sites). Both hydrogen-covered (100) terraces and (111) steps were found to be active for generating multibonded CO on (100) terraces, defects showing a higher activity, while only hydrogen-covered (111) steps were found active for generating linearly adsorbed CO. Interestingly, (100)-step sites were found to be inactive for the CO_2RR.[18] All these results are in general good agreement with Hoshi *et al.*[9,12–14]

Recently, the electroreduction of CO_2 using ionic liquids as electrolyte has attracted much attention, due to reports of high faradaic and energy efficiencies. The former effect is due to the inhibition of hydrogen evolution typical of non-aqueous electrolytes, while the latter has been attributed to the stabilisation of CO_2^- by the ionic liquid.[20] Hanc-Scherer *et al.*[21] searched for structural effects in the CO_2RR on Pt in 1-ethyl-3-methylimidazolium bis(trifluoromethysulfnyl)imide ($[C_{2mim}{}^+][NTf^-]$), also known as EMIM-NTf, by using Pt(111), Pt(100) and Pt(110). In the pure ionic liquid, Pt(111) was found to behave differently to Pt(100) and Pt(110), which was attributed to a lower activity of Pt(111) to generate CO_2^-. As a consequence, the reduction taking place on Pt(111) in the presence of CO_2 would correspond to the reductive dehydrogenation of $[C_{2mim}{}^+]$ followed by the formation of an adduct of the resulting product with CO_2, yielding a (neutral) imidazolium-2-carboxylate and H_2 (eqn (4.1)):

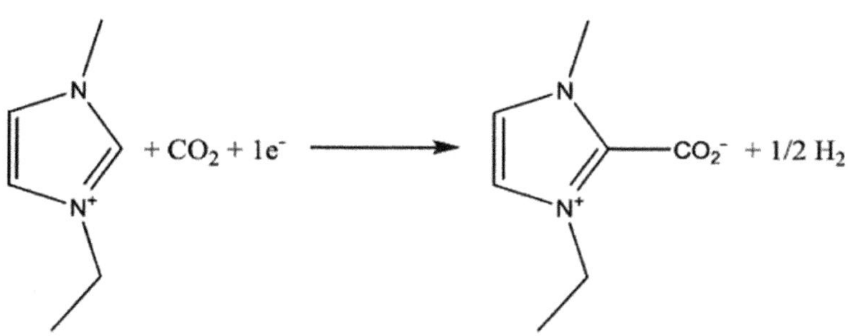

(4.1)

As in aqueous solutions, Pt(110) was found to be the most active surface, and Pt(111) the least active one. In the case of Pt(110) and Pt(100) the product of the reduction was shown to be that resulting from the coupling of the reductions of [C_{2mim}^+] and CO_2, respectively (eqn (4.2)):

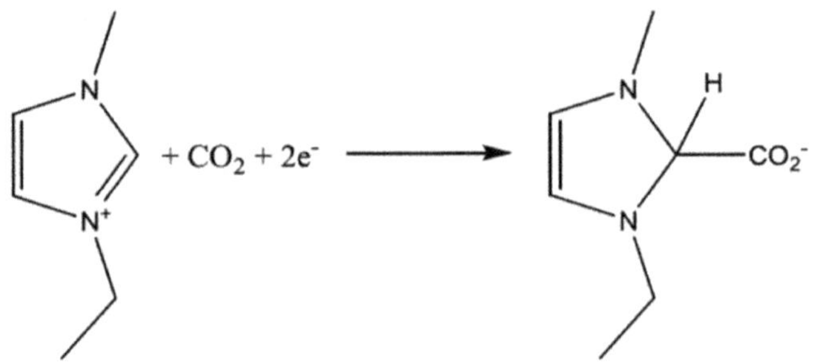

(4.2)

In order to favour the two electron reduction to CO or HCOOH, Hanc-Scherer *et al.* added [H^+][NTf^-] as a source of protons.[21] Under these conditions, HCOOH was produced upon reduction of CO_2, with Pt(110) having an activity approximately three times higher than Pt(111) and Pt(100).

4.3 Reduction of CO_2 on Ag and Au Single-crystal Electrodes

The reduction of CO_2 on silver and gold results in high selectivity towards the formation of CO due to its weak binding energy on these surfaces, which prevents it from adsorbing irreversibly and poisoning the catalyst. The reaction proceeds at relatively low overvoltage (−0.8 V *vs.* NHE at pH 7.5) with reported faradaic efficiencies as high as 87% and 91% for Ag electrodes and Au, respectively, at current densities around 4 mA cm^{-2}.[5,22]

Experimental work has shown that Ag(111), Ag(110) and Ag(100) single-crystal electrodes convert CO_2 into CO selectively at almost any overpotential.[23] Similar to the case of Pt-group metals, it was also found that the partial current density of CO on Ag(110) was remarkably higher than on Ag(111) and Ag(100) at −1.44 V. This result suggests that steps enhance the catalytic activity of CO_2 reduction to free CO. However, the effect of the surface structure seems to be less significant on Ag than on Pt single-crystal electrodes. This can be attributed to differences in the adsorption energies of CO_2 and surface intermediates, but also to the absence of adsorbed hydrogen and adsorbed anions on Ag.

To the best of our knowledge, there are no experimental studies of the CO_2RR on Au single-crystal electrodes. However, recent findings regarding the surface sensitivity of the adsorption of carbon monoxide on Au single-crystal electrodes in alkaline media suggest that the catalytic activity and selectivity towards the CO_2RR might be affected by the surface structure.[24–28] We would like to note, though, that, at the very negative potentials at which the CO_2RR proceeds, Au(111)-(1×23), hex-Au(100) and (1×2)-Au(110) reconstructed surfaces are more stable than the nominal (111), (100) and (110) planes, respectively.[29] Depending on the electrode history, the reconstructed surface present might correspond to a well-ordered, thermally induced, reconstruction to a somehow disordered, electrochemically induced, reconstruction, or to a partially reconstructed surface with an ongoing reconstruction process (this would be the case if the electrode is polarised negatively after having lifted the reconstruction at positive enough potentials). This must be taken into account when analysing eventual future work with Au single-crystal electrodes, and can be a source of irreproducibility if not adequately considered.

4.4 Reactivity and Selectivity of Cu Single-crystal Electrodes in the CO_2RR

Copper is unique in that it is the only metal that can produce significant amounts of hydrocarbons during the CO_2RR.[5] Most importantly, it is the only metal known to catalyse C–C coupling and generate ethylene, as well as C_2 and C_3 alcohols in aqueous electrolytes at ambient temperature.[5] Methane and ethylene are the main hydrocarbon products of CO_2 reduction on Cu, although they are generated at high overpotentials and, consequently, with low energy efficiency.[5] Several studies have addressed the product selectivity as a function of temperature,[30,31] electrode potential[6,7,32] and electrolyte composition.[6,33]

Extensive experiments have also demonstrated that the surface structure of copper electrodes strongly influences the activity and selectivity of the CO_2RR. Early experiments reported CH_4 and C_2H_4 as the two major gas products of the galvanostatic reduction of CO_2 at 5 mA cm^{-2} on Cu(111), Cu(110) and Cu(100).[34] These early studies showed that C_2H_4 is favourably produced on Cu(100), while Cu(111) favours the formation of CH_4. Cu(110) showed intermediate product selectivity. The authors also proposed that CO is the common reaction intermediate formed in the reduction of CO_2 to hydrocarbons and alcohols.

Further work from Hori's group using Cu(111),[35] Cu(100)[35] and stepped Cu(S)-[n(100)×(111)],[35,36] Cu(S)-[n(100)×(110)],[35,36] Cu(S)-[n(111)×(100)],[36] Cu(S)-[n(111)×(111)][36] and Cu(S)-[n(110)×(100)][36] single-crystal electrodes confirmed the favoured production of CH_4 and C_2H_4 on Cu(111) and Cu(100), respectively, and revealed that introducing (111) or (110) steps in the (100) basal plane results in the suppression of CH_4 formation and in an increase in the C_2H_4/CH_4 ratio. The highest C_2H_4 to CH_4 ratio was found for Cu(711) (*i.e.*, Cu(S)-[4(100)×(111)]).[35,36] Increasing step density in Cu(S)-[n(111)×(111)] surfaces was found to increasingly favour the formation of acetic acid, acetaldehyde and ethyl alcohol,[36] this being the first report of CH_3COOH formation during the CO_2RR on copper. The highest acetic acid yield (20%) was obtained with Cu(110) and Cu(S)-[2(111)×(111)]) electrodes. This was attributed to increased efficiency for the formation of the precursor for CH_3COOH at (111) steps adjacent to (111) terraces. It was suggested that this precursor may be sequentially reduced to CH_3COOH, CH_3CHO and C_2H_5OH. Gaseous products were predominantly found with Cu(S)-[n(110)×(100)] electrodes, increasing step density leading to increased formation of CH_4.

Kortlever *et al.* recently compared the behaviour of several Cu electrodes (polycrystalline, (111), (100), (110), (644) and (911)) in phosphate buffer solutions with pHs ranging between neutral and alkaline.[37] Particular attention was paid to solutions whose pH after CO_2 bubbling lays between 7.35 and 9.33, where >90% of the dissolved CO_2 exists as HCO_3^-. After bubbling CO_2, a reduction peak was observed only at pH ≥ 7.5, which was attributed to direct reduction of HCO_3^- to formate. Roughening the Cu surface following a procedure developed by Tang *et al.* led to a significant increase in formate production.[38] This was followed by a study using Cu single-crystal electrodes (Figure 4.3), which showed that, despite Cu(100) showing the highest overall

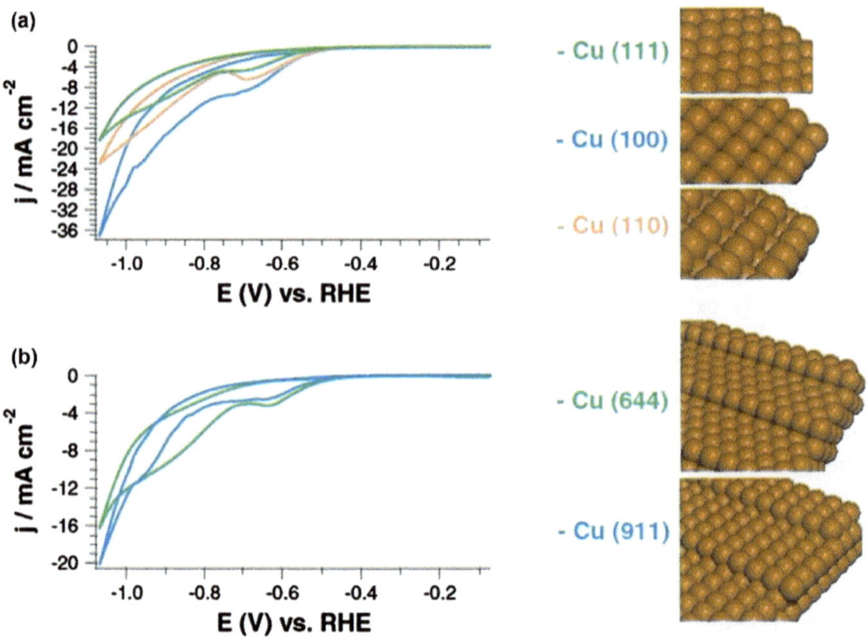

Figure 4.3 Cyclic voltammograms of CO_2 reduction in a phosphate buffer solution starting at pH 11.6 on single-crystal electrodes. Scan rate: 20 mV s^{-1}. (a) Cyclic voltammograms on Cu(111) (green), Cu(100) (blue) and Cu(110) (orange); (b) cyclic voltammograms on Cu(644) (green) and Cu(911) (blue).
Adapted from *Journal of Solid State Electrochemistry*, Electrochemical carbon dioxide and bicarbonate reduction on copper in weakly alkaline media, 17, 2013, 1843–1849, R. Kortlever, © Springer-Verlag Berlin Heidelberg 2013, with permission of Springer.[37]

catalytic activity, the other two basal planes have a higher activity for the direct reduction of bicarbonate to formate.[37] This difference was attributed to the higher activity of Cu(100) for the HER at relatively high potentials (> -0.6 V *vs.* RHE), as compared to the Cu(110) and Cu(111) surfaces. Cu(110) was the most active surface for the formation of formate, which suggested that under-coordinated sites play a relevant role in the reduction of bicarbonate to formate.[37] Although experiments with Cu(911) and Cu(644) were inconclusive regarding the role of monoatomic steps in this process, it must be taken into account that, while Cu(110) corresponds to the family of stepped surfaces with (111) terraces and (111) steps, Cu(911) and Cu(644) correspond to surfaces with (100) terraces and (111) steps, and (111) terraces and (100) steps, respectively. The possibility that a (111) step adjacent to a (111) terrace is the most active site for the reduction of HCO_3^- to $HCOO^-$ remains, hence, untested.

Encouraged by the different product selectivity of Cu(111) and Cu(100) (which favour CH_4 and C_2H_4 formation, respectively), and by Hori's proposal that the CO dimer is the common intermediate in the reduction of CO_2 to hydrocarbons and alcohols, Schouten *et al.* studied the CORR on the three basal planes, as well as on the stepped surfaces Cu(322) (Cu(S)-[5(111)×(100)]) and Cu(911) (Cu(S)-[5(100)×(111)]), in pH 7 phosphate buffer.[39] The latter are stepped surfaces with the same step density, but combining (111) terraces with (100) steps (Cu(322)), and (100) terraces with (111) steps (Cu(911)), which allowed the investigation of the different reactivities of (100) steps and (100) terraces for ethylene formation. Hori's proposal that (100) terrace sites, and not (100) steps, are active for ethylene formation on Cu was confirmed.[34,35] Experiments with ethylene oxide also provided support to the authors' previous suggestion that an oxametallacycle is the intermediate in the pathway from the CO dimer to ethylene.[40] Cu(110) showed a potential dependence for methane and ethylene formation similar to that of Cu(111), but a primary alcohol was also observed among the products.[39] The trend in CH_4/C_2H_4 selectivity on Cu single-crystal electrodes along the stereographic triangle, as resulting from the work of Schouten *et al.*, is shown in Figure 4.4.

The same authors showed later, combining cyclic voltammetry and on-line electrochemical mass spectrometry (OLEMS), that in addition to surface structure, product selectivity of the CORR and the CO_2RR is also potential and pH dependent.[41] At pH 1, CH_4 formation was found to start at -0.55 V on Cu(100), while on Cu(111) no CO_2 reduction products could be detected. By contrast, at pH 7, formation of C_2H_4 on Cu(100) starts at potentials less negative than that of CH_4, whereas on Cu(111) the onset potential is the same for both products (Figure 4.5). This led the authors to propose the existence of two possible reaction pathways to C_2H_4: (i) a pH-dependent pathway with a common intermediate for the formation of CH_4 and C_2H_4, occurring mainly on Cu(111) and (ii) a pH-independent pathway *via* a carbon monoxide dimer occurring only on Cu(100).

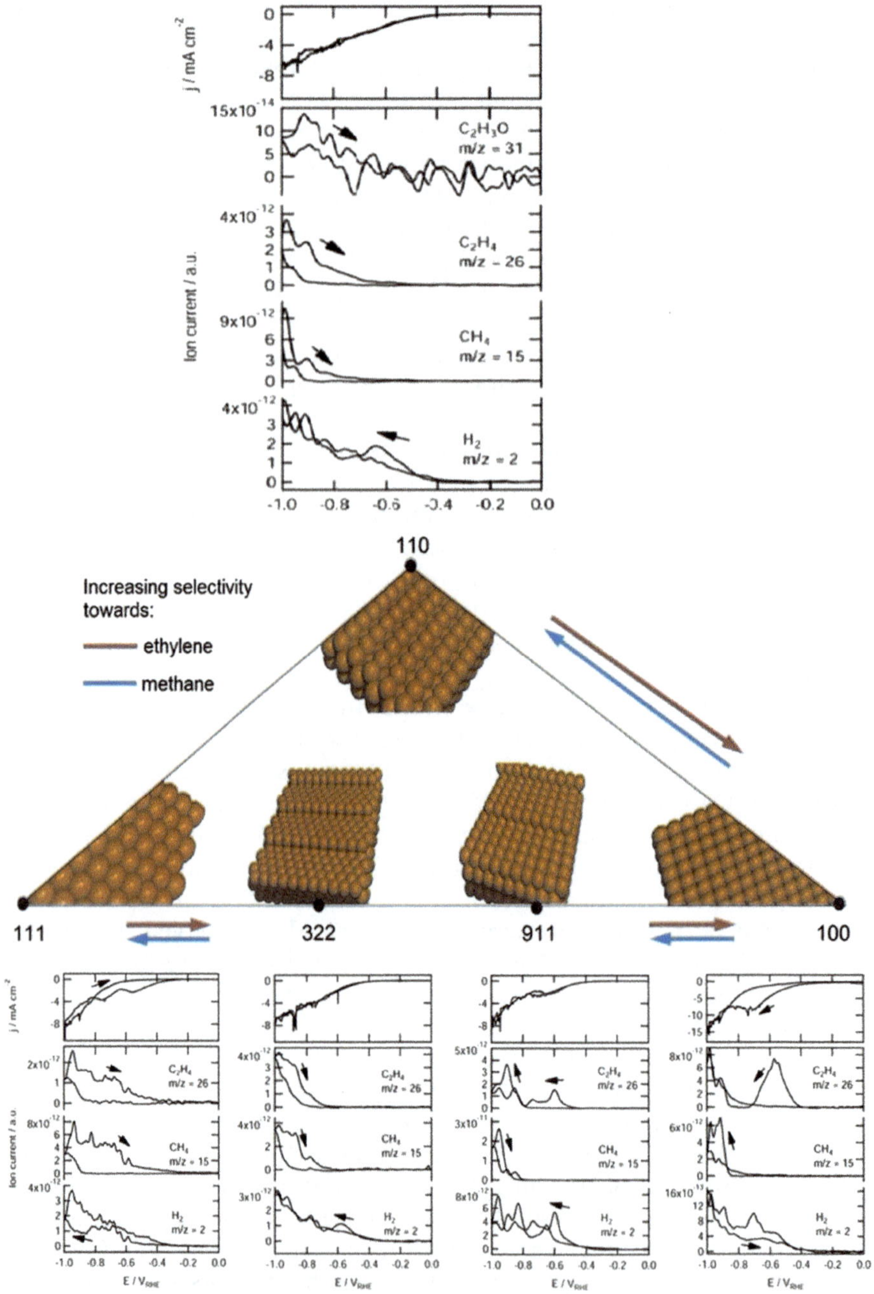

Figure 4.4 The reduction of CO in a CO-saturated (~1 mM) phosphate buffer (pH 7) on (a) Cu(911) and (b) Cu(100). The top panels show the corresponding cyclic voltammograms, and the bottom panels the associated mass fragments of volatile products measured with OLEMS. Adapted with permission from K. J. P. Schouten, E. Pérez Gallent and M. T. M. Koper, *ACS Catal.*, 2013, 3, 1292.[39] Copyright 2013 American Chemical Society.

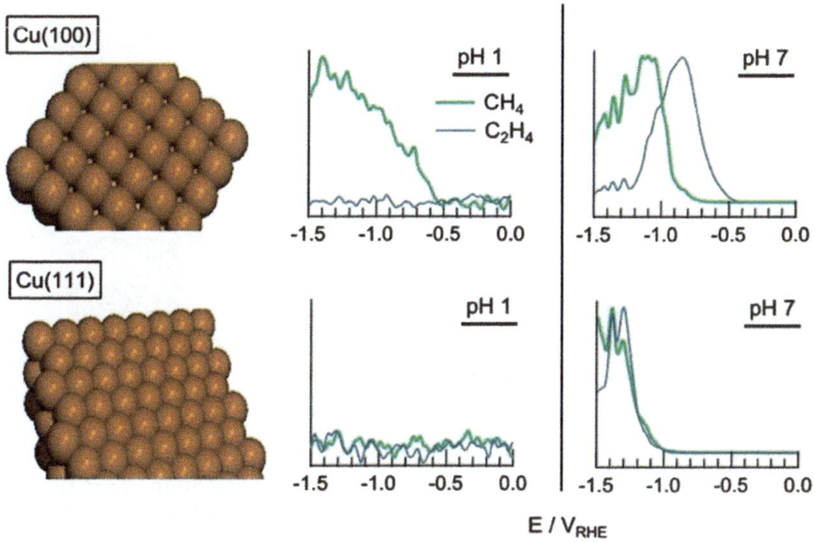

Figure 4.5 The reduction of CO_2 in 0.2 M phosphate buffers with pH 1 and 7 on Cu(111) (left) and Cu(100) (right). The formation of CH_4 ($m/z = 15$, green) and C_2H_4 ($m/z = 26$, blue) was followed using OLEMS.
Adapted from *Journal of Electroanalytical Chemistry*, **716**, K. J. P. Schouten, E. P. Galent and M. T. M. Koper, The influence of pH on the reduction of CO and CO_2 to hydrocarbons on copper electrodes, 53–57,[41] Copyright 2016,with permission from Elsevier.

Soriaga and co-workers have addressed a different, though extremely important aspect, namely the dynamic nature of the structure of copper surfaces under CO_2RR operating conditions, and the role of faceting on their activity and selectivity.[42,43] Using electrochemical scanning tunnelling microscopy (EC-STM), they showed that, after 30 min at −0.9 V *vs.* SHE in 0.1 M KOH, (111)-oriented facets are generated on polycrystalline copper surfaces. After a further 15 min at the same potentials, (100) ordered domains are generated.[42] No additional changes in the surface morphology can be observed after polarisation during more prolonged periods. This transformation was suggested to be the origin of the (100)-like selectivity of polycrystalline Cu electrodes in long-term electrolysis experiments. In a later work, the authors studied the effect of surface faceting in determining the product distribution of the CO reduction reaction (which can be considered as being embedded in the CO_2RR, as CO is thought to be the common intermediate leading to hydrocarbons and alcohols). Combining differential electrochemical mass spectrometry with STM they could show that the production of ethanol could be triggered by the formation of stepped (511) = [3(100)×(111)] facets (Figure 4.6). Formation of these (511) facets was accomplished by submitting a previously (100)-faceted polycrystalline Cu electrode (see above) to monolayer-limited $Cu \leftrightarrow Cu_2O$ oxidation–reduction cycles (ORCs).[43] Monolayer-limited ORCs did not induce such faceting on

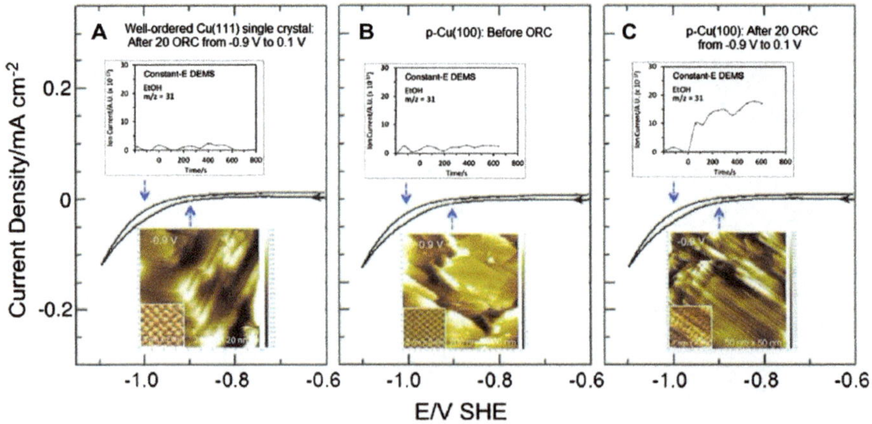

Figure 4.6 Combined (sequential) quasi-operando EC-STM and differential electrochemical mass spectrometry (DEMS) of (A) a well-defined Cu(100) single crystal after multiple ORCs, (B) a reconstructed Cu(pc)–[Cu(100)] without a prior ORC, and (C) a reconstructed Cu(pc)–[Cu(100)] after multiple ORCs, in CO-saturated 0.1 M KOH. The potentials for the DEMS and EC-STM measurements are indicated by the arrows. The DEMS signals were only for C_2H_5OH as the product. The EC-STM images were identical before and after potential excursions to −1.06 V, in the absence of CO in solution; no images could be obtained at that potential because of deleterious effects by the onset of the hydrogen evolution reaction and CO reduction.
Adapted from *Electrocatalysis*, Regulating the Product Distribution of CO Reduction by the Atomic Level Structural Modification of the Cu Electrode Surface, 7, 2016, 391–399,[43] Y.-G. Kim, A. Javier, J. H. Baricuatro and M. P. Soriaga, © Springer Science + Business Media New York 2016, with permission of Springer.

Cu(100) single-crystal electrodes, on which ethanol could not be detected either before or after the reduction/oxidation cycles. The formation of ethanol was attributed to the presence of (111) step sites adjacent to (100) terraces.

Varela *et al.* employed a different approach in the investigation of the effect of surface structure on the electrocatalytic properties of Cu electrodes, by depositing Cu overlayers on Pt(111) and Pt(211) substrates.[44] Although initially aimed at addressing the effect of both strain (the interatomic distance within pseudomorphic overlayers of Cu on Pt is larger than that on the surface of bulk Cu single-crystal electrodes) and steps sites on the electrocatalytic properties of Cu surfaces, surface reorganisation under reaction conditions exposed the Pt substrate, leading to a surface consisting of Cu islands on a Pt surface. This surface reorganisation was attributed to the formation of CO during the CO_2RR, and to the strong bond between the CO generated and Pt. As a consequence, a lower selectivity to CH_4 formation, and a higher selectivity to H_2 generation, was observed on these surfaces, as opposed to the surfaces of bulk Cu electrodes.

4.5 Computational Studies

Despite improvements in the selectivity and activity of CO_2RR electrocatalysts, the mechanism of the complex reactions involved and, in particular, how the structure of the catalyst surface impacts each one of the possible reaction pathways and steps, remains unknown. The development of new catalysts requires an in-depth understanding of the kinetics of the intrinsic steps that govern the activity and selectivity and, therefore, comprehensive information of the role of the surface structure of the catalyst in a series of sequential steps is needed. Computational studies allow exploring the feasibility of different reaction pathways, by determining the stability of possible reaction intermediates, which are often difficult or impossible to detect with experimental methods. They provide, hence, invaluable insight into the critical reaction steps, and into how to affect the activity and selectivity of CO_2RR catalysts. Obviously, computational calculations can only be performed on ideal surfaces, *i.e.*, on surfaces corresponding to those of single-crystal electrodes.

Free energy diagrams calculated by Rosen *et al.* suggest that Ag surfaces with steps and edges exhibit a significantly lower barrier for the first proton-coupled electron transfer in CO_2 reduction in comparison to the low-index surfaces (Figure 4.7).[45] A similar trend can be drawn for the CO_{ads} step, as Ag(211) exhibits a lower overall free energy change.

CO is believed to be the common intermediate in the generation of more reduced, hydrogenated products. The studies performed by Luo *et al.* suggest that the hydrogenation of CO* (where * indicates an adsorbed state) to hydroxymethylidyne (COH*) or formyl (CHO*) is a key step determining the selectivity of the reduction of CO_2 to hydrocarbons (Figure 4.8).[46] While Cu(111) favours the formation of COH*, which leads to CH_4 and C_2H_4, on Cu(100) the reaction proceeds through the formation of CHO*, and ethylene formation results from C–C coupling of two CHO* species. The potential-dependent coverage of CO* dictates the selectivity towards either COH* or CHO*, resulting in potential-dependent selectivity. At high overpotentials and high CO* coverage, the formation of COH* is preferred over CHO* on Cu(100), and methane and ethylene can be produced *via* a path similar to that on Cu(111).

Figure 4.9 shows the calculated limiting potential (U_L) as a function of the binding energy of CO for several reaction steps believed to be involved in the CO_2RR on transition metal electrodes.[47] The limiting potential is the minimum overpotential required for a given step to proceed. Consequently, the step with the most negative U_L for a specific binding energy of CO determines the theoretical overpotential of the overall reaction. As can be seen, the competitive path from adsorbed CO to COH* or CHO* dominates the theoretical overpotential in most of the metals. On those metals with moderate CO binding energy, like Pd, Ni and Cu, the CHO* route is more favoured. By contrast, on Pt, which binds CO more strongly, the COH* route is preferred. The vertical line at -0.5 eV shows the binding energy at which

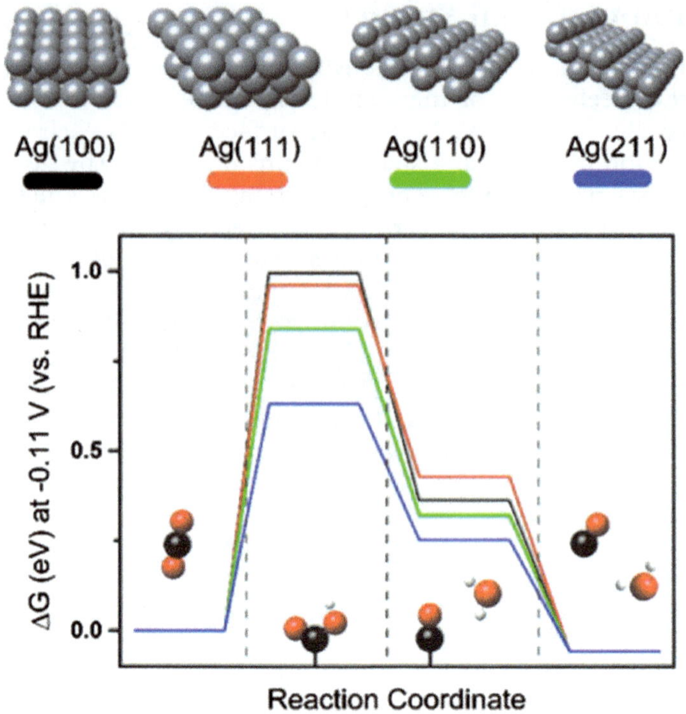

Figure 4.7 Free energy diagrams for the electrochemical reduction of CO_2 to CO on flat (Ag(100) and Ag(111)) and edge (Ag(221) and Ag(110)) surfaces. The first two steps include a simultaneous proton/electron transfer, with the final molecular surface configuration at each step depicted on the bottom of the graph. Values of ΔG correspond to an applied potential of -0.11 V *vs.* RHE. Sphere colours: white, H; black, C; red, O; silver, Ag.
Reprinted with permission from J. Rosen, G. S. Hutchings, Q. Lu, S. Rivera, Y. Zhou, D. G. Vlachos and F. Jiao, *ACS Catal.*, 2015, 5, 4293.[45]

adsorbed CO is in equilibrium with gaseous CO at a partial pressure of 1%. CO is expected to desorb at weaker binding energies. Au and Ag lie on the right side of this line so, once formed, CO will desorb from the surface of these metals. Catalysts effective for alcohol or hydrocarbon production are required to catalyse the protonation of adsorbed CO to adsorbed CHO or COH efficiently, and must simultaneously exhibit poor activity for the competitive HER. Cu electrodes appear at the top of the volcano, but still require a high overpotential for the CO to CHO hydrogenation step.

Along these lines, Shi *et al.* determined the limiting potential for the CORR on the (111) and (211) (*i.e.*, 3(111)×(100)) surfaces of several fcc transition metals using density functional theory (DFT)-based calculations.[51] As shown in Figure 4.10, the theoretical overpotential for the reduction of CO to *COH is similar on Au(211) and Cu(111). Thus, the presence of (100) steps

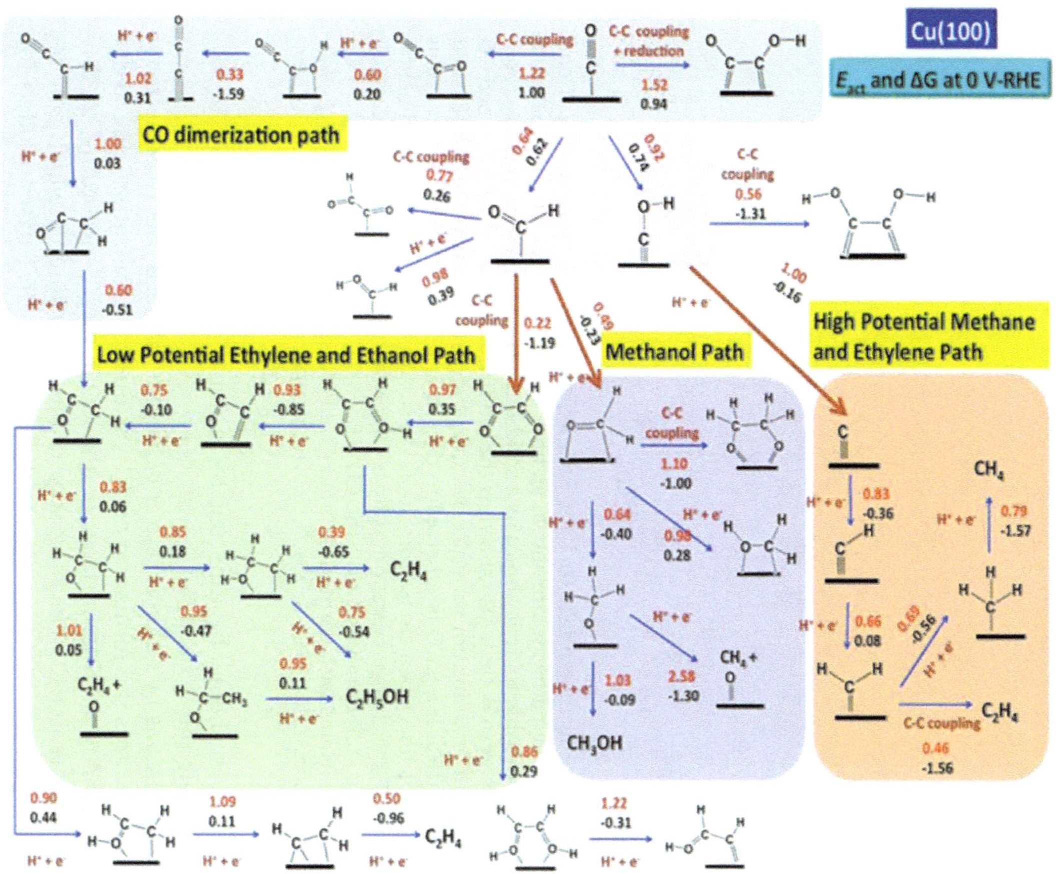

Figure 4.8 Overview of the reaction networks on Cu(100) examined starting from adsorbed CO (CO*). For every reaction step, two values are given. The values given in red are barriers (E_{act}) at 0 V *vs.* RHE, and the values given in black are reaction energies (ΔG) at 0 V *vs.* RHE. Reprinted with permission from W. Luo, X. Nie, M. J. Janik and A. Asthagiri, *ACS Catal.*, 2016, **6**, 219.[46] Copyright 2016 American Chemical Society.

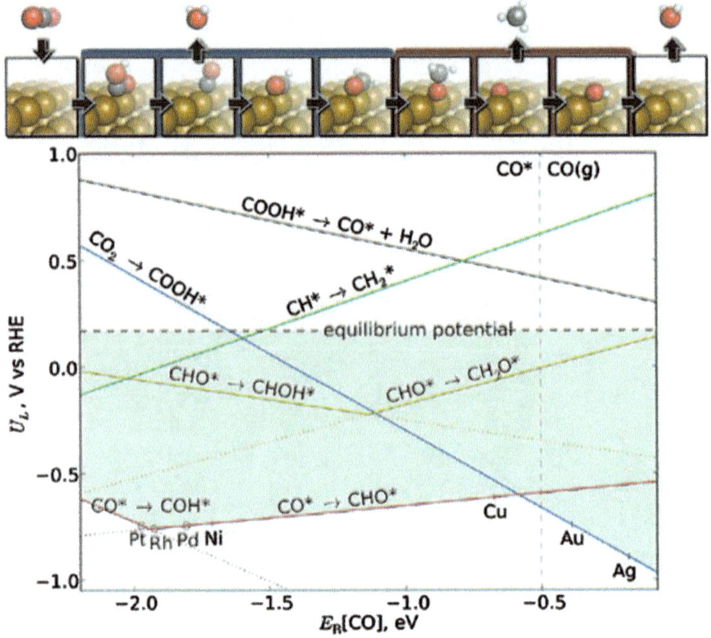

Figure 4.9 Limiting potentials (U_L) for competitive reaction mechanisms. Competitive pathways are shown as lines of the same colour, and the more favourable route is shown as a solid line, while the less favourable route is shown dotted. Also shown is the binding energy at which adsorbed CO is in equilibrium with gaseous CO at a partial pressure of 1% (dashed vertical line). CO is expected to desorb at binding energies weaker than this line. Open circles correspond to surfaces on which the experimental HER exchange current density is larger than 10^{-4} A cm^{-2},[48–50] on which the HER dominates over the CO_2RR.
Reprinted with permission from A. A. Peterson and J. K. Nørskov, *J. Phys. Chem. Lett.*, 2012, **3**, 251.[47] Copyright 2012 American Chemical Society.

on Au surfaces should be expected to promote the formation of hydrocarbons during the CO_2RR.

DFT calculations by Calle-Vallejo *et al.* suggest instead that the selectivity of Cu(100) towards the formation of C_2H_4 is due to stabilization, due to the square symmetry of this surface, of the CO dimer.[40] According to this mechanism, the CO–CO coupling step is rate determining, which can explain some experimental observations that previous DFT-based models for C–C coupling in the CO_2RR and CORR cannot reproduce.[52,53] It can also explain why, at low overpotentials, only C_2 products are observed on Cu(100),[34,35] while other models predict the simultaneous formation of CH_4 and C_2H_4, as well as pH-independent C_2H_4 formation on the SHE scale.[41] The relevance of CO–CO coupling in the CO_2RR pathway leading to C_2 products has also been recognised by Nørskov's group, which has very recently published a DFT study on the effect of strain, coverage and electric

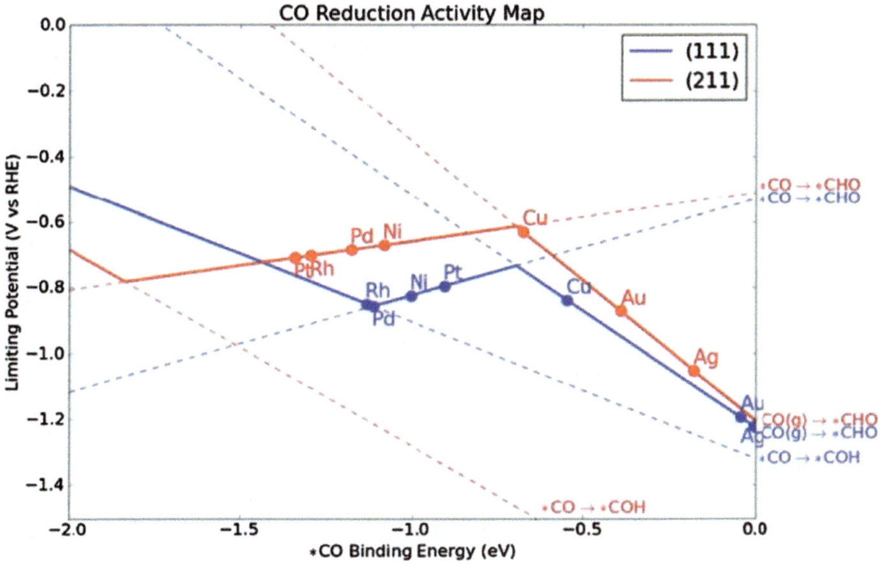

Figure 4.10 Map of the activity of (111) and (211) surfaces of different metals for the reduction of CO to hydrocarbons, based on the limiting-potential concept. The dashed lines correspond to elementary steps, while blue and red bold lines highlight the more favourable paths for (111) and (211) facets, respectively.
Reproduced from ref. 51 with permission from the PCCP Owner Societies.

field on the energetics of the CO–CO coupling step on Cu(100), (111) and (211).[54] According to this work, the electric field at potentials more negative than the pzc (*i.e.*, under reaction conditions at medium and high pH) makes CO dimerization facile on all three surfaces studied, with Cu(100) having the lowest activation barrier. All these works suggest that dimerization of adsorbed CO must play a relevant role in the formation of C_2 products in the CO_2RR. Since the formation of these final products must involve the insertion of H in the CO dimer, surface hydrogenation steps must also play a relevant role in the overall process.[54]

4.6 Relevant Considerations and Future Challenges

In order to determine structure–reactivity relationships in electrocatalysis, the preparation of reproducibly well-ordered and clean surfaces, and the use of ultra-pure solutions, is mandatory. Otherwise, the interpretation of the results might lead to misleading conclusions. During the 1980s and 1990s, experiments with Pt and Au single-crystal electrodes demonstrated the importance of the pre-treatment in obtaining clean and well-ordered surfaces for reproducible results.[55–64] Electropolishing pre-treatments have been proposed for the preparation of Cu single-crystal electrodes. However, very little is known about the effect of these pre-treatments in the roughness and the reconstruction of high index surfaces.[65,66] Taking into account the work by Magnussen and colleagues,[67–69] and the recent results by Soriaga and co-workers,[42] the understanding of the transformation of Cu(*hkl*) surfaces under CO_2RR conditions seems to be a significant topic to address. Adapting methods developed for the preparation *via* annealing in oxygen-free atmosphere of clean and well-ordered single-crystal electrodes of metals more reactive than Pt and Au[70] might provide a route to better control of the surface state. This is surprisingly an approach that, as far as we know, has not been attempted yet.

Barely any work has been done with alloy single-crystal electrodes. Metal alloys allow tuning the activity and selectivity through a combination of electronic and geometric effects, but very little is known on the effect of the surface structure of these systems.[71–77] Previous experimental and theoretical work on the ORR and CO electro-oxidation has revealed the importance of the surface structure of alloy systems on their electrocatalytic activity and selectivity.[78–80] It is then anticipated that changes in the surface structure of alloys with different compositions will also result in changes in their CO_2RR activity and selectivity.

Further information on catalytic activity and selectivity might also by obtained by tuning the surface properties *via* surface modification of single-crystal electrodes. The presence of irreversibly adsorbed species, like cyanide,[81,82] organic molecules[83] and inorganic adatoms[84–86] might change the electronic structure or the distribution of surface atomic ensembles, affecting the energetics of the adsorption of intermediate species and/or blocking some of the possible reaction pathways. The CO_2RR on Pt single-crystal electrodes modified with metal adatoms has been explored with promising results. Sanchez-Sanchez *et al.* have reported that Bi–Pt(111), Te–Pt(111), and Sb–Pt(100) electrodes show a significant increase in activity for CO_2 reduction in neutral bicarbonate buffer solution.[87]

Over the last few decades, the role of surface strain in catalysis and electrocatalysis has attracted the interest of the scientific community. Such strained surfaces, obtained by deposition of metal overlayers onto a single-crystal substrate, allow for fine tuning of the catalytic activity and selectivity of electrochemical reactions. The tensile strain or compressive stress in the overlayers results in changes in the electronic structure of the metal surface and, therefore, in the adsorption energy of intermediate species. Studies of the CO_2RR on Cu@Pt flat electrodes,[88] Cu@Au(111),[89] Au@Cu nanocubes[71] and Pd@Au nanoparticles[72,73] have demonstrated the capabilities and versatility of this approach. However, a better understanding of the electronic and structural properties of these overlayers on single-crystal electrodes is still required.

4.7 Conclusions

Despite the enormous amount of work devoted to understanding the CO_2RR on metal electrodes over the past 15 years, the reaction mechanism and other significant questions with the potential to impact our ability to develop more active and selective electrocatalysts remain unclear. We have discussed the surface structure parameters that influence the catalytic activity of metal surfaces for the CO_2RR, mainly in aqueous media. The work reviewed here has shown that the geometry of both step and terrace sites affects the kinetics and the reaction pathway followed.

The ability of copper and copper alloy electrodes to convert CO_2 into hydrocarbons is strongly pushing research on understanding the mechanism of the CO_2RR on these systems. In order to develop materials with optimal catalytic activity and selectivity major efforts and further investments in this area are needed. Some of the approaches being currently considered in order to optimize the binding energies of reaction intermediates like *CO, and to mitigate parasite reactions like the HER include (i) the development and understanding of metal alloys with preferential orientation,[90,91] (ii) the modification of electrode surfaces with ad-atoms[84,87] and (iii) the preparation of strained overlayers.[44,71–73,88] Finally, the long term durability of efficient catalysts and the changes in their selectivity as a function of time remains one of the main open questions in the immediate future.

References

1. J. Clavilier, D. Armand, S. G. Sun and M. Petit, *J. Electroanal. Chem.*, 1986, **205**, 267.
2. J. Solla-Gullón, F. J. Vidal-Iglesias, P. Rodríguez, E. Herrero, J. M. Feliu, J. Clavilier and A. Aldaz, *J. Phys. Chem. B*, 2004, **108**, 13573.
3. M. Duca, P. Rodriguez, A. I. Yanson and M. T. M. Koper, *Top. Catal.*, 2014, **57**, 255.
4. Y. Hori, H. Wakebe, T. Tsukamoto and O. Koga, *Electrochim. Acta*, 1994, **39**, 1833.
5. Y. Hori, in *Modern Aspects of Electrochemistry*, ed. C. Vayenas, R. White and M. Gamboa-Aldeco, Springer, New York, 2008, vol. 42, p. 89.
6. Y. Hori, A. Murata and R. Takahashi, *J. Chem. Soc., Faraday Trans. 1*, 1989, **85**, 2309.
7. A. A. Peterson, F. Abild-Pedersen, F. Studt, J. Rossmeisl and J. K. Nørskov, *Energy Environ. Sci.*, 2010, **3**, 1311.
8. R. Kortlever, J. Shen, K. J. P. Schouten, F. Calle-Vallejo and M. T. M. Koper, *J. Phys. Chem. Lett.*, 2015, **6**, 4073.
9. N. Hoshi and Y. Hori, *Electrochim. Acta*, 2000, **45**, 4263.
10. N. Hoshi, H. Ito, T. Suzuki and Y. Hori, *J. Electroanal. Chem.*, 1995, **395**, 309.
11. N. Hoshi, M. Noma, T. Suzuki and Y. Hori, *J. Electroanal. Chem.*, 1997, **421**, 15.
12. N. Hoshi, E. Sato and Y. Hori, *J. Electroanal. Chem.*, 2003, **540**, 105.
13. N. Hoshi, T. Suzuki and Y. Hori, *J. Electroanal. Chem.*, 1996, **416**, 61.
14. N. Hoshi, T. Suzuki and Y. Hori, *Electrochim. Acta*, 1996, **41**, 1647.

15. N. Hoshi, T. Uchida, T. Mizumura and Y. Hori, *J. Electroanal. Chem.*, 1995, **381**, 261.
16. A. Rodes, E. Pastor and T. Iwasita, *J. Electroanal. Chem.*, 1994, **369**, 183.
17. A. Rodes, E. Pastor and T. Iwasita, *J. Electroanal. Chem.*, 1994, **373**, 167.
18. A. Rodes, E. Pastor and T. Iwasita, *J. Electroanal. Chem.*, 1994, **377**, 215.
19. B. Z. Nikolic, H. Huang, D. Gervasio, A. Lin, C. Fierro, R. R. Adzic and E. Yeager, *J. Electroanal. Chem.*, 1990, **295**, 415.
20. B. A. Rosen, A. Salehi-Khojin, M. R. Thorson, W. Zhu, D. T. Whipple, P. J. A. Kenis and R. I. Masel, *Science*, 2011, **334**, 643.
21. F. A. Hanc-Scherer, M. A. Montiel, V. Montiel, E. Herrero and C. M. Sanchez-Sanchez, *Phys. Chem. Chem. Phys.*, 2015, **17**, 23909.
22. Y. Hori, A. Murata, K. Kikuchi and S. Suzuki, *J. Chem. Soc., Chem. Commun.*, 1987, 728.
23. N. Hoshi, M. Kato and Y. Hori, *J. Electroanal. Chem.*, 1997, **440**, 283.
24. P. Rodriguez, Y. Kwon and M. T. M. Koper, *Nat. Chem.*, 2012, **4**, 177.
25. P. Rodríguez, A. A. Koverga and M. T. M. Koper, *Angew. Chem., Int. Ed.*, 2010, **49**, 1241.
26. P. Rodriguez, N. Garcia-Araez, A. Koverga, S. Frank and M. T. M. Koper, *Langmuir*, 2010, **26**, 12425.
27. P. Rodriguez, N. Garcia-Araez and M. T. M. Koper, *Phys. Chem. Chem. Phys.*, 2010, **12**, 9373.
28. P. Rodriguez, J. M. Feliu and M. T. M. Koper, *Electrochem. Commun.*, 2009, **11**, 1105.
29. D. M. Kolb, *Prog. Surf. Sci.*, 1996, **51**, 109.
30. H. De Jesús-Cardona, C. del Moral and C. R. Cabrera, *J. Electroanal. Chem.*, 2001, **513**, 45.
31. Y. Hori, K. Kikuchi, A. Murata and S. Suzuki, *Chem. Lett.*, 1986, **15**, 897.
32. K. P. Kuhl, E. R. Cave, D. N. Abram and T. F. Jaramillo, *Energy Environ. Sci.*, 2012, **5**, 7050.
33. Y. Hori, R. Takahashi, Y. Yoshinami and A. Murata, *J. Phys. Chem. B*, 1997, **101**, 7075.
34. Y. Hori, H. Wakebe, T. Tsukamoto and O. Koga, *Surf. Sci.*, 1995, **335**, 258.
35. Y. Hori, I. Takahashi, O. Koga and N. Hoshi, *J. Phys. Chem. B*, 2002, **106**, 15.
36. Y. Hori, I. Takahashi, O. Koga and N. Hoshi, *J. Mol. Catal. A: Chem.*, 2003, **199**, 39.
37. R. Kortlever, K. H. Tan, Y. Kwon and M. T. M. Koper, *J. Solid State Electrochem.*, 2013, **17**, 1843.
38. W. Tang, A. A. Peterson, A. S. Varela, Z. P. Jovanov, L. Bech, W. J. Durand, S. Dahl, J. K. Nørskov and I. Chorkendorff, *Phys. Chem. Chem. Phys.*, 2012, **14**, 76.
39. K. J. P. Schouten, E. Pérez Gallent and M. T. M. Koper, *ACS Catal.*, 2013, **3**, 1292.
40. F. Calle-Vallejo and M. T. M. Koper, *Angew. Chem., Int. Ed.*, 2013, **52**, 7282.

41. K. J. P. Schouten, E. Pérez Gallent and M. T. M. Koper, *J. Electroanal. Chem.*, 2014, **716**, 53.
42. Y.-G. Kim, J. H. Baricuatro, A. Javier, J. M. Gregoire and M. P. Soriaga, *Langmuir*, 2014, **30**, 15053.
43. Y.-G. Kim, A. Javier, J. H. Baricuatro and M. P. Soriaga, *Electrocatalysis*, 2016, **7**, 391.
44. A. S. Varela, C. Schlaup, Z. P. Jovanov, P. Malacrida, S. Horch, I. E. L. Stephens and I. Chorkendorff, *J. Phys. Chem. C*, 2013, **117**, 20500.
45. J. Rosen, G. S. Hutchings, Q. Lu, S. Rivera, Y. Zhou, D. G. Vlachos and F. Jiao, *ACS Catal.*, 2015, **5**, 4293.
46. W. Luo, X. Nie, M. J. Janik and A. Asthagiri, *ACS Catal.*, 2016, **6**, 219.
47. A. A. Peterson and J. K. Nørskov, *J. Phys. Chem. Lett.*, 2012, **3**, 251.
48. S. Trasatti, *J. Electroanal. Chem.*, 1972, **39**, 163.
49. J. O. M. Bockris and A. K. Reddy, *Modern Electrochemistry*, Springer, 1970, vol. 2, p. 623.
50. D. Eberhardt, E. Santos and W. Schmickler, *J. Electroanal. Chem.*, 1999, **461**, 76.
51. C. Shi, H. A. Hansen, A. C. Lausche and J. K. Nørskov, *Phys. Chem. Chem. Phys.*, 2014, **16**, 4720.
52. J. H. Montoya, A. A. Peterson and J. K. Nørskov, *ChemCatChem*, 2013, **5**, 737.
53. X. Nie, M. R. Esopi, M. J. Janik and A. Asthagiri, *Angew. Chem., Int. Ed.*, 2013, **52**, 2459.
54. R. B. Sandberg, J. H. Montoya, K. Chan and J. K. Nørskov, *Surf. Sci.*, 2016, **654**, 56.
55. J. Clavilier, D. Armand and B. L. Wu, *J. Electroanal. Chem.*, 1982, **135**, 159.
56. J. Clavilier, R. Faure, G. Guinet and R. Durand, *J. Electroanal. Chem.*, 1980, **107**, 205.
57. V. Climent and J. Feliu, *J. Solid State Electrochem.*, 2011, **15**, 1297.
58. L. A. Kibler, A. Cuesta, M. Kleinert and D. M. Kolb, *J. Electroanal. Chem.*, 2000, **484**, 73.
59. D. Kolb, *J. Solid State Electrochem.*, 2011, **15**, 1391.
60. D. M. Kolb, *Angew. Chem., Int. Ed.*, 2001, **40**, 1162.
61. X. Gao, A. Hamelin and M. J. Weaver, *Phys. Rev. Lett.*, 1991, **67**, 618.
62. X. Gao, A. Hamelin and M. J. Weaver, *Phys. Rev. B*, 1991, **44**, 10983.
63. X. Gao, A. Hamelin and M. J. Weaver, *J. Chem. Phys.*, 1991, **95**, 6993.
64. X. Gao, A. Hamelin and M. J. Weaver, *Phys. Rev. B*, 1992, **46**, 7096.
65. G. M. Brisard, E. Zenati, H. A. Gasteiger, N. Markovic and P. N. Ross, *Langmuir*, 1995, **11**, 2221.
66. K. J. P. Schouten, Y. Kwon, C. J. M. van der Ham, Z. Qin and M. T. M. Koper, *Chem. Sci.*, 2011, **2**, 1902.
67. M. R. Vogt, A. Lachenwitzer, O. M. Magnussen and R. J. Behm, *Surf. Sci.*, 1998, **399**, 49.
68. H. Matsushima, A. Taranovskyy, C. Haak, Y. Gründer and O. M. Magnussen, *J. Am. Chem. Soc.*, 2009, **131**, 10362.

69. H. Matsushima, C. Haak, A. Taranovskyy, Y. Grunder and O. M. Magnussen, *Phys. Chem. Chem. Phys.*, 2010, **12**, 13992.
70. A. Cuesta, L. A. Kibler and D. M. Kolb, *J. Electroanal. Chem.*, 1999, **466**, 165.
71. J. Monzo, Y. Malewski, R. Kortlever, F. J. Vidal-Iglesias, J. Solla-Gullon, M. T. M. Koper and P. Rodriguez, *J. Mater. Chem. A*, 2015, **3**, 23690.
72. D. Plana, J. Florez-Montaño, V. Celorrio, E. Pastor and D. J. Fermin, *Chem. Commun.*, 2013, **49**, 10962.
73. J. J. L. Humphrey, D. Plana, V. Celorrio, S. Sadasivan, R. P. Tooze, P. Rodríguez and D. J. Fermín, *ChemCatChem*, 2016, **8**, 952.
74. J. Christophe, T. Doneux and C. Buess-Herman, *Electrocatalysis*, 2012, **3**, 139.
75. H. Mistry, R. Reske, P. Strasser and B. Roldan Cuenya, *Catal. Today*, 2017, **288**, 30.
76. R. Kortlever, I. Peters, S. Koper and M. T. M. Koper, *ACS Catal.*, 2015, **5**, 3916.
77. Z. Xu, E. Lai, Y. Shao-Horn and K. Hamad-Schifferli, *Chem. Commun.*, 2012, **48**, 5626.
78. V. Stamenkovic, M. Arenz, B. B. Blizanac, K. J. J. Mayrhofer, P. N. Ross and N. M. Markovic, *Surf. Sci.*, 2005, **576**, 145.
79. V. Stamenkovic, B. Fowler, B. S. Mun, G. Wang, P. N. Ross, C. A. Lucas and N. M. Markovic, *Science*, 2007, **315**, 493.
80. V. Stamenkovic, T. J. Schmidt, P. N. Ross and N. M. Markovic, *J. Phys. Chem. B*, 2002, **106**, 11970.
81. A. Cuesta, *ChemPhysChem*, 2011, **12**, 2375.
82. M. Escudero-Escribano, G. J. Soldano, P. Quaino, M. E. Zoloff Michoff, E. P. M. Leiva, W. Schmickler and Á. Cuesta, *Electrochim. Acta*, 2012, **82**, 524.
83. B. Genorio, D. Strmcnik, R. Subbaraman, D. Tripkovic, G. Karapetrov, V. R. Stamenkovic, S. Pejovnik and N. M. Markovic, *Nat. Mater.*, 2010, **9**, 998.
84. I. Kerbach, V. Climent and J. M. Feliu, *Electrochim. Acta*, 2011, **56**, 4451.
85. E. Herrero, J. M. Feliu and A. Aldaz, *J. Electroanal. Chem.*, 1994, **368**, 101.
86. E. Herrero, A. Fernandez-Vega, J. M. Feliu and A. Aldaz, *J. Electroanal. Chem.*, 1993, **350**, 73.
87. C. M. Sánchez-Sánchez, J. Souza-Garcia, E. Herrero and A. Aldaz, *J. Electroanal. Chem.*, 2012, **668**, 51.
88. R. Reske, M. Duca, M. Oezaslan, K. J. P. Schouten, M. T. M. Koper and P. Strasser, *J. Phys. Chem. Lett.*, 2013, **4**, 2410.
89. A. Eilert, F. S. Roberts, D. Friebel and A. Nilsson, *J. Phys. Chem. Lett.*, 2016, 7, 1466.
90. H. A. Hansen, C. Shi, A. C. Lausche, A. A. Peterson and J. K. Nørskov, *Phys. Chem. Chem. Phys.*, 2016, **18**, 9194.
91. Z. P. Jovanov, H. A. Hansen, A. S. Varela, P. Malacrida, A. A. Peterson, J. K. Nørskov, I. E. L. Stephens and I. Chorkendorff, *J. Catal.*, 2016, **343**, 215.

CHAPTER 5

Homogeneous M(bpy)(CO)$_3$X and Aromatic N-heterocycle Catalysts for CO_2 Reduction

MITCHELL C. GROENENBOOM, KARTHIKEYAN SARAVANAN AND JOHN A. KEITH*

Department of Chemical and Petroleum Engineering, Swanson School of Engineering University of Pittsburgh, 804 Benedum Hall, 3700 O'Hara Street, Pittsburgh, PA 15261, USA
*Email: jakeith@pitt.edu

5.1 Introduction

Sustainable, efficient and economical CO_2 utilization as a chemical feedstock addresses two potentially catastrophic problems facing humanity.[1–3] First, it would stem the accumulation of anthropogenic CO_2, which is correlated with severe weather patterns[4] and global climate change[5] that bring severe economic consequences. Second, it would alleviate the global dependence on petroleum for transportation fuels and petrochemical feedstocks while allowing greater flexibility in how finite petroleum reserves are used. Industrially scalable routes to convert CO_2 to liquid fuels (*e.g.* methanol) are greatly desired but not yet available. The US Energy Information Administration projects that more 32% of the energy used by OECD nations will come from liquid fuels until 2040.[6] There is a massive barrier for completely abandoning petroleum based liquid fuels for other energy

Energy and Environment Series No. 21
Electrochemical Reduction of Carbon Dioxide: Overcoming the Limitations of Photosynthesis
Edited by Frank Marken and David Fermin

Published by the Royal Society of Chemistry, www.rsc.org

sources, so chemical processes that regenerate fuels and petrochemicals from post-combustion CO_2 are greatly desired.

Regenerating petroleum *via* carbon-neutral[7] CO_2 recycling processes requires that anthropogenic CO_2 be captured[8] and then converted.[3,9] Synthesizing chemicals and fuels from coal, natural gas, or renewable forms of carbon inevitably requires H_2, which is most economically generated by steam reforming fossil fuels like coal or natural gas. Unfortunately, these processes release CO_2. Producing H_2 from water electrolysis itself does not release CO_2, but this requires electricity that is typically generated from burning coal or natural gas. Entirely solar-driven processes[10] may hopefully soon bring sustainable and economical carbon-neutral solar fuels, but doing so requires improved fundamental understanding of how to activate and convert CO_2.

Converting CO_2 into useful products requires substantial amounts of energy. Possible schemes to do so range from utilizing bacterial microorganisms,[11] molten salts,[12,13] formate and CO dehydrogenase enzymes[14,15] enzyme surrogate models,[16,17] molecular homogeneous catalysts,[15,18–23] industrial thermal processes,[24] solar-powered metal oxide reactors,[25] and electrochemical methods in traditional[26] or solid oxide[27] devices as well as continuous flow electrolyzers.[28] Though these processes are diverse, at an atomic level, CO_2 conversion fundamentally requires energetically efficient hydrogentations under different chemical environments. An atomic-level understanding of how different H-transfer steps occur would likely provide fundamental insight into how such processes can be better engineered.

Many consider electrochemical CO_2 reduction a promising avenue to investigate. Unfortunately, the electricity consumed by these processes is often generated by burning fossil fuels. To design more sustainable and environmentally friendly processes, we can utilize photovoltaic devices to generate carbon-neutral electricity from sunlight that can in turn be used to drive CO_2 reduction.[29] A common concept in electrochemistry is the standard redox potential, E, defined as the free energy change in an electrochemical reaction, ΔG, divided by the number of electrons transferred, n, and Faraday constant, F:

$$E = -\Delta G/nF. \tag{5.1}$$

Figure 5.1 lists several relevant standard redox potentials involved in CO_2 reduction and exemplifies why such processes are challenging. First, Figure 5.1(2) shows that adding one electron to CO_2 requires more than 1.9 eV (43.8 kcal mol^{-1}, since the redox potential corresponds to the lower limit of the actual reaction barrier). Second, while less-negative redox potentials are associated with simultaneously transferring multiple electrons and protons, such reactions carry an unfavorable entropy penalty, necessitating higher overpotentials. (Note that electrochemical reaction barriers govern activation overpotentials defined within the Butler–Volmer equation.) However, activation overpotentials nowadays are also often referred to as the extra thermodynamic energy, relative to the thermodynamic

$CO_2 + e^-$	$\longrightarrow$	$CO_2^{\bullet-}$	-1.90 V	(2)
$CO_2 + 2\,H^+ + 2\,e^-$	$\longrightarrow$	HCOOH	-0.43 V	(3)
$CO_2 + 2\,H^+ + 2\,e^-$	$\longrightarrow$	$CO + H_2O$	-0.52 V	(4)
$CO_2 + 4\,H^+ + 4\,e^-$	$\longrightarrow$	$HCHO + H_2O$	-0.48 V	(5)
$CO_2 + 6\,H^+ + 6\,e^-$	$\longrightarrow$	$CH_3OH + H_2O$	-0.38 V	(6)
$CO_2 + 8\,H^+ + 8\,e^-$	$\longrightarrow$	$CH_4 + 2\,H_2O$	-0.25 V	(7)
$2\,CO_2 + 12\,H^+ + 12\,e^-$	$\longrightarrow$	$CH_3CH_2OH + 3\,H_2O$	-0.33 V	(8)
$2\,CO_2 + 12\,H^+ + 12\,e^-$	$\longrightarrow$	$C_2H_4 + 4\,H_2O$	-0.34 V	(9)
$2\,H^+ + 2\,e^-$	$\longrightarrow$	H_2	-0.41 V	(10)

Figure 5.1 Standard redox potentials referenced to the standard hydrogen electrode (SHE) in aqueous electrolyte solutions (pH = 7) at 25 °C; taken from ref. 26 and 32.

equilibrium potential, required to make all sequential reaction intermediates downhill in energy. Though less-rigorous, this approximation is often heuristically valid.[30–32] Lastly, since electrochemical reduction involves proton and electron transfers, CO_2 reduction must compete with the hydrogen evolution reaction (Figure 5.1(10)). To maximize their efficiency, CO_2 electro-reduction catalysts must have the lowest possible overpotential while also inhibiting pathways that result in H_2 generation.

Another consideration is the energy required to convert CO_2 relative to the energy that is stored within the final reduced products. Often times the sunlight driven processes are energetically efficient, but they also have small thermodynamic driving forces to form products, and thus reaction rates are slow. High overpotential processes, on the other hand, have much higher driving forces for product formation but require much more energy.

In each case, there is knowledge that can be gained by studying catalytic reaction pathways, particularly by using first principles quantum chemistry. With these tools, reaction energies and barrier heights for elementary reactions can be predicted with reasonable degrees of accuracy. By linking together elementary pathways, we can better understand full reaction mechanisms, and obtain insights into how to make slow reaction steps occur faster and with less energy.

The intent of this chapter is to provide a contemporary review of two different electrochemical processes for CO_2 reduction, primarily from an atomic scale perspective. In the first section, we discuss reaction mechanisms involving homogeneous inorganic complexes that function as catalysts for CO_2 reduction. These complexes, first designed by Jean-Marie Lehn and co-workers in the 1980s, remain of interest to researchers due to their versatility in carrying out CO_2 reduction in different conditions and with different metal centers. In the second section, we discuss CO_2 reduction with aromatic N-heterocycles, a subject first reported by Andrew Bocarsly and co-workers in the mid-1990s. We will discuss our own contributions in these areas and outline outstanding questions and future directions.

5.2 CO_2 Reduction with Re and Mn Complexes

5.2.1 Background

Homogeneous catalysis is a highly active area of research for CO_2 reduction. Modern synthetic techniques allow inorganic complexes to be designed with specific functionalities that enable the transfer of electrons and protons to CO_2. Of particular interest are complexes designed with biomimetic functionalities that can transfer hydrides and/or protons in energetically efficient reaction pathways.[33] Although these complexes are rarely stable or active enough to function as industrially scalable catalysts, well-designed moieties could be integrated into material catalysts for enhanced reactivities.

Among the various catalysts studied, *fac*-Re(bpy-R)$(CO)_3$X (where bpy-R = 4,4-disubstituted-2,2-bipyridine and X = anionic ligand or solvent with counteranion) complexes reduce CO_2 to CO with extremely high faradaic efficiencies. Understanding the atomic scale mechanism of such efficient processes is important for fundamentally understanding artificial photosynthesis. Additionally, CO is a valuable component of syngas and can be further reduced to liquid fuels (*e.g.* methanol). There are at least two well-studied mechanisms for CO_2 reduction to CO using this Re catalyst.[34] One is a low overpotential (energetically efficient) but slow one-electron pathway. The other is a high overpotential (energetically inefficient) but rapid two-electron pathway.

In the one-electron process, Re(bpy-R)$(CO)_3$X undergoes a one-electron reduction to form the active species that catalyzes the reductive disproportionation of two CO_2 molecules to CO and CO_3^{2-}. The absence of current enhancement at the first reductive potential in cyclic voltammetry experiments indicates the slow rates of this reaction. Similarly, the photochemical reduction of CO_2 using this complex is also slow (only 5–12 turnovers per hour) and follows a similar pathway. The two-electron mechanism, on the other hand, has rapid kinetics (10–100 turnovers per second) but is limited by a very high overpotential that is only a few hundred millivolts away from the one-electron reduction potential of CO_2 (Figure 5.1(2)). Thus, this reaction's overpotential is far too high for practical applications. While a serious impediment, the high overpotentials can likely be avoided by utilizing synthetically modified ligand backbones and semiconductor electrodes.[35,36] The atomic scale reaction pathway seems to take place through two subsequent one-electron transfers to Re(bpy-R)$(CO)_3$X complexes (where R = 4,4-(H or *t*Bu)-2,2-bipyridine).

Interestingly, the rate for CO_2 reduction is dramatically increased in the presence of higher concentrations of Brønsted acids.[36,37] This is counter-intuitive since higher Brønsted acid concentrations would normally be expected to increase H_2 production. This indicates a particularly selective catalyst for CO_2 reduction.[38] Despite $[Re(bpy\text{-}R)(CO)_3]^-$ anions being extensively characterized,[36,39,40] the mechanism behind the two-electron process has remained elusive for many years and hindered the design of other catalysts that might make use of this rapid process with a low overpotential.

5.2.2 Re Complex Reaction Pathways

Several quantum chemistry studies have modeled the reaction mechanisms for these rhenium complexes. In the early 2000s, Fujita and Muckerman studied the electronic structure of Re(bpy)(CO)$_3$ complexes and used Kohn–Sham density functional theory (DFT) to understand the energetics of its dimerization process. Time-dependent DFT calculations on [Re(bpy)(CO)$_3$]$_2$ and [Re(bpy)(CO)$_3$]$^-$ identified the origin of UV–vis absorption bands in THF solvents.[43] More recent studies by Muckerman reported yet another feasible pathway for the one-electron process.[44,45]

Keith *et al.* studied the mechanism of electrocatalytic reduction of CO_2 by Re(bpy)(CO)$_3$Cl in acetonitrile solvents with added proton sources.[42] They used standard reduction potentials, pK_a values, reaction free energies, and activation barriers calculated with DFT to elucidate the electrochemical reaction mechanism. The calculated one-electron standard redox potentials for Re(bpy)(CO)$_3$Cl (**1**) to form [Re(bpy)(CO)$_3$Cl]$^-$ (**3**) and then subsequently reduce **3** to form [Re(bpy)(CO)$_3$]$^-$ (**2**) + Cl^- were found to be −1.50 V and −1.68 V *vs.* SCE, respectively. These values agree reasonably well with the experimental standard redox potentials of −1.34 V and −1.73 V *vs.* SCE, respectively.[38] The difference in standard redox potentials between the DFT calculations and experimental results were found to decrease to −1.41 and −1.74 V *vs.* SCE, respectively when an explicit counterion (in this case K^+) was added to the calculation of the neutral and anionic complexes. This signified to the authors that the quantum chemical model benefited from the inclusion of a counterion that would be expected to exist in the local solvation environment of an electrolyte.

Two distinct pathways branch off after the formation of complex **2**, one that would result in the formation of H_2 and the other resulting in the formation of CO. A possible intermediate to form H_2 involves proton addition to **2** generating Re(bpy)(CO)$_3$H (**4**) followed by a one-electron reduction to form [Re(bpy)(CO)$_3$H]$^-$ (**5**). The pathway leading to the formation of CO could involve the addition of CO_2 to **3** to form [Re(bpy)(CO)$_3$(CO_2)]$^-$ (**6**), followed by a proton addition to yield Re(bpy)(CO)$_3$(CO_2H) (**7**). Notably, DFT calculations found that **6** was not stable in the gas phase, but in the presence of a counterion and/or using molecular geometry optimized using a continuum solvation method, the complex was stable. This signified that the reaction environment plays a key role in determining which reaction intermediates are energetically feasible.

According to DFT calculations summarized in Figure 5.2, **4** is 1.0 kcal mol^{-1} more stable than **7**. **4** also has a substantially higher affinity for protons when compared to 7 (as evidenced by the pK_a of **4** being seven pK_a units higher than that of 7). The overall thermodynamics for the two pathways are slightly in favor of H_2 formation over the production of CO. However,

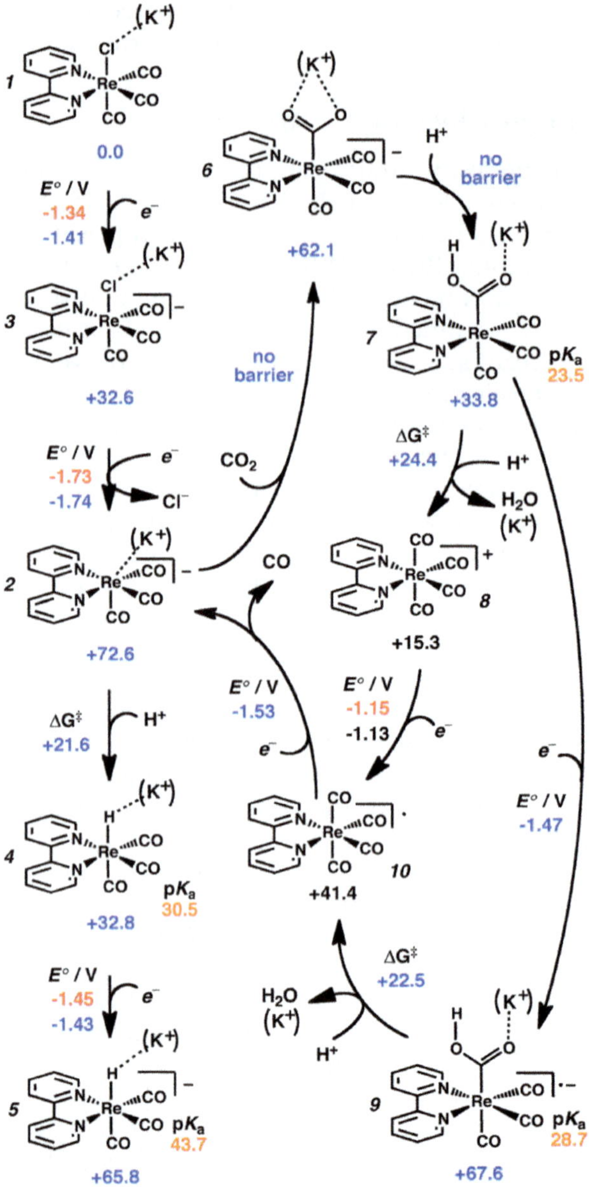

Figure 5.2 Reaction free energies (labeled below each complex and referenced to complex **1-K**) and barrier heights (ΔG) are reported in kcal mol^{-1}. Data colors: red = experimentally obtained reduction potentials; black = calculated data with no counterion present; blue = calculated data involving an explicit K^+ counterion interacting with the Re complex; orange = calculated pK_as obtained after applying a linear correction to reduce systematic errors in acetonitrile pK_as (similar to that done in ref. 41).
Reprinted with permission from J. A. Keith *et al.*, *J. Am. Chem. Soc.*, 2013, **135**, 15823–15829.[42]

reaction barriers must also be accounted for to identify the selectivity of the reaction. The barrier calculated by DFT for the protonation of **2** is $+10.1\ \text{kcal mol}^{-1}$ larger than the barrier for the $2+CO_2 \rightarrow 7$ pathway when MeOH is used as the Brønsted acid. Thus, while H_2 formation is the thermodynamic product, the barrier is substantially higher than that needed for the catalyst to form CO.

7 is known to be isolatable and relatively stable in solution. DFT calculations showed that two different pathways for the formation of CO are energetically accessible. In the first pathway, **7** can become protonated by a Brønsted acid to form complex **8** and H_2O, and this process is downhill in energy. **8** would then become reduced *via* a one-electron process to form the reduced complex **10**. In the other pathway, **7** could also undergo a one-electron reduction to form **9** with a standard redox potential of −1.41 V. The anion radical **9** would then readily become protonated by a Brønsted acid to form **10**. From **10**, CO can easily be removed after a one-electron reduction with a relatively low standard redox potential, −1.53 V. Accounting for all these data, the DFT predicted pathway at experimentally applied potentials has **2** adding CO_2, then a proton, then an electron, at which point a Brønsted acid must associate with **9** to catalyze rate-determining C–O bond breaking (Figures 5.3 and 5.4). Figure 5.5 shows transition state geometries for these processes and the participatory role of counterions in some of the reaction steps.

5.2.3 Mn Complex Pathways

Riplinger and Carter performed additional mechanistic studies of related Mn complexes.[46] Despite having superior activities and stability, the high cost and scarcity associated with second and third row transition metals prohibits their use in large scale applications. However, Mn is much more abundant than Re, making it economically suitable for practical scale-up. Compared to the Re complexes (**3** = −1.34 V, **2** = −1.73 V *vs.* SCE), the analogous Mn complexes had somewhat less negative standard redox potentials (−1.26 V and −1.50 V *vs.* SCE, respectively).[47]

The active intermediate in CO_2 reduction by Mn and Re complexes is formed *via* different pathways involving one-electron and two-electron reduction products (Figure 5.6). As discussed in the previous section, the Re complex (**1-Re**) undergoes a one-electron reduction to form a reduced species (**2X-Re**), which then loses its Cl^- ligand upon the second reduction in order to form the active catalyst (**3**). The Mn complex takes a different pathway where **1-Mn** is preferentially reduced to **2-Mn**, loses the Br^- ligand, and is reduced from **2-Mn** to **3-Mn**.

One other significant difference between the Mn and Re complexes is that the one-electron product of the former (**2-Mn**) forms a dimer (**2D-Mn**) more readily than the latter. The dimerization of Mn inhibits CO_2 reduction as it increases the overpotential to form the catalytically active **3-Mn**. Recent studies by Wishart, Ishitani, and their respective coworkers have shown that

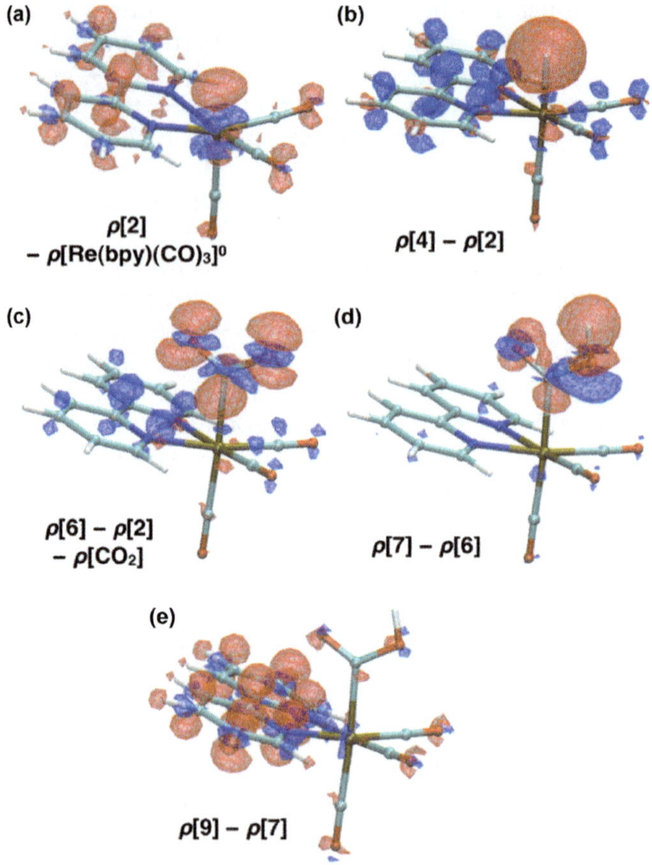

Figure 5.3 Density difference plots denoting the polarization that occurs upon adding electrons or binding molecules within the complexes. Red signifies increased charge density while blue signifies decreased charge density. Note that (a) shows that the HOMO orbital of **2** is delocalized within the π^* system of the bpy ligand; (b) and (c) show that binding H^+ or CO_2, respectively, to the Re center causes electron transfer from the bpy; (d) shows that adding a proton to **6** does not influence the electronic structure of the bpy, but (e) shows that the HOMO orbital of 7 is delocalized on the bpy ligand.
Reprinted with permission from J. A. Keith *et al.*, *J. Am. Chem. Soc.*, 2013, **135**, 15823–15829.[42] Copyright (2013) American Chemical Society.

the dimer is catalytically active in the photochemical reduction of CO_2 to formic acid.[48,49] Furthermore, functionalization of the bpy ligand with electron-donating groups, such as methyl groups, has improved the electrocatalytic reduction rates of CO_2 to CO.[50] These catalytic reactions are slower than those of the two-electron reduced complex, **3-Mn.**

Using the combined DFT/microkinetics approach, it was found that the Mn and Re catalysts take different catalytic routes to forming CO at their respective minimum operating potentials (−1.45 and −1.76 V *vs.* SCE for

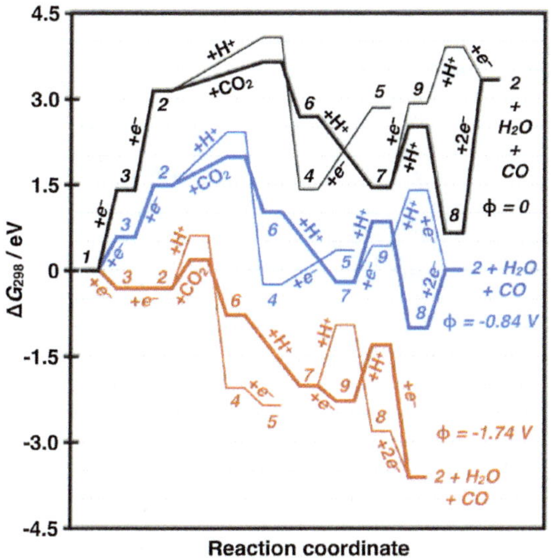

Figure 5.4 Potential dependent reaction mechanisms for the two-electron reduction of CO_2. The reaction proceeds rapidly at negative applied potentials that cause steps **2**→**3** to be downhill in energy.
Reprinted with permission from J. A. Keith *et al.*, *J. Am. Chem. Soc.*, 2013, **135**, 15823–15829.[42] Copyright (2013) American Chemical Society.

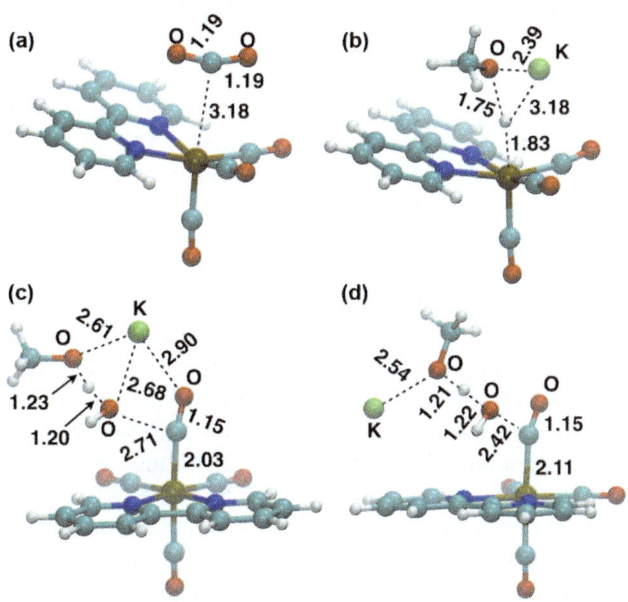

Figure 5.5 Transition state geometries for Re-catalyzed CO_2 reduction obtained using DFT. Labeled distances are in angstrom units.
Reprinted with permission from J. A. Keith *et al.*, *J. Am. Chem. Soc.*, 2013, **135**, 15823–15829.[42] Copyright (2013) American Chemical Society.

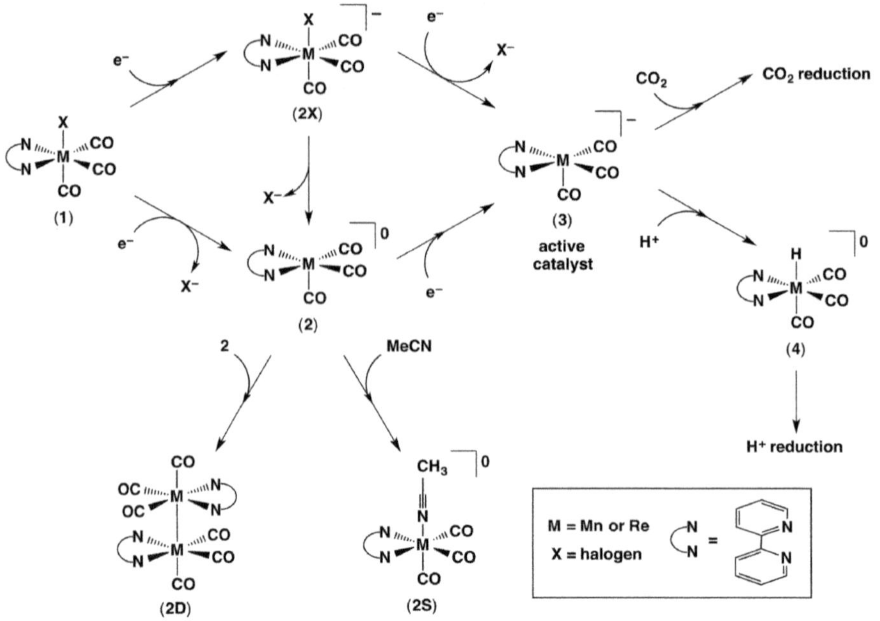

Figure 5.6 Overview of the possible products after one- and two-electron reductions of $Mn(bpy)(CO)_3Br$ (1-Mn) and $Re(bpy)(CO)_3Cl$ (1-Re).
Reprinted with permission from C. Riplinger *et al.*, *J. Am. Chem. Soc.*, 2014, **136**, 16285–16298.[46] Copyright (2014) American Chemical Society.

Mn and Re), as shown in Figure 5.7. For both catalysts, the first step involves CO_2 binding to the active catalyst complex (**3**), followed by proton addition. At this stage, one of the CO bonds has weakened in preparation for C–O bond cleavage due to the transfer of an electron from the bpy-ligand. The next step in the mechanism depends on the applied potential and the catalyst.

The Mn catalyst can operate through different pathways depending on the operating potential. At −1.4 V *vs.* SCE, C–O bond cleavage from **6** takes place by protonation of the OH group in **6-Mn** to form water and **7-Mn,** followed by another reduction and loss of an axial CO ligand to form **2-Mn** (which is further reduced to regenerate active catalyst **3-Mn**). However, at this potential, no catalysis is observed for Re catalysts since they form the two-electron reduced active complex only at −1.74 V *vs.* SCE. At −1.8 V *vs.* SCE, both the Mn and Re catalysts proceed *via* the same catalytic mechanism. After forming **6**, reduction of the bpy ligand produces **8**. The subsequent protonation of the –OH group initiates C–O bond cleavage, resulting in formation of either $[Mn(bpy)(CO)_3]^0$ (**2-Mn**) and release of CO or $[Re(bpy)(CO)_4]^0$ (**2CO-Re**), respectively. **2CO-Re** can undergo ligand exchange with an acetonitrile molecule at low CO concentrations ([CO]<1 mM). An additional one-electron reduction of the resulting intermediates regenerates the active

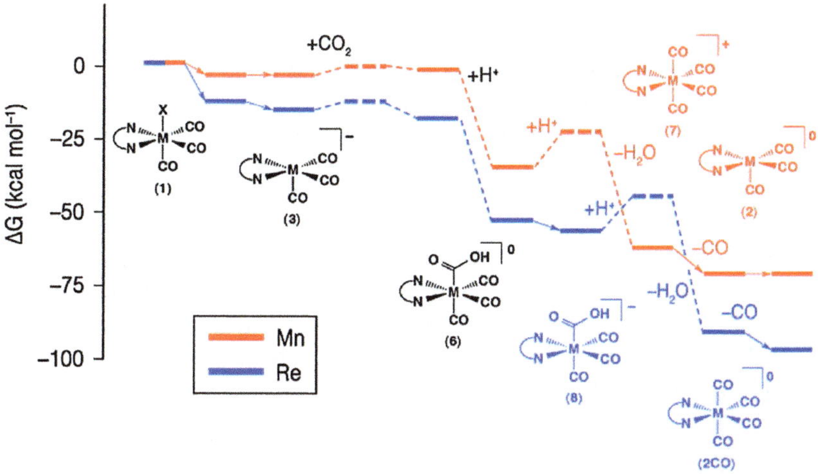

Figure 5.7 Reaction pathways for the catalytic reaction cycle for the Mn and Re catalysts. Reaction intermediates are depicted with solid lines and transition states with dashed lines. Arrows indicate electron transfer steps. The mechanisms are calculated at the least negative operating potentials for each catalyst (approximately −1.45 and −1.76 V *vs.* SCE for Mn and Re, respectively).
Reprinted with permission from C. Riplinger *et al.*, *J. Am. Chem. Soc.*, 2014, **136**, 16285–16298.[46] Copyright (2014) American Chemical Society.

catalyst (**3**). Goddard and co-workers later found that by replacing the bipyridine ligand in the Mn complex with bipyrimidine, the reduction potential required to convert **6** to **8** is less negative by 0.5 V and predicted the required overpotential to be nearly 0.25 V lower.[51]

The high selectivity of Re and Mn complexes towards CO_2 reduction over H_2 evolution arises from the high barrier of binding a proton when compared to binding a CO_2 to the active complex (**3**). However, CO_2 binding to **3-Mn** is endergonic while it is exergonic for **3-Re**. The process becomes thermodynamically favorable if **3-Mn** gets protonated by a weak acid, explaining the necessity of a weak Brønsted acid for the Mn-catalysts to be active. Riplinger and Carter also studied the role of weak Brønsted acids in this process.[52] They found that turnover frequencies of Mn and Re complexes depend on the type of Brønsted acid used, and that the Mn catalyst exhibits no catalytic turnover without Brønsted acid for the reason discussed above. Furthermore, catalysis using some weak acids require more negative applied potential or higher acid concentrations to protonate CO_2 or stabilize CO_2 binding. The reaction rates of each system vary with the acid used and achieve maximum rates with different acids. In the absence of Brønsted acids, the solvent acetronitrile can act as the proton donor for Re catalysts but prohibits the cleavage of C–O bonds in systems containing the Mn complexes.

5.2.4 Subsequent Studies

Later studies on M(bpy)$(CO)_3$X complexes (M = Re, Mn) have focused on improving the rate of electrocatalytic reduction of CO_2 by changing the metal center and ligands of the metal complexes. Tetracarbonyl molybdenum and tungsten complexes [M(bpy-R)$(CO)_4$ (R = H, M = Mo, W; R = *t*Bu, M = Mo, W)] were found to be competent catalysts for CO_2 reduction through a two-electron reduction process but they possess slower rates than [M(bpy)$(CO)_3$X].[53] A study on how the methyl substituent positions in the bpy ligand impacts catalytic activity showed that the catalytic current response is decreased when a methyl substituent occupies $n = 3$ and 5 on the bpy ligand Re(n,n-dimethyl-bpy)$(CO)_3$Cl due to steric hindrance disrupting optimal charge transfer in the catalytic cycle.[54] Other ligand modification studies on Mn complexes include a bulky bpy ligand (6,6-dimesityl-2,2-bipyridine),[55] a non-aromatic α diimine ligand[56] and N-heterocyclic carbene (NHC-pyridine) based Mn complexes [MnBr(*N*-methyl-*N*-2-pyridylbenzimidazol-2-ylidene)$(CO)_3$].[57]

Interestingly, modification of the bipyridine ligand with methyl acetamidomethyl groups at the 4 and 4′ positions altered the dominant electrocatalytic response in acetonitrile. This change was attributed to the combination of two complexes to form a hydrogen bonded dimer.[59,60] An alternate mechanism involving the bimolecular Re complex involved a reductive disproportion reaction that generated CO and CO_3^{2-} from two equivalents of CO_2. These interesting results have led others to study heterobimetallic systems comprised of both Re and Mn complexes (as shown in Figure 5.8), which have been shown to result in a modest enhancement of the observed current as a result of a cooperative catalytic effect.[58]

Recently, Grice and Saucedo explored the electrochemistry of group 6 $M(CO)_6$ (M = Cr, Mo, or W) and group 7 $M_2(CO)_{10}$ (M = Mn or Re) metal carbonyl complexes and studied their electrocatalytic CO_2 reduction ability.[61] The study also focused on determining if the group 7 compounds were capable of CO_2 reduction without the use of the "non-innocent" ligands. Group 6 $M(CO)_6$ catalysts showed a significant increase in current densities under a CO_2 atmosphere, which corresponds to catalytic reduction of CO_2. On the other hand, group 7 complexes were inactive for CO_2 reduction, indicating their necessity for non-innocent ligands to facilitate CO_2 reduction. Adding protons did not improve the ability of group 7 catalysts to reduce CO_2, and inhibited the CO_2 reduction ability of the group 6. The distribution of products obtained varied with the presence of water with only CO and carbonate being formed at anhydrous conditions. The authors attribute the observed difference between the activities of group 6 and group 7 complexes in CO_2 reduction to the nuclear charges. A group 6 $[M(CO)_5]^{2-}$ species will have a negative charge that experiences a lesser net nuclear charge than that of the comparable group 7 species and therefore will be a more effective reducing agent as the electrons are more readily removed from the complex.

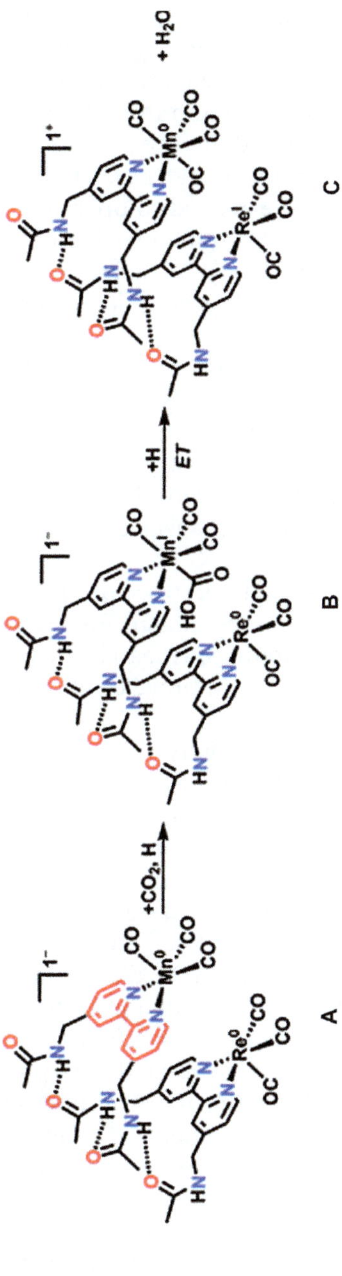

Figure 5.8 Proposed mechanism of CO_2 reduction by the hydrogen-bonded heterobimetallic active species A. Reproduced from ref. 58 with permission from The Royal Society of Chemistry.

5.3 Aromatic N-heterocycle Promoted Processes

5.3.1 Background

Many CO_2 conversion processes only operate with high overpotentials. Again, this is usually attributed to the high reduction potential to form $CO_2^{\bullet -}$ from CO_2 (Figure 5.1(2)) or the energy required to regenerate reaction sites on the catalyst by removing reaction intermediates (*i.e.*, CO or CHO).[62] Brønsted acids are often included in these systems to provide protons for proton-coupled electron transfers (PCET). PCET mechanisms are less endoergic,[63–65] and thus can be expected to operate at lower reaction overpotentials. Unfortunately, these protons also increase hydrogen evolution reaction activity and thus lower the overall selectivity and efficiency for CO_2 reduction. An ideal process would have a low overpotential and a high selectivity for CO_2 reduction.

The past decade has seen numerous experimental reports of CO_2 reduction occurring at low overpotentials with high faradaic efficiencies using aromatic N-heterocycles (ANHs) in aqueous solutions. Bocarsly and co-workers first reported this chemistry using electrolytes containing pyridinium with hydrogenated Pd electrodes.[66] Later, the chemistry was revisited using pyridinium[67,68] and imidazolium[69] to promote CO_2 reduction on several types of metal electrode.

While some reports have claimed not to see products,[70,71] others have,[72,73] raising the question of what the mechanism for these processes might be. There has been a significant amount of research towards using pyridine and other ANH containing molecules to improve electrochemical CO_2 reduction with metal electrodes.

It has been argued that this chemistry must have a surface dependence because ANH-promoted CO_2 reduction was reported to occur on Pt but not glassy carbon electrodes.[74] However, MacDonnell and co-workers[75] used homogeneous photochemical cells containing $[Ru(phen = phenanthroline)_3]^{2+}$ chromophores to reduce CO_2 to methanol in the presence of pyridinium. Furthermore, Dyer has reported ^{13}C-labeled experiments resulting in methanol using mercaptopteridine ANH molecules,[76] though interpretations of these results have recently been questioned by Tard and Savéant.[70]

Portenkichner *et al.* observed CO_2 reduction to methanol in the presence of pyridine and pyridazine on platinum electrodes.[72] They also note that no methanol was formed when only acetic acid was present, suggesting that pyridine plays an integral role in the reaction and serves as more than a proton source. Similar studies by Rybchenko,[73] Yang,[77] Chernyshova,[78] and Lee[79] have also reported CO_2 reduction to formate or methanol in the presence of pyridine, pyridine embedded into platinum electrodes, pyridine based polymers wrapped around copper electrodes, and pyridoxine (vitamin B-6, also an ANH molecule) at various faradaic efficiencies and overpotentials. However a number attempts to obtain similar results with analogous studies have found that these ANH molecules appear to only increase the rate of the hydrogen evolution.[70,71,80,81] Currently, there is no consensus on how ANH molecules serve in these reactions, but on the basis of work by Portenkirchner *et al.*, it appears the ANH molecules are playing a role beyond that of simply being a Brønsted acid. Understanding why ANH molecules might cause lower overpotentials and higher faradic efficiencies would provide helpful design principles for improved renewable energy catalysts.

5.3.2 Theoretical Studies of Homogeneous Mechanisms

Several computational groups have attempted to elucidate the role of ANH molecules in CO_2 electroreduction reactions with quantum chemistry calculations. This section discusses homogeneous reaction mechanism studies that do not explicitly account for reactions taking place at electrode surfaces. The impact of the electrode surface will be discussed later.

The first computational study related to ANH chemistry utilized DFT to calculate reaction energetics and HOMO orbitals of several proposed intermediate states.[67] Later, Tossell calculated additional thermodynamic energetics including pK_as and standard redox potentials, which showed that the one-electron reduction potential of pyridine and protonated pyridine were both significantly more negative (−2.90 and −1.44 V *vs.* SCE, respectively) than the reported experimental conditions (−0.58 V *vs.* SCE).[82] Tossell also calculated a series of ANH–CO_2 complexes, an example of two such complexes is shown in Figure 5.9. Binding CO_2 in this complex is energetically uphill for pyridine (complex (a) in Figure 5.9), but it is energetically more favorable with other ANH molecules (such as imidazole and 1,5,7-triaza-bicyclo[4.4.0]dec-5-ene (TBD)). The complexes also form spontaneously if either pyridine or CO_2 has been reduced by one electron (complex (b) in Figure 5.9). By calculating one-electron reduction potentials for these ANH–CO_2 complexes (−1.44 to −1.76 V *vs.* SHE) he showed that they were easier to reduce than CO_2 alone (−2.16 V *vs.* SCE). While imidazole and TBD more readily formed complexes with CO_2, the redox potentials of those complexes were significantly more negative (and less favorable) than that of the pyridine–CO_2 complexes. These calculations supported the claim that ANH molecules could potentially serve as CO_2 reduction catalysts.

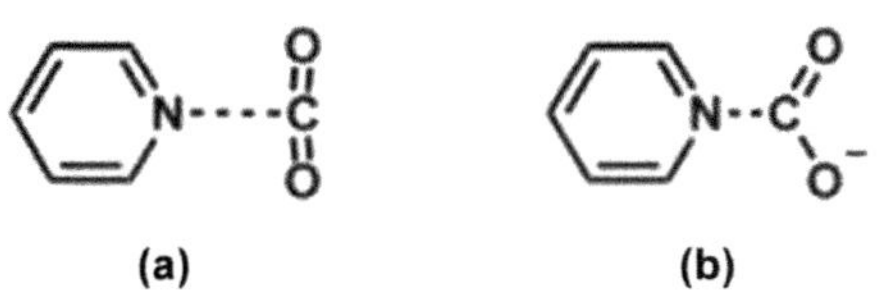

Figure 5.9 Example ANH–CO_2 complexes studied by Tossell. (a) A neutral pyridine and CO_2 complex. CO_2 remains linear, and the complex has a N–C length bond of 2.79 Å. (b) A negatively charged pyridine and CO_2 complex. CO_2 becomes bent, and the N–C bond length decreases to 1.46 Å.

Other mechanisms, such as those proposed by the Carter group and the Musgrave group considered pyridinium as a CO_2 reducing agent. The conclusions of these studies from the two groups differed in the extent to which the ANH molecule reduces CO_2. In 2012, calculations by Keith and Carter[83] reiterated that the one-electron reduction potential of pyridinium (−1.37 V *vs.* SCE) to form the pyridinyl radical ($PyH^•$) as well as the ANH–CO_2 complexes calculated by Tossell were significantly more negative than the experimental reduction potential (−0.58 V *vs.* SCE) reported by Bocarsly.[67] This extremely negative reduction potential means that it is very unlikely that $PyH^•$ would participate in CO_2 reduction unless at very high applied potentials or in the presence of photolysis conditions.[83,84] However, other ANH molecules, such as doubly protonated 4,4′-bipyridine, have less negative one-electron reduction potentials (−0.37 V *vs.* SCE) and could be more active reduction catalysts.

Keith and Carter later predicted that the reduction event observed at −0.58 V *vs.* SCE on Pt electrodes may actually correspond to the two-electron reduction of pyridine to dihydropyridine, see Figure 5.10.[85] By calculating the energy of various protonated and reduced states of pyridine they created a molecular Pourbaix diagram, as shown in Figure 5.11. Pourbaix diagrams are electrochemical phase diagrams that show the most stable state for a molecule (or material) at different applied potentials and pH values. These can be used to predict which form of reduced pyridine would be most stable near the reaction conditions. The Pourbaix diagram revealed that Py, PyH^+, and 1,4-dihydropyridine all have similar chemical potentials near the experimental conditions for CO_2 reduction. 1,4-dihydropyridine closely resembles the active moiety in NADH, one of nature's most active redox catalysts. This suggests that these Py species might reduce CO_2 through some type of coupled proton–electron transfer, or a biomimetic proton–hydride transfer reaction. These Pourbaix diagram triple-points may also serve as descriptors of the electrochemical conditions where ANH molecules are most active as proton/hydride transfer agents. Marjolin and Keith later continued in this direction to show that several different ANH molecules also have two-electron reduced species with standard redox potentials close to those for CO_2 reduction.[86] While this methodology produces useful thermodynamic descriptors, accurate reaction barriers would still be needed to fully understand this proposed mechanism.

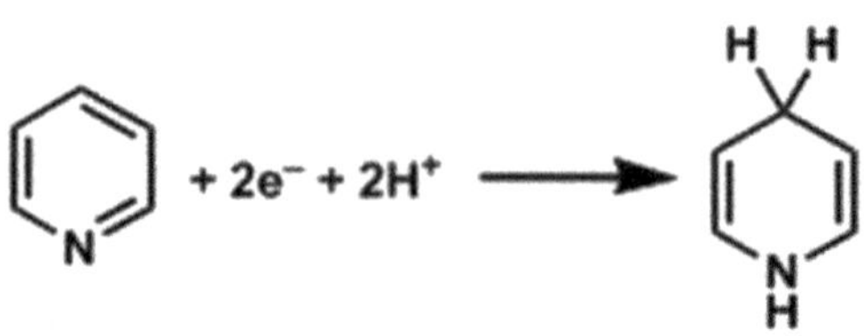

Figure 5.10 The reduction of pyridine into para-dihydropyridine, a pyridine derivative that Keith and Carter suggested as being responsible for CO_2 reduction.

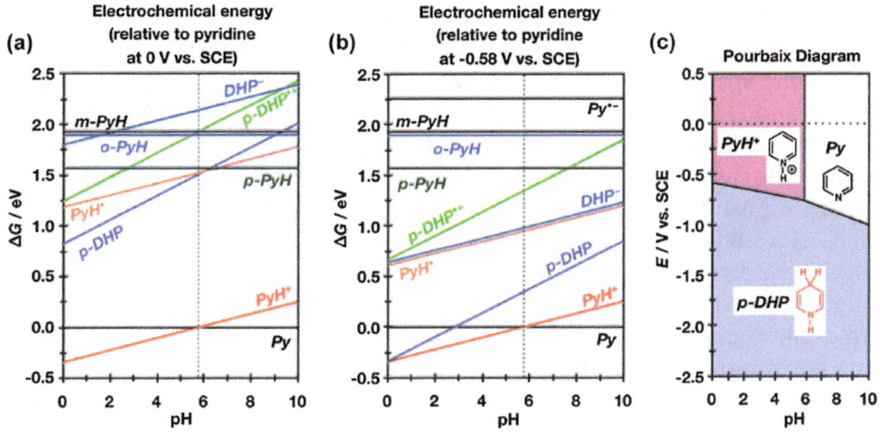

Figure 5.11 Relative electrochemical energies referenced to the energy of Py at an SCE potential of (a) 0 V and (b) −0.58 V *vs.* the SCE for pyridine species in solution at different pH. (The dotted line denotes the calculated p*K*a for PyH^+.) (c) Pourbaix diagram depicting the most thermodynamically stable species at a given pH and electrode potential. Calculation data here used high level (U)CCSD(T)-F12/aug-cc-pVTZ-F12 calculations.
Reproduced from ref. 85 with permission from The Royal Society of Chemistry.

Musgrave and coworkers presented data supporting a one-electron pyridine assisted CO_2 reduction mechanism in 2013.[87] Calculations on a Pt surface proposed that $PyH^\bullet$ can be formed, and that this radical can react with CO_2 to form $PyCOOH^\bullet$ through an inner-sphere electron transfer, as was originally proposed by Bocarsly.[67] While the reaction to form $PyCOOH^\bullet$ is energetically uphill, they report that the reaction barrier is substantially lower when the electron transfer is accompanied by a proton transfer across a water chain from the aqueous solvent. Because $PyH^\bullet$ has a very high pK_a (approximately 29 as reported by Keith and Carter[88]), the proton transfer must likely be coupled with the electron transfer if it is to occur. Similar proton relay mechanisms were reported by Siegbahn.[89] This proton relay can be significantly more energetically favorable than the direct reaction of $PyH^\bullet$ and CO_2 to produce PyH^+ and $CO_2^{\bullet-}$. While the formation of $PyCOOH^\bullet$ is energetically favorable, their mechanism does not account for the high energetic cost to form $PyH^\bullet$.

In 2014 Musgrave and coworkers[90,91] presented another pyridine assisted CO_2 reduction mechanism that utilizes 1,2-dihydropyridine to reduce CO_2. The argument was that although 1,2-dihydropyridine was less stable than the 1,4-dihydropyridine species proposed by Keith and Carter, the sequential proton and electron transfer steps to make it were more energetically accessible. Again, their reaction mechanism assumes the facile formation of $PyH^\bullet$, which has a very negative calculated redox potential, as previously mentioned, and very short lifetime in solutions.[83] They stress that there are several routes to generate PyH•, such as photochemical production (which we address below). The proton and electron addition to PyH• to form dihydropyridine species are much more favorable and occur at easily achievable pH (4.1) and electrode potentials (0.11 V *vs.* SCE), respectively. Once formed, they report barrier heights for elementary steps to reduce CO_2 into methanol through a series of coupled proton–hydride transfers with reaction barriers between 6 and 20 kcal mol^{-1} and quite exoergic reaction free energies between -15.5 and -36.7 kcal mol^{-1}.

Other experimental studies have presumed that pyridine assisted CO_2 reduction occurs through the PyH• species. MacDonnell and coworkers reported photochemical catalytic CO_2 reduction to formate and methanol using a ruthenium(II) trisphenanthroline chromophore and pyridine.[92] They achieved 76 and 0.15 turnovers per Ru for formate and methanol, respectively. It was presumed that once the PyH• was formed, it could proceed to reduce CO_2 by mechanisms similar to those originally postulated by Bocarsly and coworkers. PyH• can also be generated through photolysis and then react with CO_2 to form a PyH–$COO^{\bullet -}$ complex, as demonstrated by Colussi and coworkers.[84] Although the mechanism of generating PyH• is different in photolysis experiments, there is indication that high energy PyH• may result in PyH–$COO^{\bullet -}$ complexes.

5.3.3 ANH Reactions on Surfaces

We now turn to discuss studies of pyridine/ANH assisted CO_2 reduction processes that explicitly studied the role of electrode surfaces. Batista and coworkers were the first to report reaction pathways consisting of a proton coupled hydride transfer (PCHT) for ANH chemistry.[93] Their mechanisms involved the formation of a metal hydride on the platinum surface that then transferred to CO_2 while a nearby pyridinium ion donates a proton to result in CO_2 reduction into formic acid (HCOOH). The surface hydrogen could then regenerate by the one-electron reduction of a pyridinium ion near the electrode surface to regenerate a pyridine and a surface bound hydrogen atom. This process has a free energy barrier of 13 kcal mol^{-1}, and is predicted to occur at $E^0 = -0.72$ V *vs.* SCE, in good agreement with the original experimental redox potential reported for this process. In addition to providing protons during CO_2 reduction, pyridinium ions help establish a high proton concentration near the electrode surface necessary for the PCHT reaction. They later predicted that imidazole would exhibit similar electrochemical properties and facilitate CO_2 reduction with the same mechanism.[94] While their predicted redox potentials are in good agreement with the experimental measurements[95] for both pyridine and imidazole, CO_2 reduction products were not observed in the presence of acetic acid[72] (a weak acid with a pK_a similar to pyridine), as one might predict if pyridine only served as a proton shuttle, as suggested by this mechanism.

Bélanger and coworkers reported surface dependence of pyridinium reduction in the context of CO_2 reduction by using cyclic voltammetry on different metal surfaces.[96] However, Lucio and Shaw reported that gold electrodes behave differently than the earlier work with platinum electrodes.[71] They observed an irreversible reduction wave that they attributed to the one-electron reduction of pyridinium to the PyH$^•$ at −1.0 V *vs.* Ag/AgCl. They also noted an increase in reduction current when CO_2 and pyridine were present in solution, but no CO_2 reduction products. This is consistent with work by Savéant, which showed that this current enhancement was likely due to carbonic acid catalyzing the hydrogen evolution reaction.[74]

Keith and Carter have also reported calculations for ANH assisted CO_2 reduction on models for GaP photoelectrodes.[97,98] Their calculations predicted that proton and pyridinium ion reduction will be energetically unfavorable except at very negative electrode potentials, but the two-electron/two-proton reduction of pyridine to dihydropyridine on the GaP surface should be thermodynamically feasible ($E^0 = -0.63$ to -0.71 V *vs.* SCE). Studies by Bocarsly using GaP photoelectrodes reported very high faradaic efficiencies for CO_2 reduction. Keith and Carter suggested that the standard reduction potential of 1,4-dihydropyridine is not that different to that obtained in aqueous solution and thus remains similar to that needed to electrochemically convert CO_2 into a variety of products. This contrasts with previous arguments that the illuminated p-GaP electrode can produce pyridnyl radicals.[87]

To address this point, Lessio and Carter reported that the transfer of photoexcited electrons to pyridinium from the GaP electrode as well as pyridinium adsorption to the GaP surface were not energetically favorable.[99] The conduction band minimum of the GaP electrode lies too low in energy to transfer electrons to PyH$^+$. Investigating alternative mechanisms showed that reducing pyridinium to pyridine and an adsorbed hydrogen atom was more likely to occur than direct pyridinium reduction. This further supports a mechanism involving more than one-electron reductions. Koel and coworkers used STM to experimentally probe the spatial positions of the LUMO of pyridine adsorbed to a GaP surface. This technique can determine atomic sites most susceptible to nucleophilic attack. Figure 5.12 compares the experimental STM to simulated data from DFT, and a simpler DFT side view. The STM data clearly identifies the sites that would be susceptible to nucleophilic attack from a surface hydrogen to form an adsorbed dihydropyridine. While not conclusive proof of this mechanism, it lends further support to a mechanism involving a dihydropyridine like species that has been recently investigated with computational theory[100] and experiment.[101] As with the homogeneous studies, insight from high quality studies of barrier heights will likely be needed.

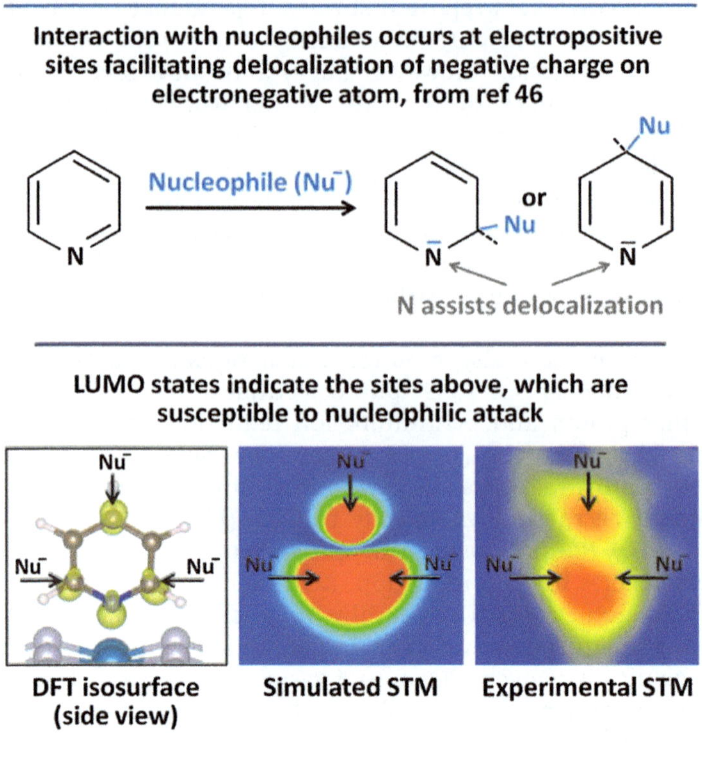

Figure 5.12 Combined experimental and DFT results showing that by spatially resolving the LUMO, the STM images predict the sites susceptible to nucleophilic attack[102] on adsorbed pyridine by adsorbed hydrides and protons from solution to produce 1,2- and 1,4-dihydropyridine, as described in ref. 98.
Reprinted with permission from C. X. Kronawitter *et al.*, *J. Phys. Chem. C*, 2015, **119**, 28917–28924.[103] Copyright (2015) American Chemical Society.

5.4 Conclusions

We have reviewed two different processes for CO_2 reduction whose mechanistic studies are still rich with interesting facets. For CO_2 reduction with inorganic complexes, progress has been made understanding high overpotential but very fast reactions, and efforts are underway to reduce the overpotentials while attempting to remain on fast and pathways selective for CO_2 reduction. These studies focus on designing and selecting ligands for inorganic complexes that improve their selectivity towards CO_2 reduction, rate of reaction, and stability in these reactive environments.

The ANH studies present a tantalizing process where energetically efficient CO_2 reduction to form methanol occurs, but there still is a lack of clarity how reaction mechanisms proceed. DFT calculations have indicated that one-electron standard redox potentials to form pyridinyl radicals are very negative, and other pathways are being pursued. In all cases, there is an important need to account for reaction barriers accurately as well as obtain spectroscopic characterization of intermediates.

Acknowledgements

This work was financially supported by the Department of Chemical & Petroleum Engineering at the University of Pittsburgh, the R. K. Mellon Foundation. Acknowledgment is made to the donors of The American Chemical Society Petroleum Research Fund (#55595-DNI4) for support of this research.

References

1. G. A. Olah, A. Goeppert and G. K. S. Prakash, in *Beyond Oil and Gas: The Methanol Economy*, Wiley-VCH Verlag GmbH & Co. KGaA, 2009, pp. 77–89.
2. M. Aresta, in *Carbon Dioxide as Chemical Feedstock*, Wiley-VCH Verlag GmbH & Co. KGaA, 2010, pp. 1–13.
3. E. A. Quadrelli, G. Centi, J.-L. Duplan and S. Perathoner, *ChemSusChem*, 2011, **4**, 1194–1215.
4. M. A. Bender, T. R. Knutson, R. E. Tuleya, J. J. Sirutis, G. A. Vecchi, S. T. Garner and I. M. Held, *Science*, 2010, **327**, 454–458.
5. H. Le Treut, R. Somerville, U. Cubasch, Y. Ding, C. Mauritzen, A. Mokssit, T. Peterson and M. Prather, in *Climate Change 2007: The Physical Science Basis. Contribution of Working Group I to the Fourth Assessment Report of the Intergovernmental Panel on Climate Change*, 2007.
6. *International Energy Outlook 2016, May 2016, U.S. Energy Information Administration, DOE/EIA-0484(2016),* tech. rep.
7. G. A. Olah, G. K. S. Prakash and A. Goeppert, *J. Am. Chem. Soc.*, 2011, **133**, 12881–12898.
8. K. Z. House, A. C. Baclig, M. Ranjan, E. A. van Nierop, J. Wilcox and H. J. Herzog, *Proc. Natl. Acad. Sci. U. S. A.*, 2011, **108**, 20428–20433.
9. Y. Izumi, *Coord. Chem. Rev.*, 2013, **257**, 171–186.
10. N. S. Lewis and D. G. Nocera, *Proc. Natl. Acad. Sci. U. S. A.*, 2006, **103**, 15729–15735.
11. M. W. Keller, G. J. Schut, G. L. Lipscomb, A. L. Menon, I. J. Iwuchukwu, T. T. Leuko, M. P. Thorgersen, W. J. Nixon, A. S. Hawkins, R. M. Kelly and M. W. W. Adams, *Proc. Natl. Acad. Sci. U. S. A.*, 2013, **110**, 5840–5845.
12. S. Licht, B. Wang and H. Wu, *J. Phys. Chem. C*, 2011, **115**, 11803–11821.
13. H. Yin, X. Mao, D. Tang, W. Xiao, L. Xing, H. Zhu, D. Wang and D. R. Sadoway, *Energy Environ. Sci.*, 2013, **6**, 1538–1545.
14. T. W. Woolerton, S. Sheard, Y. S. Chaudhary and F. A. Armstrong, *Energy Environ. Sci.*, 2012, **5**, 7470–7490.

15. A. M. Appel, J. E. Bercaw, A. B. Bocarsly, H. Dobbek, D. L. DuBois, M. Dupuis, J. G. Ferry, E. Fujita, R. Hille, P. J. A. Kenis, C. A. Kerfeld, R. H. Morris, C. H. F. Peden, A. R. Portis, S. W. Ragsdale, T. B. Rauchfuss, J. N. H. Reek, L. C. Seefeldt, R. K. Thauer and G. L. Waldrop, *Chem. Rev.*, 2013, **113**, 6621–6658.
16. J.-H. Jeoung and H. Dobbek, *Science*, 2007, **318**, 1461–1464.
17. T. Spatzal, M. Aksoyoglu, L. Zhang, S. L. A. Andrade, E. Schleicher, S. Weber, D. C. Rees and O. Einsle, *Science*, 2011, **334**, 940.
18. Y. Oh and X. Hu, *Chem. Soc. Rev.*, 2013, **42**, 2253–2261.
19. C. Costentin, M. Robert and J.-M. Savéant, *Chem. Soc. Rev.*, 2013, **42**, 2423–2436.
20. E. E. Benson, C. P. Kubiak, A. J. Sathrum and J. M. Smieja, *Chem. Soc. Rev.*, 2009, **38**, 89–99.
21. A. J. Morris, G. J. Meyer and E. Fujita, *Acc. Chem. Res.*, 2009, **42**, 1983–1994.
22. Y. Tamaki, T. Morimoto, K. Koike and O. Ishitani, *Proc. Natl. Acad. Sci. U. S. A.*, 2012, **109**, 15673–15678.
23. J.-M. Savéant, *Chem. Rev.*, 2008, **108**, 2348–2378.
24. S. Wang, G. Q. Lu and G. J. Millar, *Energy Fuels*, 1996, **10**, 896–904.
25. G. P. Smestad and A. Steinfeld, *Ind. Eng. Chem. Res.*, 2012, **51**, 11828–11840.
26. Y. Hori, in *Modern Aspects of Electrochemistry*, ed. C. G. Vayenas, R. E. White and M. E. Gamboa-Aldeco, Springer New York, New York, NY, 2008, pp. 89–189.
27. Z. Zhan, W. Kobsiriphat, J. R. Wilson, M. Pillai, I. Kim and S. A. Barnett, *Energy Fuels*, 2009, **23**, 3089–3096.
28. B. Kim, S. Ma, H.-R. M. Jhong and P. J. Kenis, *Electrochim. Acta*, 2015, **166**, 271–276.
29. K. Branker, M. Pathak and J. Pearce, *Renewable Sustainable Energy Rev.*, 2011, **15**, 4470–4482.
30. J. K. Nørskov, J. Rossmeisl, A. Logadottir, L. Lindqvist, J. R. Kitchin, T. Bligaard and H. Jónsson, *J. Phys. Chem. B*, 2004, **108**, 17886–17892.
31. J. Greeley and J. K. Nørskov, *Surf. Sci.*, 2007, **601**, 1590–1598.
32. J. Greeley and J. K. Nørskov, *J. Phys. Chem. C*, 2009, **113**, 4932–4939.
33. A. M. Appel, J. E. Bercaw, A. B. Bocarsly, H. Dobbek, D. L. DuBois, M. Dupuis, J. G. Ferry, E. Fujita, R. Hille, P. J. A. Kenis, C. A. Kerfeld, R. H. Morris, C. H. F. Peden, A. R. Portis, S. W. Ragsdale, T. B. Rauchfuss, J. N. H. Reek, L. C. Seefeldt, R. K. Thauer and G. L. Waldrop, *Chem. Rev.*, 2013, **113**, 6621–6658.
34. B. P. Sullivan, C. M. Bolinger, D. Conrad, W. J. Vining and T. J. Meyer, *J. Chem. Soc., Chem. Commun.*, 1985, 1414–1416.
35. B. Kumar, J. M. Smieja and C. P. Kubiak, *J. Phys. Chem. C*, 2010, **114**, 14220–14223.
36. J. M. Smieja, E. E. Benson, B. Kumar, K. A. Grice, C. S. Seu, A. J. M. Miller, J. M. Mayer and C. P. Kubiak, *Proc. Natl. Acad. Sci. U. S. A.*, 2012, **109**, 15646–15650.

37. K.-Y. Wong, W.-H. Chung and C.-P. Lau, *J. Electroanal. Chem.*, 1998, **453**, 161–170.
38. J. M. Smieja and C. P. Kubiak, *Inorg. Chem.*, 2010, **49**, 9283–9289.
39. E. E. Benson, K. A. Grice, J. M. Smieja and C. P. Kubiak, *Polyhedron*, 2013, **58**, 229–234.
40. E. E. Benson and C. P. Kubiak, *Chem. Commun.*, 2012, **48**, 7374.
41. J. T. Muckerman, J. H. Skone, M. Ning and Y. Wasada-Tsutsui, *Biochim. Biophys. Acta*, 2013, **1827**, 882–891.
42. J. A. Keith, K. A. Grice, C. P. Kubiak and E. A. Carter, *J. Am. Chem. Soc.*, 2013, **135**, 15823–15829.
43. E. Fujita and J. T. Muckerman, *Inorg. Chem.*, 2004, **43**, 7636–7647.
44. J. Agarwal, B. C. Sanders, E. Fujita, H. F. Schaefer III, T. C. Harrop and J. T. Muckerman, *Chem. Commun.*, 2012, **48**, 6797.
45. J. Agarwal, E. Fujita, H. F. Schaefer and J. T. Muckerman, *J. Am. Chem. Soc.*, 2012, **134**, 5180–5186.
46. C. Riplinger, M. D. Sampson, A. M. Ritzmann, C. P. Kubiak and E. A. Carter, *J. Am. Chem. Soc.*, 2014, **136**, 16285–16298.
47. M. Bourrez, F. Molton, S. Chardon-Noblat and A. Deronzier, *Angew. Chem., Int. Ed.*, 2011, **50**, 9903–9906.
48. D. C. Grills, J. A. Farrington, B. H. Layne, S. V. Lymar, B. A. Mello, J. M. Preses and J. F. Wishart, *J. Am. Chem. Soc.*, 2014, **136**, 5563–5566.
49. H. Takeda, H. Koizumi, K. Okamoto and O. Ishitani, *Chem. Commun.*, 2014, **50**, 1491–1493.
50. M. Bourrez, M. Orio, F. Molton, H. Vezin, C. Duboc, A. Deronzier and S. Chardon-Noblat, *Angew. Chem., Int. Ed.*, 2013, **53**, 240–243.
51. Y. C. Lam, R. J. Nielsen, H. B. Gray and W. A. Goddard, *ACS Catal.*, 2015, **5**, 2521–2528.
52. C. Riplinger and E. A. Carter, *ACS Catal.*, 2015, **5**, 900–908.
53. M. L. Clark, K. A. Grice, C. E. Moore, A. L. Rheingold and C. P. Kubiak, *Chem. Sci.*, 2014, **5**, 1894.
54. S. A. Chabolla, E. A. Dellamary, C. W. Machan, F. A. Tezcan and C. P. Kubiak, *Inorg. Chim. Acta*, 2014, **422**, 109–113.
55. M. D. Sampson and C. P. Kubiak, *J. Am. Chem. Soc.*, 2016, **138**, 1386–1393.
56. Q. Zeng, J. Tory and F. Hartl, *Organometallics*, 2014, **33**, 5002–5008.
57. J. Agarwal, T. W. Shaw, C. J. Stanton, G. F. Majetich, A. B. Bocarsly and H. F. Schaefer, *Angew. Chem., Int. Ed.*, 2014, **53**, 5152–5155.
58. C. W. Machan and C. P. Kubiak, *Dalton Trans.*, 2016, **45**, 15942–15950.
59. C. W. Machan, S. A. Chabolla, J. Yin, M. K. Gilson, F. A. Tezcan and C. P. Kubiak, *J. Am. Chem. Soc.*, 2014, **136**, 14598–14607.
60. C. W. Machan, J. Yin, S. A. Chabolla, M. K. Gilson and C. P. Kubiak, *J. Am. Chem. Soc.*, 2016, **138**, 8184–8193.
61. K. A. Grice and C. Saucedo, *Inorg. Chem.*, 2016, **55**, 6240–6246.
62. A. A. Peterson, F. Abild-Pedersen, F. Studt, J. Rossmeisl and J. K. Nørskov, *Energy Environ. Sci.*, 2010, **3**, 1311–1315.
63. J. M. Mayer, *Annu. Rev. Phys. Chem.*, 2004, **55**, 363–390.

64. D. R. Weinberg, C. J. Gagliardi, J. F. Hull, C. F. Murphy, C. A. Kent, B. C. Westlake, A. Paul, D. H. Ess, D. G. McCafferty and T. J. Meyer, *Chem. Rev.*, 2012, **112**, 4016–4093.
65. S. Hammes-Schiffer, *Acc. Chem. Res.*, 2009, **42**, 1881–1889.
66. G. Seshadri, C. Lin and A. B. Bocarsly, *J. Electroanal. Chem.*, 1994, **372**, 145–150.
67. E. B. Cole, P. S. Lakkaraju, D. M. Rampulla, A. J. Morris, E. Abelev and A. B. Bocarsly, *J. Am. Chem. Soc.*, 2010, **132**, 11539–11551.
68. E. E. Barton, D. M. Rampulla and A. B. Bocarsly, *J. Am. Chem. Soc.*, 2008, **130**, 6342–6344.
69. A. B. Bocarsly, Q. D. Gibson, A. J. Morris, R. P. LEsperance, Z. M. Detweiler, P. S. Lakkaraju, E. L. Zeitler and T. W. Shaw, *ACS Catal.*, 2012, **2**, 1684–1692.
70. J.-M. Savéant and C. Tard, *J. Am. Chem. Soc.*, 2016, **138**, 1017–1021.
71. A. J. Lucio and S. K. Shaw, *J. Phys. Chem. C*, 2015, **119**, 12523–12530.
72. E. Portenkirchner, C. Enengl, S. Enengl, G. Hinterberger, S. Schlager, D. Apaydin, H. Neugebauer, G. Knr and N. S. Sariciftci, *ChemElectroChem*, 2014, **1**, 1543–1548.
73. S. I. Rybchenko, D. Touhami, J. D. Wadhawan and S. K. Haywood, *ChemSusChem*, 2016, **9**, 1660–1669.
74. C. Costentin, J. C. Canales, B. Haddou and J.-M. Savéant, *J. Am. Chem. Soc.*, 2013, **135**, 17671–17674.
75. D. J. Boston, C. Xu, D. W. Armstrong and F. M. MacDonnell, *J. Am. Chem. Soc.*, 2013, **135**, 16252–16255.
76. D. Xiang, D. Magana and R. B. Dyer, *J. Am. Chem. Soc.*, 2014, **136**, 14007–14010.
77. H.-P. Yang, S. Qin, H. Wang and J.-X. Lu, *Green Chem.*, 2015, **17**, 5144–5148.
78. S. Ponnurangam, C. M. Yun and I. V. Chernyshova, *ChemElectroChem*, 2016, **3**, 74–82.
79. J. H. Lee, S. J. Lauw and R. D. Webster, *Electrochem. Commun.*, 2016, **64**, 69–73.
80. A. G. Peroff, E. Weitz and R. P. Van Duyne, *Phys. Chem. Chem. Phys.*, 2016, **18**, 1578–1586.
81. C. Costentin, J. C. Canales, B. Haddou and J.-M. Savéant, *J. Am. Chem. Soc.*, 2013, **135**, 17671–17674.
82. J. Tossell, *Comput. Theor. Chem.*, 2011, **977**, 123–127.
83. J. A. Keith and E. A. Carter, *J. Am. Chem. Soc.*, 2012, **134**, 7580–7583.
84. F. Riboni, E. Selli, M. R. Hoffmann and A. J. Colussi, *J. Phys. Chem. A*, 2015, **119**, 4433–4438.
85. J. A. Keith and E. A. Carter, *Chem. Sci.*, 2013, **4**, 1490–1496.
86. A. Marjolin and J. A. Keith, *ACS Catal.*, 2015, **5**, 1123–1130.
87. C.-H. Lim, A. M. Holder and C. B. Musgrave, *J. Am. Chem. Soc.*, 2013, **135**, 142–154.
88. J. A. Keith and E. A. Carter, *J. Chem. Theory Comput.*, 2012, **8**, 3187–3206.
89. P. E. M. Siegbahn, *J. Phys. Chem.*, 1996, **100**, 14672–14680.

90. C.-H. Lim, A. M. Holder, J. T. Hynes and C. B. Musgrave, *J. Am. Chem. Soc.*, 2014, **136**, 16081–16095.
91. C.-H. Lim, A. M. Holder, J. T. Hynes and C. B. Musgrave, *J. Phys. Chem. Lett.*, 2015, **6**, 5078–5092.
92. D. J. Boston, C. Xu, D. W. Armstrong and F. M. MacDonnell, *J. Am. Chem. Soc.*, 2013, **135**, 16252–16255.
93. M. Z. Ertem, S. J. Konezny, C. M. Araujo and V. S. Batista, *J. Phys. Chem. Lett.*, 2013, **4**, 745–748.
94. K. Liao, M. Askerka, E. L. Zeitler, A. B. Bocarsly and V. S. Batista, *Top. Catal.*, 2015, **58**, 23–29.
95. E. L. Zeitler, M. Z. Ertem, J. E. Pander, Y. Yan, V. S. Batista and A. B. Bocarsly, *J. Electrochem. Soc.*, 2015, **162**, H938–H944.
96. E. Lebègue, J. Agullo, M. Morin and D. Bélanger, *ChemElectroChem*, 2014, **1**, 1013–1017.
97. J. A. Keith, A. B. Muñoz-García, M. Lessio and E. A. Carter, *Top. Catal.*, 2015, **58**, 46–56.
98. J. A. Keith and E. A. Carter, *J. Phys. Chem. Lett.*, 2013, **4**, 4058–4063.
99. M. Lessio and E. A. Carter, *J. Am. Chem. Soc.*, 2015, **137**, 13248–13251.
100. T. P. Senftle, M. Lessio and E. A. Carter, *ACS Cent. Sci.*, 2017, **3**, 968–974.
101. C. X. Kronawitter, Z. Chen, P. Zhao, X. Yang and B. E. Koel, *Catal. Sci. Technol.*, 2017, 7, 831–837.
102. J. A. Joule and K. Mills, in *Heterocyclic Chemistry at a Glance*, John Wiley and Sons, Ltd, 2012, pp. 33–47.
103. C. X. Kronawitter, M. Lessio, P. Zahl, A. B. Muñoz-García, P. Sutter, E. A. Carter and B. E. Koel, *J. Phys. Chem. C*, 2015, **119**, 28917–28924.

CHAPTER 6

DFT Modelling Tools in CO_2 Conversion: Reaction Mechanism Screening and Analysis

LUIS MIGUEL AZOFRA*[a,b] AND CHENGHUA SUN*[a,c]

[a] ARC Centre of Excellence for Electromaterials Science (ACES), School of Chemistry, Faculty of Science, Monash University, Clayton, VIC 3800, Australia; [b] Current affiliation: KAUST Catalysis Center (KCC), King Abdullah University of Science and Technology (KAUST), Thuwal 23955-6900, Saudi Arabia; [c] Faculty of Science, Engineering & Technology, Swinburne University of Technology, Hawthorn, VIC 3122, Australia
*Email: luis.azoframesa@kaust.edu.sa; chenghuasun@swin.edu.au

6.1 Introduction

Attending to the International Union of Pure and Applied Chemistry (IUPAC) Gold Book's definition of catalyst, this is "*a substance that increases the rate of a reaction without modifying the overall standard Gibbs energy change*".[1] The process in which a catalyst is involved receives the name of catalysis, and, depending on whether the catalytic process takes place in the same phase, or on the contrary, occurs at or near the interphase that separates two different phases, catalysis can be classified into homogeneous and heterogeneous, respectively.

Energy and Environment Series No. 21
Electrochemical Reduction of Carbon Dioxide: Overcoming the Limitations of Photosynthesis
Edited by Frank Marken and David Fermin

Published by the Royal Society of Chemistry, www.rsc.org

In this regard, Scheme 6.1 shows a schematic representation of a chemical reaction in terms of the Gibbs free energy in which reactants (R) evolve to products (P) through a first-order transition state (TS; one imaginary frequency). As indicated in (a), the presence of a catalyst decreases the activation barrier ($G_c^{\neq}$) with respect the non-catalysed process ($G_{nc}^{\neq}$), and therefore increases the rate constant of the reaction by virtue of the Arrhenius equation.[2,3] Since the catalyst participates in the process as a promoter of the reaction but it is regenerated once reactants reach products, the reaction energy (G_R) is, by definition, not affected. At the end, the negative value of the Gibbs free energy variation between products and reactants leads to a spontaneous/exergonic process for this specific instance. By contrast, (b) represents the case of a chemical reaction in which the activation barrier increases ($G_i^{\neq} > G_{ni}^{\neq}$) in the presence of what can be considered a 'negative' catalyst. This behaviour is known as inhibition, that is, a process in which "*a substance known as inhibitor decreases the rate of a reaction without modifying the overall standard Gibbs energy change*". Finally and as happened in (a), G_R remains unchanged when compared with the non-inhibited and inhibited cases, however, its positive value leads to a non-spontaneous/endergonic process for this specific example.

However, why is the concept of catalysis so important? The answer is clear. If the introduction of a catalyst during a chemical reaction decreases the activation barrier from, for instance, 1.00 to 0.25 eV, this supposes an energy saving of three quarters; the economic consequences are therefore evident, and if this energy comes from fossil fuel sources, it also means the production of just a quarter of the greenhouse gases that otherwise would have been completely emitted into the atmosphere.

Despite the fact that inhibition processes are also of paramount importance for many chemical applications, we will just focus on those chemical reactions in which $G_{nc}^{\neq} > G_c^{\neq}$ (Scheme 6.1(a)), that is, catalysis. Scheme 6.2 presents the case of a multi-step catalysed chemical reaction

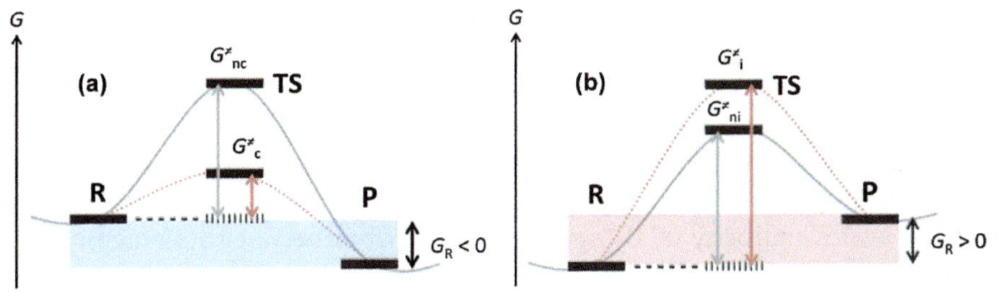

Scheme 6.1 Schematic representation of a chemical reaction in terms of the Gibbs free energy: (a) catalysed and non-catalysed; and (b) non-inhibited and inhibited. R, TS, and P refer to reactants, transition state, and products, respectively. $G^{\neq}$ and G_R refer to the Gibbs free activation and reaction energies, in each case.

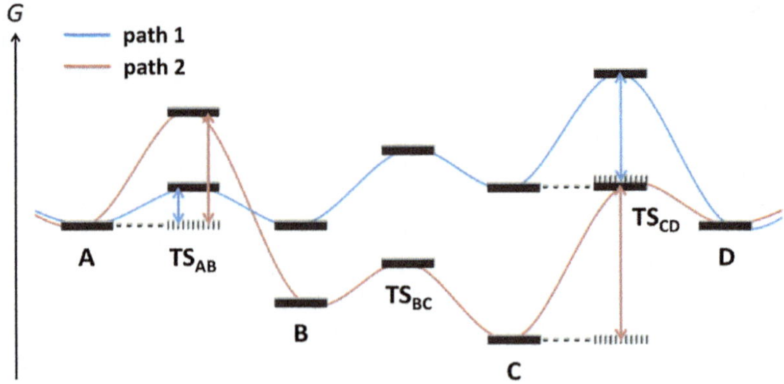

Scheme 6.2 Schematic Gibbs free energy diagram for a catalysed **A**→**D** chemical reaction following two mechanistic routes: path 1 (in blue) and path 2 (in red).

in which **A** (reactant) evolves to **D** (product) prior to passing from the **B** and **C** intermediate species and their respective TS. The use of a hypothetical catalyst leads to two different mechanistic routes, named as path 1 and path 2. First of all, it seems that an inconsistency arises with respect the formal concept of catalysis. That is, why are the reaction energies for obtaining the **B** and **C** intermediate species different when comparing both reaction mechanisms? The concept 'intermediate species' precisely refers to chemical entities existing as transitional minima or metastable structures. Since the nature of the species–catalyst interactions are different based on when they occur, therefore different thermodynamics can be seen in such cases. However, both paths obey the principle of the non-modified overall standard Gibbs energy change between products (**D**) and reactants (**A**).

Thus, at the beginning of the reaction process, **A** is converted into **B** once it has passed the respective $\mathbf{TS_{AB}}$. Despite $\mathbf{B_2}$ being thermodynamically more stable than $\mathbf{B_1}$, kinetics reveal that the activation barrier for $\mathbf{TS_{AB,\,1}}$ is lower than the one demanded when going through $\mathbf{TS_{AB,\,2}}$. This entails a majority of $\mathbf{B_1}$ intermediate species being obtained despite $G(\mathbf{B_2}) - G(\mathbf{B_1}) < 0$, *i.e.* the reaction takes place *via* the minimum energy path, being path 1 in this circumstance. Against this fact, we say that the reaction is governed by a kinetic control. Alternatively, giving the hypothetical case in which $G(\mathbf{TS_{AB,\,1}}) \geq G(\mathbf{TS_{AB,\,2}})$, undoubtedly thermodynamic control will be imposed and a majority of $\mathbf{B_2}$ species will be obtained at this stage of the reaction.

Finally, the reaction follows with obtaining $\mathbf{C_1}$ and $\mathbf{D_1}$ and their respective $\mathbf{TS_{BC,\,1}}$ and $\mathbf{TS_{CD,\,1}}$. However, since the activation barrier for the evolution of $\mathbf{C_1}$ up to $\mathbf{TS_{CD,\,1}}$ is the largest of the activation Gibbs free energies, this indicates that precisely this step becomes the limiting one of the whole process.

One of the key objectives of the quantum mechanical modelling of many-body (or *N*-body) systems is precisely the calculation of the energy. The information that can be extracted from this and its directly related properties (structural parameters, harmonic vibrational frequencies, orbitals/bands,† dipole moment, *etc.*) allows the hypothesising of the factors that explain the stability and reactivity of the chemical species and the role that the introduction of catalysts plays in such properties, *inter alia*. Among the variety of quantum mechanical methods at the disposal of the theoreticians' community, well-resolved density functional theory (DFT)[4] is, undoubtedly, one of the most powerful, and, therefore, widely used in the literature. In this regard:

1. DFT is based on the electron density, a physical observable.
2. DFT scales $O(N^3)$, which is much better than *ab initio* post-Hartree–Fock second-order Møller–Plesset perturbation theory [MP2, $O(N^4)$],[5] and extremely more superior than coupled clusters with singles and doubles [CCSD, $O(N^6)$],[6] for instance.
3. Multiple algorithms and approximations have been developed/implemented to decrease the time that DFT demands; this overcomes the bottleneck imposed by the $O(N^3)$ scaling, especially when the size of the system exceeds a certain number of atoms (as from a hundred).
4. On the market (commercial), and also under academic and GNU licenses, there are several tens of codes for DFT modelling.
5. And, what could be understood as a limitation of DFT, that is, the inexistence of an accurate and universal functional to the external potential, it is supplied with a wide range of general and specific functionals and the respective vast literature has been applied, even compared, to a multitude of systems.

Hence, it is not surprising to find that the DFT methodology is implemented in fields such as material sciences, and, by extension, in the *in silico* study of catalytic processes. In this regard, quantum mechanical modelling (but especially DFT due to the aforementioned strong points), allows the proposition of candidates ('initial guesses' from the theoretical formalism) to be evaluated before they are synthesised. This benefits experimentalists, who can start from those promising 'targets' previously identified by computational calculations; this can dramatically improve the success rate of materials development, and means, in the majority of cases, the saving of time and economic resources, besides offering an explanatory and complementary point-of-view of the events that occur at the atomic level. Thus, in the present chapter, we offer a comprehensive guide on DFT modelling of the catalytic carbon dioxide (CO_2) conversion mechanism into hydrocarbon compounds through the electrochemical approach.

---

†Formally, an orbital is a mathematical function that is an exact solution to the Schrödinger equation for a hydrogen-like system (H, He^+, Li^{2+}, $H_2{}^+$, *etc.*). However, the term has been improperly extended to many-body systems being widely accepted by the scientific community.

6.2 Insights into the Electrochemical CO_2 Conversion Reaction Mechanisms

Even after more than four decades of research, the development of realistic machinery for the catalytic conversion of CO_2 into hydrocarbon compounds still remains a 'hot topic'. This impact has its base in the implications that CO_2 has in the energy, environmental, biological, or industrial fields, among others. Since the beginning of the first industrial revolution to today, humanity has used non-renewable sources as the main form of energy production through the combustion of fossil fuels. In this regard, CO_2 has been a major product, and according to the National Oceanic and Atmospheric Administration (NOAA) Earth System Research Laboratory[7] its concentration in the atmosphere has reached a concerning level of 400 ppm with expectations of further rises at a rate of 2 ppm per year.[8] As result of fierce industrial activity, the role of CO_2—a natural gas with important implications for life—is clearly compromised by its consequences of the intensified greenhouse effect and climate change.[9,10] Thus, the reverse of this process is profiled as one 'green' alternative for energy consumption.[11] In this regard:

1. CO_2 is a natural source found in large quantities as a component of air.
2. The re-burning of hydrocarbon fuels as result of CO_2 conversion produces 'clean' energy since the balance of greenhouse emissions is exactly zero, always assuming that the energy provided during the conversion process comes from renewable sources.
3. The 'green' fuels generated through CO_2 conversion are easily manageable and transportable.

In nature, CO_2 is an abundant gas with a very high stability. Form the structural point-of-view, CO_2 is a linear molecule ($D_{\infty h}$ group of symmetry) with in-equilibrium R(C=O) distances equal to 1.162 Å. From an electrostatic perspective, CO_2 exhibits one electrostatic potential minimum on the 0.001 au electron density iso-surface (van der Waals iso-surface in atoms) along the C=O direction of each terminal O atom, and a set of multiple maxima surrounding the central C atom (see Figure 6.1(a)). These stationary points in the potential have been estimated to be −0.46 and 1.20 eV, respectively, at the MP2/aug-cc-pVDZ computational level, and correspond to the classical 'rabbit ear' lone pairs on O (O_{lp}) and the π-holes surrounding the sp-hybridised C of CO_2 in each case.[12]

This information revealed by the molecular electrostatic potential (MEP)[13] on the 0.001 au electron density iso-surface is of great significance since it indicates the potential role that these moieties would play when interacting with complementary moieties from partner molecules. In this regard, it advertises that O_{lp} will act as electron donor moieties with electropositive binding sites *via* presumably weak $O{=}C{=}O_{lp}\cdots X(\delta^+)$ interactions, while stronger contacts can occur with electronegative binding sites through $X(\delta^-)\cdots C(\pi\text{-hole})$ interactions, and acting C(π-hole) as an electron acceptor

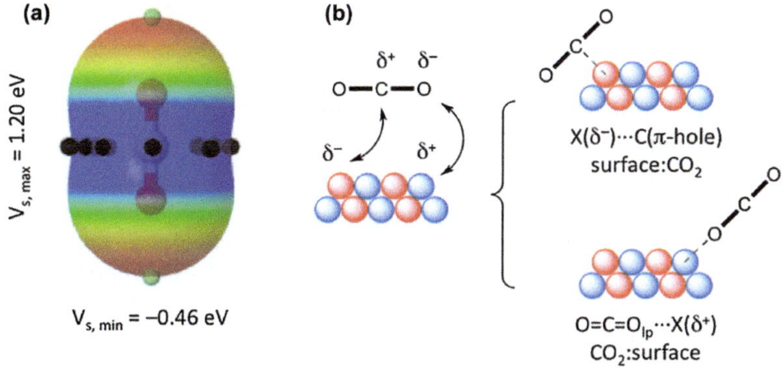

Figure 6.1 (a) MEP (±0.015 au iso-contour) on the 0.001 au electron density iso-surface for the isolated molecule of CO_2 calculated at the MP2/aug-cc-pVDZ computational level (WFA-SAS program).[14] From red to blue colours, from more negative to more positive potentials are indicated. Green and black spheres represent minima (O_{lp}) and maxima (π-holes on C), respectively. (b) Potential binding sites between CO_2 and a hypothetical surface catalyst indicating the cases in which CO_2 acts as electron acceptor or donor, respectively.

(see Figure 6.1(b)). Undoubtedly, among the variety of interactions that might take place, those of electrostatic nature will be always the most intense, although a significant dose of dispersion could also come into play.

The study of the interaction of CO_2 with partner substances is of paramount relevance since these so-called 'partners' could be the catalyst. Obviously, and as a prerequisite to start the catalytic conversion process, this cannot occur if there is no physicochemical contact between the substrate (CO_2, in this case) and the catalyst. In this sense, there are several lines of research exclusively dedicated to the description and development of novel and improved techniques for the enhancement of CO_2 adsorption as well as the search for new materials or approaches in which CO_2 could be chemisorbed[15–17] or captured.[18–20]

Once CO_2 is fixed (physi- or chemisorbed) on the catalytic surface, the reduction process takes places in successive elementary electrochemical reactions, that is, one proton (H^+) and one electron (e^-) are added to CO_2 or the immediately preceding species. Attending to the even number of H^+/e^- pairs transferred along the whole process, different hydrocarbon compounds can be obtained: carbon monoxide (CO) or formic acid (HCOOH, two), formaldehyde (H_2CO, four), methanol (CH_3OH, six), or methane (CH_4, eight). Although the possibility of the highly efficient production of CO and/or HCOOH is of great importance for chemical research, from a practical point-of-view, the low energy storage capacity of these less-reduced compounds do not make them good candidates for use in 'green' fuels technology; the amount of energy that can be obtained *via* combustion will be larger when the final product has a higher number of C–H covalent bonds and, in that sense, more reduced compounds are much more desirable.

However, which factors affect the major production of one hydrocarbon compound against the others? And, before this, how? Scheme 6.3 describes in a very clear way the different paths that isolated CO_2 can follow when reduced into CO, and HCOOH, H_2CO, CH_3OH, and CH_4 hydrocarbon compounds. Each one of these steps occurs from elementary electrochemical reactions in which one given species gains one H^+/e^- pair to further produce the hydrogenated intermediate species/product. In this sense, for the first electrochemical hydrogenation of CO_2, two options appear as separated mechanisms: on the one hand, leading to the $HOCO^\bullet$ intermediate species (eqn (6.1)), *i.e.* H is linked to one of the two O atoms from CO_2; on the other, the $OCHO^\bullet$ intermediate species is postulated (eqn (6.2)), H being linked on the C atom of CO_2:

$$CO_2 + H^+ + e^- \rightarrow HOCO^\bullet \tag{6.1}$$

$$CO_2 + H^+ + e^- \rightarrow OCHO^\bullet \tag{6.2}$$

As is clearly shown in Scheme 6.3, and obviating inter-conversion processes, the further mechanistic route depends on which radical is produced. That is, during a second H^+/e^- pair gain on the $HOCO^\bullet$ radical, three reductions might take place: (i) electrochemical hydrogenation on the previously hydrogenated O atom to reach CO and a released H_2O

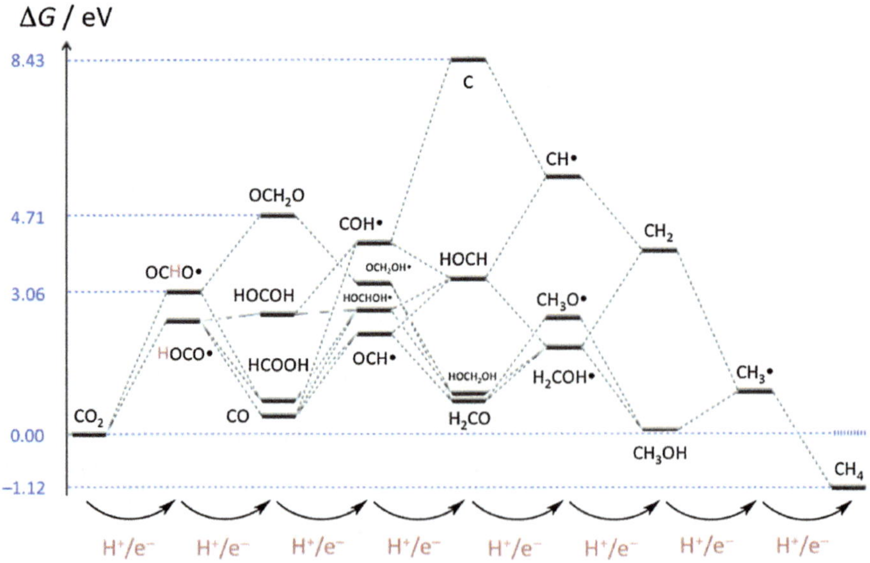

Scheme 6.3 Schematic Gibbs free energy diagram *vs.* SHE (just thermodynamics, in eV) for the electrochemical CO_2 reduction into CO, and HCOOH, H_2CO, CH_3OH, and CH_4 hydrocarbon compounds, in the gas phase and at mild conditions ($T = 298.15$ K), calculated at the MP2/aug-cc-pVTZ computational level. (Calculated through the facilities provided by the Gaussian09 package, revision D.01.)[21]

molecule; (ii) electrochemical hydrogenation on the C atom to reach HCOOH; and (iii) electrochemical hydrogenation on the non-reacted O atom to reach HOCOH. However, if the second H^+/e^- pair gain occurs on the OCHO• radical, HCOOH can be also obtained if it occurs on one of the O atoms or OCH_2O if it occurs on the previously hydrogenated C atom. Obviously, HCOOH is a common product, while the CO product (plus a released H_2O molecule) and the HOCOH intermediate species can only come from the HOCO• radical, or, alternatively, OCH_2O from OCHO•.

From here, it is found that the reaction evolves increasingly through a considerably complicated mechanism. Then, a series of multiple pathways can be postulated, as for instance, the two indicated in Scheme 6.4 and coming into play on the catalytic surface:

Thus, once the adsorbed CO_2 species is hydrogenated to reach the HOCO• radical, CO (with the release of a H_2O molecule), a HCO• radical, H_2CO, and a CH_3O• radical are obtained along the second, third, fourth and fifth H^+/e^- pair gains, respectively. In this context, and giving a hypothetical catalytic surface in which red and pink balls represent electronegative and positive binding sites, a set of specific interactions might take place. Strong $C^\bullet\cdots X(\delta^+)$ connections between HCO• and the surface are expected, or weak H-bonds between H_2CO and the $X(\delta^-)$ moiety from the material. These questions are not of trivial nature, since the specific interactions between substrate and surface could, and in fact can, define the minimum energy path. It is for that reason that, *in praxis*, delving into the interactions and the conformational exploration is of crucial importance for realistic modelling, and the conclusions which can be drawn from the thermodynamics and kinetics analysis derived from that.

Additionally, during the sixth to eighth H^+/e^- pair gains, two different mechanistic routes can be seen: on the one hand, a CH_3O• radical could be reduced by formation of CH_3OH (Scheme 6.4(6a)), to finally reach CH_4 prior passing to the CH_3• intermediate species (and the release of the second H_2O molecule); on the other, it is hypothesised an alternative path in which CH_4 is firstly produced (Scheme 6.4(6b)) being the catalytic surface as O-deposited, and producing the second H_2O molecule at the end of the process.

From a pragmatic point-of-view, some strategic points deserve mentioning in this context:

1. The isolated transformation of isolated CO_2 exhibits very large thermodynamic impediments. For example, the first elementary electrochemical reduction that could lead to the formation of the HOCO• and OCHO• radicals exhibit Gibbs free reaction energies (at MP2/aug-cc-pVTZ level) of 2.41 and 3.06 eV *vs.* SHE, respectively. It is obvious that the introduction of the catalyst is precisely based on a decrease in these impediments, describing a smoother reaction route. It is for that reason that an adequate and smart idea as an initial approach is the modelling of just the clean surface, fixed CO_2, and adsorbed HOCO• and OCHO• radical states in order to test the Gibbs

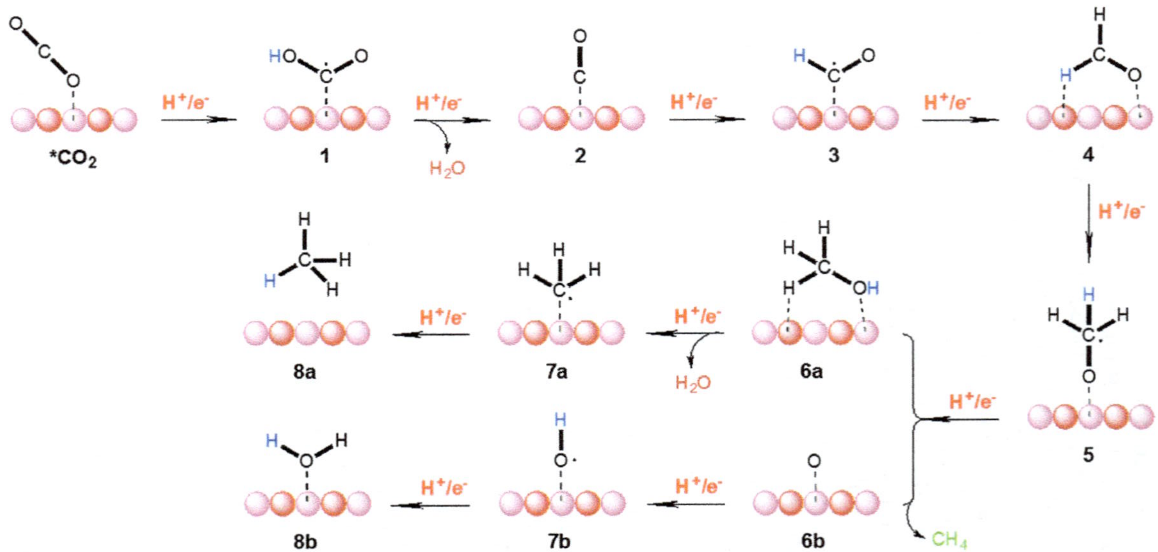

Scheme 6.4 Schematic representation for the electrochemical CO_2 conversion into the CH_4 mechanism through the CO pathway and following two different mechanistic routes from the fifth H^+/e^- pair gain. Red and pink balls represent electronegative and positive binding sites for a hypothetical catalytic surface material, respectively. Introductions of new H^+/e^- pairs are highlighted in blue.

free binding energy of CO_2 and the reaction Gibbs free energies due to the first electrochemical hydrogenation steps. If the test results are positive, that is, if a low reaction energy is obtained, it would be essential to estimate the activation barriers for both $HOCO^•$ and $OCHO^•$ radical formation.

2. Strongly related to this, it is well known that the electro-reduction of isolated CO_2 demands a very high potential, −1.90 V *vs.* NHE (eqn (6.3)),[22] and the first electrochemical hydrogenation represents the largest impediment, limiting the step of the overall process. This is also very usual when catalysed, and, once located, which is the step corresponding to the minimum energy path ($HOCO^•$ *vs.* $OCHO^•$ radicals), the mechanism is greatly simplified.

$$CO_2 + e^- \rightarrow CO_2^{•-}, \qquad \varepsilon^0 = -1.90 \text{ V } vs. \text{ NHE.} \tag{6.3}$$

3. Also, one might assume that we will not find, except in special cases, Gibbs free energies (both reaction and activation values) much larger during the successive H^+/e^- pair gains. This leads us to conclude that a comprehensive, accurate prediction of the thermochemistry during the early CO_2 fixation and first electrochemical hydrogenation steps generally represents a good approximation of the limitations imposed by the selected material.

Finally, very close potentials are seen for the CO and hydrocarbon compounds obtained during the CO_2 reduction (eqn (6.4)–(6.8)). However, the catalytic surface imposes the selectivity. Thus, following the minimum energy path can be hypothesised by the major product that would be obtained.

$$CO_2 + 2H^+ + 2e^- \rightarrow CO + H_2O, \qquad \varepsilon^0 = -0.61 \text{ V} \tag{6.4}$$

$$CO_2 + 2H^+ + 2e^- \rightarrow HCOOH, \qquad \varepsilon^0 = -0.53 \text{ V} \tag{6.5}$$

$$CO_2 + 4H^+ + 4e^- \rightarrow H_2CO + H_2O, \qquad \varepsilon^0 = -0.48 \text{ V} \tag{6.6}$$

$$CO_2 + 6H^+ + 6e^- \rightarrow CH_3OH + H_2O, \qquad \varepsilon^0 = -0.38 \text{ V} \tag{6.7}$$

$$CO_2 + 8H^+ + 8e^- \rightarrow CH_4 + 2H_2O, \qquad \varepsilon^0 = -0.24 \text{ V.} \tag{6.8}$$

6.3 Thermochemistry and Chemical Kinetics in Electrochemical Reactions[‡]

In theoretical chemistry, there are three widely used energy quantities that provide very useful, specific information for the understanding of the

[‡]A basic understanding of fundamental concepts in quantum mechanics, physical chemistry and statistical mechanics is assumed.

physicochemistry during a chemical (catalysed or not) reaction. We refer to the electronic energy (E), the enthalpy (H), and the Gibbs free energy (G). Negative changes in E, H, and G lead to favoured, exothermic, and spontaneous/exergonic processes, respectively, while positive values indicate non-favoured, endothermic, and non-spontaneous/endergonic processes, respectively. From the fundamental relations of thermodynamics it is inferred that:

$$G^0 = H^0 - TS \tag{6.9}$$

where G^0 and H^0 refer to the standard Gibbs free energy and enthalpy, and T and S denote temperature and entropy. At temperatures greater than 0 K and constant pressure (for mild conditions, $T = 298.15$ K), enthalpy can be expressed in terms of H^0 and the heat capacity, C_P:

$$H = H^0 + \int C_P \,\mathrm{d}T. \tag{6.10}$$

In addition, the entropy term can be expressed as the sum of the translational, rotational, vibrational, and electronic contributions as to:

$$S = S_t + S_r + S_v + S_e. \tag{6.11}$$

And therefore:

$$G = H^0 + \int C_P \,\mathrm{d}T - T(S_t + S_r + S_v + S_e). \tag{6.12}$$

In addition to these formal quantities, two additional correction terms can be added to eqn (6.12): on the one hand, and of intrinsic nature, the vibrational zero-point energy (ZPE); on the other hand, extrinsic dispersion (D) corrections (the explicit formulae of D depends on the method employed):

$$G = H^0 + \int C_P \,\mathrm{d}T - T(S_t + S_r + S_v + S_e) + \mathrm{ZPE} + D \tag{6.13}$$

$$\mathrm{ZPE} = \frac{1}{2}\sum h\nu_i \tag{6.14}$$

ν_i being each one of the vibrational frequencies. Thus, the Gibbs free energy variation (relative Gibbs free energy) between two different states, 2 and 1, for a given system, can be expressed as follows:

$$\begin{aligned}\Delta G_{21} = G_2 - G_1 = H_2^0 &+ \int C_{P,2} \,\mathrm{d}T - T(S_{t,2} + S_{r,2} + S_{v,2} + S_{e,2}) \\ &+ \mathrm{ZPE}_2 + D_2 - H_1^0 - \int C_{P,1} \,\mathrm{d}T + T(S_{t,1} + S_{r,1} + S_{v,1} + S_{e,1}) \\ &- \mathrm{ZPE}_1 - D_1\end{aligned} \tag{6.15}$$

or simply:

$$\Delta G_{21} = G_2 - G_1 = \Delta H_{21}^0 + \Delta \int (C_P)_{21} \,\mathrm{d}T - T\Delta S_{21} + \Delta \mathrm{ZPE}_{21} + \Delta D_{21}. \tag{6.16}$$

For the specific case of the heterogeneous catalytic electrochemical CO_2 conversion mechanisms, three different energy changes are usually calculated. The first one refers to the CO_2 adsorption on the surface, and it is also known as binding energy. Thus, the chemical equation representing the adsorption of $CO_2(g)$ (or to simplify, just CO_2) on a given surface (denoted as '*') is expressed as:

$$* + CO_2 \rightarrow {}^*CO_2. \tag{6.17}$$

This, which is formally a chemical reaction, exhibits a Gibbs free energy variation between the product (adsorbed CO_2) and reactants (clean surface plus isolated CO_2) representing the aforementioned binding Gibbs free energy. If it is positive, the adsorption is spontaneously produced, *i.e.* no energy injection is needed. By contrast, if it is negative, it means that fixation is non-spontaneous and additional energy (pressure) is required to enhance such contact. Thus:

$$\Delta G = G_b = G({}^*CO_2) - G(^*) - G(CO_2) \tag{6.18}$$

where:

$$G({}^*CO_2) = H^0({}^*CO_2) + \int C_P({}^*CO_2)\,dT - T[S_t({}^*CO_2) + S_r({}^*CO_2) + S_{v,}({}^*CO_2) + S_e({}^*CO_2)] + \text{ZPE}({}^*CO_2) + D({}^*CO_2) \tag{6.19}$$

$$G(^*) = H^0(^*) + \int C_P(^*)\,dT - T[S_t(^*) + S_r(^*) + S_{v,}(^*) + S_e(^*)] + \text{ZPE}(^*) + D(^*) \tag{6.20}$$

$$G(CO_2) = H^0(CO_2) + \int C_P(CO_2)\,dT - T[S_t(CO_2) + S_r(CO_2) + S_{v,}(CO_2) + S_e(CO_2)] + \text{ZPE}(CO_2) + D(CO_2). \tag{6.21}$$

For these last equations, some approximations can be applied:

1. At the fundamental electronic level: $S_e \approx 0$.
2. For gases,[§] translational, rotational, and vibrational entropy terms have contributions that might not be neglected. Therefore: $S = S_t + S_r + S_v$.
3. For solids and adsorbates, both $S_t \approx 0$ and $S_r \approx 0$. Therefore: $S = S_v$.
4. Since $\int C_P\,dT$ is almost negligible and $\Delta \int C_P\,dT \approx 0$, no thermal corrections for the enthalpy can be taken into account.

Finally, eqn (6.19)–(6.21) can be approximated as:

$$G({}^*CO_2) = H^0({}^*CO_2) - TS_{v,}({}^*CO_2) + \text{ZPE}({}^*CO_2) + D({}^*CO_2) \tag{6.22}$$

$$G(^*) = H^0(^*) - TS_{v,}(^*) + \text{ZPE}(^*) + D(^*) \tag{6.23}$$

$$G(CO_2) = H^0(CO_2) - T[S_t(CO_2) + S_r(CO_2) + S_{v,}(CO_2)] + \text{ZPE}(CO_2) + D(CO_2). \tag{6.24}$$

This can be also extended to the calculation of the Gibbs free reaction energy once ${}^*CO_2$ starts to be reduced through the gain of a set of H^+/e^- pairs. The

[§]See foundations in *Quantum Chemistry*, D. A. McQuarrie and M. Hanson, Macmillan Education, 2nd revised edition, 2007.

chemical equation for the first H^+/e^- pair gain leading to the $*OCHO^\bullet$ radical species obeys:

$$*CO_2 + H^+ + e^- \rightarrow *OCHO^\bullet. \quad (6.2)$$

In view of what was previously discussed concerning the calculation of the binding Gibbs free energy, it seems obvious how $G(*OCHO^\bullet)$ and $G(*CO_2)$ can be expressed as sum of the enthalpy, entropy, ZPE, and dispersion terms. However, a fundamental question arises: which is the Gibbs free energy of the $H^+/e^-(aq)$ (or just H^+/e^-) pair? In this context, the chemical potential of the H^+/e^- pair has the half value of the chemical potential of the dihydrogen molecule, $H_2(g)$, (or just H_2, see eqn (6.25)) when working at standard hydrogen electrode (SHE) conditions, *i.e.* $f(H_2) = 101\,325$ Pa, and $U = 0$ V, being $f(H_2)$, and U the fugacity of H_2 and the external potential applied, respectively. This is known as the proton-coupled electron transfer (PCET) approach:

$$\mu(H^+/e^-) = \tfrac{1}{2}\,\mu(H_2). \quad (6.25)$$

Then:

$$\begin{aligned}\Delta G = G_R &= G(*OCHO^\bullet) - G(*CO_2) - G(H^+ + e^-)\\ &= G(*OCHO^\bullet) - G(*CO_2) - \tfrac{1}{2}\,G(H_2). \quad (6.26)\end{aligned}$$

Ultimately, the reaction Gibbs free energy to any step of the electrochemical CO_2 conversion mechanism, where n is the number of H^+/e^- pairs transferred and m the number of $H_2O(g)$ (or just H_2O) molecules released, if applicable, results in:

$$\begin{aligned}\Delta G = G_R = G(*CO_{2-m}H_{n-2m}) + m\ G(H_2O) - G(*) - G(CO_2)\\ -n/2\ G(H_2). \quad (6.27)\end{aligned}$$

Obviously, for $n = 0$ and $m = 0$, $G_R = G_b$, that is, the binding Gibbs free energy.

Some observations deserve mentioning in this context:

1. Eqn (6.25) can be only applied in the Gibbs free energy regime.
2. Despite eqn (6.25) being delimited to SHE conditions, pH corrections can be applied by inclusion of the 2.303 RT pH quantity. (For pH = 0, natural hydrogen electrode (NHE) conditions, it is obvious that 2.303 RT pH = 0.)
3. The effect of an additional external potential can be added to the calculation of the Gibbs free energy values by inclusion of the $-eU$ term. At SHE/NHE conditions, $U = 0$ V.
4. As solution of the Schrödinger equation, quantum mechanical modelling codes usually provide the electronic energy. Formally, the electronic energy only refers to the electronic part (kinetic and potential electronic energies), however, the nuclear part (just as function of the nuclei positions attending to the Born–Oppenheimer approximation) and the thermal corrections to the electronic energy are assumed to be included in the 'electronic energy' expression.

5. The relation between the electronic energy and the enthalpy, taking into account the assumption previously commented on about E, is:

$$H = E + k_B T \tag{6.28}$$

therefore $\Delta H = \Delta E$. It is for this reason that, *in praxis*, ΔH is directly estimated as ΔE.

Finally, the computation of electrochemical activation energies or barriers is a challenging topic in which important efforts have been recently reported by Janik and co-workers[23,24] in the development of a strong, founded methodology in which an inner-sphere Marcus mechanism is assumed. In classical chemical reactions, the TS is located with a zero charge–electron balance during the chemical process. In the case of elementary electrochemical reactions, a chemical species A, fixed on a catalytic surface (then, *A), gains one proton (H^+) from the bulk medium and one electron (e^-) travelling through the catalyst, in order to produce an adsorbed *AH• intermediate state:

$$^*A + H^+ + e^- \rightarrow {}^*AH^\bullet. \tag{6.29}$$

Given the obvious difficulties with the location of the TS in such an elementary electrochemical reaction, Janik and co-workers proposed a location of the analogous hydrogenation (chemical) reaction in which both the *A and *H• species are fixed on the surface and react in order to produce the *AH• intermediate state:

$$^*A + {}^*H^\bullet \rightarrow {}^*AH^\bullet. \tag{6.30}$$

Once located, the TS for this chemical reaction and the TS for the electro-reduction of *A can be assumed as identical at one specific electrode potential, U_0, *i.e.* U_0 equates the energy of the $^*A + {}^*H^\bullet$ reactants to $^*A + H^+ + e^-$, as shown in the schematic parabolic curves approximated by the Marcus theory (Scheme 6.5):

Finally and attending to the Butler–Volmer theory, the Gibbs free activation barrier of the electro-reduction of *A, $G^{\neq}$, can be approximated as:

$$G^{\neq}(U) = G^{\neq 0} + \beta F(U - U_0) \tag{6.31}$$

where $G^{\neq 0}$ is the activation barrier for the hydrogenation reaction of *A, β the symmetry coefficient (and approximated to 0.5), F the Faraday constant, and U the applied electrode potential.

6.4 In Practice

In the present section, practical aspects for the well-resolved DFT modelling of the catalysed electrochemical mechanisms of CO_2 will be treated. In this regard, a few points deserve overall consideration before starting:

1. As fundamental requisite for the existence of catalysis, it is evident that a ‘catalyst’ should be considered. But how to know whether a material

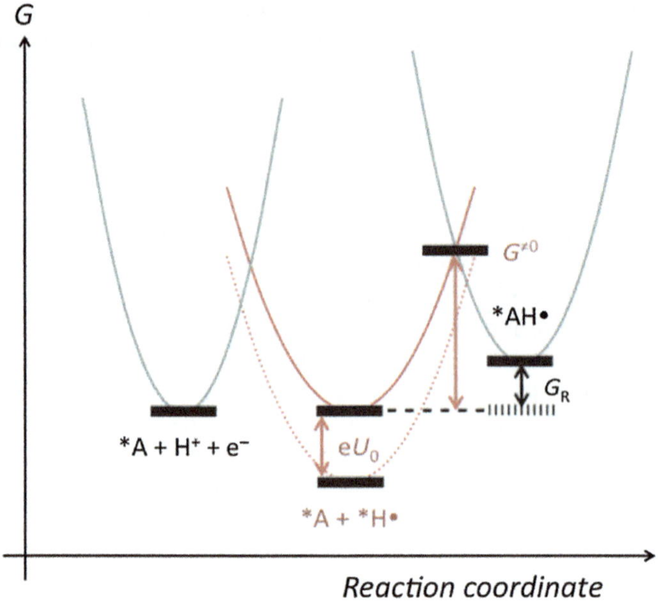

Scheme 6.5 Schematic representation of Marcus curves for the $*A+H^{+}+e^{-}$, $*A+*H^{\bullet}$, and $*AH^{\bullet}$ species.

will act as a catalyst or not? Through an analysis of specific elements, such as the atoms' connectivity, potential electrostatic binding-points, or selective reactivity, amongst others, it is possible to propose promising candidates.

2. The selection of the methodology plays a key role. An inadequate theoretical treatment might lead to unrealistic outcomes, either false or positive-negative results.
3. Since the theoretical DFT modelling is carried out thanks to the facilities provided by a software program, it is essential to have a broad knowledge of the function, settings, and theoretical basis with which such a program works.
4. Last, but not least, in many of the cases the use of a significant amount of computational resource is required. In this sense, it should be noted that most of the calculations demand computational time that simply exceeds the limits of a personal computer, even of small clusters used for the modelling of small-size systems or dedicated to performing tests.

Answering these questions:

1. As a catalyst, our choice is the (001) surface of anatase titanium dioxide, TiO_2.
2. As a methodology, we use well-resolved DFT through the generalised gradient approximation (GGA) with the Perdew–Burke–Ernzerhof (PBE)

functional[25] and employ explicit dispersion corrections through the use of the D3 method with the standard parameters programmed by Grimme and co-workers.[26,27]

3. As a simulation program, we employ the facilities provided by the Vienna *Ab-Initio* Simulation Package (VASP, version 5.3.5).[28–31]
4. All calculations were carried out at the National Computational Infrastructure (NCI), which is supported by the Australian Government.

Thus, anatase $TiO_2(001)$ is constituted, in such specific surface, by two kind of O atoms: two-fold or coordinated (labelled as 2c-O in Figure 6.2) and three-fold (3c-O) atoms, being the first the highest reactive (since 3c-O atoms are electronically saturated) for its plausible interactions with positive moieties from partner molecules. Also, all Ti atoms are in the form of five-fold (5c-Ti) atoms, presenting the ability to be potentially bound with negative moieties from partner molecules.

Before carrying out any modelling, the catalyst cell should be build and optimised. For the case of materials with crystal structure, it is possible and convenient to use experimental data provided by the various crystallography databases, many of them of with free access. In the case at hand (see Figure 6.2), the unit cell of anatase TiO_2 was downloaded, expanded, and cut in order to produce a $3\times3\times5$ super-cell of anatase TiO_2 ($Ti_{45}O_{90}$) in which the (001) surface pointed towards the *OZ* axis, and the (100) surface pointed towards the *OX* and *OY* axes.

Once the super-cell is built, it is required to proceed with the optimisation of the lattice parameters defined throughout the $\boldsymbol{a}$, $\boldsymbol{b}$, and $\boldsymbol{c}$ vectors. As a practical hint, it is quite desirable to centre the *OX* and *OY* positions in order to avoid change in the relative positions of the atoms as well as to prevent the same for the *OZ* positions moving the super-cell a sufficiently reasonable amount with respect the $z=0$ plane.

The cell parameters for anatase $TiO_2(001)$ constitute a triclinic (3D) lattice system, being the cell angles $\alpha=\beta=\gamma=90^\circ$ and the $|\boldsymbol{a}|$ and $|\boldsymbol{b}|$ cell length

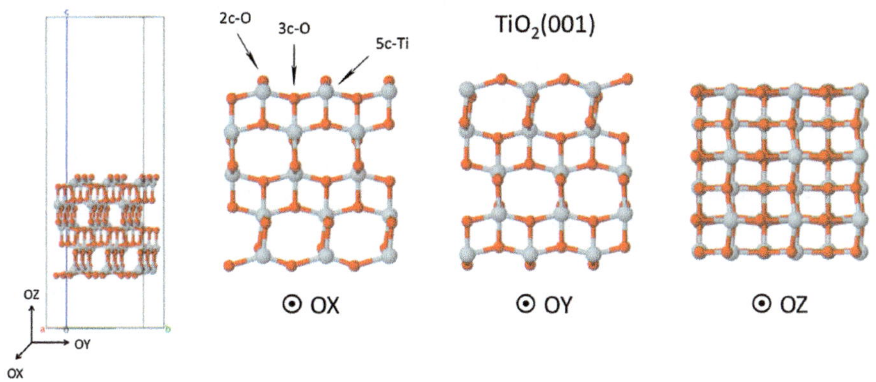

Figure 6.2 From left to right: (001) surface (*OZ*) for a $3\times3\times5$ super-cell of TiO_2 ($Ti_{45}O_{90}$) and its *OX*, *OY*, and *OZ* perspectives.

modules equal to 11.328 Å for this specific case. Since the (001) surface is expressed through the *OZ* axis, it represents a cutting in the crystal structure. For this reason, it is imposed that $|\boldsymbol{c}| = 35$ Å, and a vacuum distance of around 25 Å is supposed to avoid interactions between periodic images.

In the VASP environment, four files need to be prepared:[32]

1. POSCAR: containing the cell parameters, atom types and their quantity, and direct lattice coordinates (eventually, also forces if data were provided in a previous calculation).
2. POTCAR: containing the plane-wave pseudo-potentials. In the case at hand, projected augmented-wave (PAW) PBE pseudo-potentials: Ti (PAW_PBE Ti 08Apr2002) and O (PAW_PBE O 08Apr2002) for TiO_2, and also C (PAW_PBE C 08Apr2002) and H (PAW_PBE H 15Jun2001), which will be required later for the modelling of the CO_2 conversion.
3. KPOINTS: containing, in this case, representative $3\times3\times1$ k-points following the Monkhorst–Pack scheme and having been tested before with a larger set of k-points to make sure that there were no significant changes in the calculated energies.
4. INCAR: containing the settings for the modelling. Here, well-resolved DFT through: (i) a GGA-PBE functional; (ii) plane-wave cut-off energy of 450 eV;[33,34] (iii) energy and force convergence limits equal to 10^{-4} eV per atom and $|0.02|$ eV Å^{-1}, respectively; (iv) Gaussian smearing for the f_{nk} partial occupancies set for each wavefunction; (v) 0.02 eV for the width of the smearing; (vi) a conjugate gradient algorithm; (vii) 0.2 fs for the scaling constant of the forces; (viii) explicit D3 dispersion corrections; and (ix) the rest of the by-default settings.

Once the required accuracy was reached, that is, once it was optimised, VASP provides a total electronic energy ('TOTEN' in the VASP formalism) of −1194.74 eV. This optimisation is just the optimised structure for a given lattice parameters and does not mean that it is the optimum. The lattice parameters should be optimised as well, and for that reason, eqn (6.32) to (6.34) should be applied in order to search those lattice parameters that lead to the minimum energy. The χ factor just expands (if $\chi > 1$) or contracts (if $\chi < 1$) the cell. It is assumed that at the beginning, $\chi = 1.00$ (at the top of the INCAR file) was used. Then, with an accuracy in the hundredths, we can obtain the TOTEN when $\chi = 1.01$ (expansion) and $\chi = 0.99$ (contraction) but always keeping the same value of $|\boldsymbol{c}|$ to proceed with a consistent comparison:

$$\boldsymbol{a} = \chi\boldsymbol{a_x} + \chi\boldsymbol{a_y} + \chi\boldsymbol{a_z} = \chi\boldsymbol{a_x} = \chi \cdot 11.328\boldsymbol{i} \quad (6.32)$$

$$\boldsymbol{b} = \chi\boldsymbol{b_x} + \chi\boldsymbol{b_y} + \chi\boldsymbol{b_z} = \chi\boldsymbol{b_y} = \chi \cdot 11.328\boldsymbol{j} \quad (6.33)$$

$$\boldsymbol{c} = \chi\boldsymbol{c_x} + \chi\boldsymbol{c_y} + \chi\boldsymbol{c_z} = \chi\boldsymbol{c_z} = 35\boldsymbol{k}, \text{ in all cases.} \quad (6.34)$$

Table 6.1 TOTEN (in eV) along the optimisation of the lattice parameters of the $Ti_{45}O_{90}$ super-cell of $TiO_2(001)$.

	χ	TOTEN
↑	1.01	−1194.74
→	1.00	−1195.31
↓	**0.99**	−1195.52
↑	0.98	−1195.45

Results (see Table 6.1) indicate that TOTEN($\chi=1.01$)>TOTEN($\chi=1.00$), and TOTEN($\chi=0.99$)<TOTEN($\chi=1.00$). This leads us to conclude that it is still required to contract the cell up to find the minimum electronic energy. Finally, since TOTEN($\chi=0.98$)>TOTEN($\chi=0.99$), it is evident that for $\chi=0.99$ the optimised lattice parameters can be found, being $|\boldsymbol{a}|=|\boldsymbol{b}|=11.215$ Å and $|\boldsymbol{c}|=35$ Å.

Here, a fundamental question still arises: does the methodology employed represent well the crystal structure and properties of $TiO_2(001)$? As can be seen, the optimised lattice parameters and, then, the optimised structure from this turn out to be very close to the crystal structure, with similar O–Ti covalent distances between the experimentally measured and the theoretically predicted, and exhibiting a good agreement of the band gap when a HSE06 (Heyd–Scuseria–Ernzerhof functional)[35–37] single point calculation over the PBE optimised geometry was carried out. Thus, the methodology can be considered as validated.

The next step consists in the analysis of the CO_2 interactions on the $TiO_2(001)$ surface as fundamental prerequisite to start the catalytic process. In this regard, it seems evident that there is no catalysis if both surface and substrate do not present physicochemical contact. Since the calculation of the Gibbs free binding energy previously shown in eqn (6.18) consists in the subtraction of the Gibbs free energy of the CO_2:$TiO_2(001)$ complex minus the Gibbs free energy of the isolated moieties, their calculation is required.

At the present time, we have obtained the TOTEN and the associated optimised lattice parameters and structure of the clean $TiO_2(001)$ surface. How to calculate, *in praxis*, the Gibbs free energy? For this purpose, it is required a second calculation with the following details:[32]

1. POSCAR: the previously optimised structure contained in the CONTCAR file.
2. POTCAR: the same as in the previous case.
3. KPOINTS: since it is necessary carry out a vibrational/phonon frequency analysis, this is only possible to do for Γ points. Once done, a set of vibrational/phonon frequencies appear and, eventually, also a set of imaginary low values. They are associated with vibrational motions in the margins/limits of the cell and as result of the cutting to Γ points.

4. INCAR: in this case, a single-point calculation will be performed with the following parameters: (i) along the ionic and electronic relaxations, the same parameters can be used, although a more accurate precision setting can be chosen; (ii) if dispersion corrections were previously carried out, they should also be included here; and (iii) for the calculation of vibrational/phonon frequencies, the finite differences method is applicable in order to determine the second derivatives throughout the Hessian matrix. It is important to note that the number of iterations done to reach a self-consistent field (SCF) should be lower than the maximum number of iterations programmed. Otherwise, the results can develop into energy artefacts/inaccuracies because the SCF has not been reached. Also, it is recommended that a minimum of 6–8 iterations per alteration during the finite differences calculation should be carried out.

From these sets of calculations the Γ points can be extracted: the total electronic energy (and then, enthalpy), which directly includes the dispersion corrections, the entropy contributions, and the ZPE. However, there are different strategies concerning the selection of TOTEN. Some authors include the thermal and vibrational terms to obtain the Gibbs free energy from the TOTEN with the largest set of *k*-points, others choose the value directly obtained from the Γ points. In this case, we are working with the second case. As an important hint, it is quite important to take into account that, in the first case, the TOTEN that should be selected is the last one before reaching the required accuracy, while in the second it should be the first one after reaching the SCF since it is a single-point calculation. It is true that for an *N*-atoms system, there appear $3 \cdot 2 \cdot N+1$ iterations with their respective TOTEN. As we stated before, the first one corresponds to the single point calculation, while the rest, $3 \cdot 2 \cdot N$, carry out the alteration of each atom in the three directions (positive and negative) to evaluate the energy corresponding to each $3N$ vibrational/phonon frequency.

Table 6.2 gathers the energy quantities for the isolated H_2 and CO_2 molecules. Also, it is quite common see authors who prefer to use data from the optimised geometries of these small molecules, while others proceed with single-point calculations over the experimentally measured geometries, which are very well known due to the high precision of the detection methods in the gas phase. Also, we are working with the second case.

Table 6.2 Energy quantities for the isolated H_2 and CO_2 molecules. Experimental values from the NIST database (http://cccbdb.nist.gov): H_2 (*g*): $d_{HH} = 0.741$ Å, $\nu_1(\Sigma_g) = 4401$ cm^{-1}; CO_2 (*g*): $d_{CO} = 1.162$ Å, $A_{OCO} = 180.0°$, $\nu_1(\Sigma_g) = 1333$ cm^{-1}, $\nu_2(\Sigma_u) = 2349$ cm^{-1}, $\nu_3(\Pi_u) = 667$ cm^{-1}.

Species	$E+D$	$-TS_t$	$-TS_r$	$-TS_v$	ZPE	G
H_2 (*g*)	−6.77	−0.35	−0.04	0.00 (0)	0.27	−6.89
CO_2 (*g*)	−22.92	−0.47	−0.17	−0.00 (5)	0.27	−23.29

Unlike the previous cases in which the initial geometries are almost trivial, searching for the optimised $CO_2:TiO_2(001)$ complexes is much more complicated since it requires an initial approach in order to propose initial structures (guesses in the theoretical formalism) as close to the real minima. This not only facilitates the convergence process and saves computational time, but also could produce real but secondary minima in stability. The consequences of this are evident and the experience here usually plays an important role.

Thus, since 5c-Ti atoms are electron-poor moieties that exhibit one electropositive vacancy, it seems reasonable that this can interact with the O lone pairs (O_{lp}) from CO_2, which are electron-rich moieties. This means plausible OCO$\cdots$Ti interactions being CO_2 physisorbed on the surface. Also, another option is that 2c-O atoms in $TiO_2(001)$, which are more reactive than 3c-O ones, could interact with the π-holes from C in CO_2 in order to physically fix both moieties through an O$\cdots CO_2$ interaction. Alternatively, a secondary OCO$\cdots$Ti interaction could take place in this case. Finally, the last option is that CO_2 could be chemisorbed on the surface, being in this case strong interactions in which both moieties are closer to being covalently bound than in the form of weak interactions, as in the two previous assumptions.

Effectively, such initial guesses were good representative structures of the real minima, as shown in Figure 6.3. Concerning the stability analysis, while

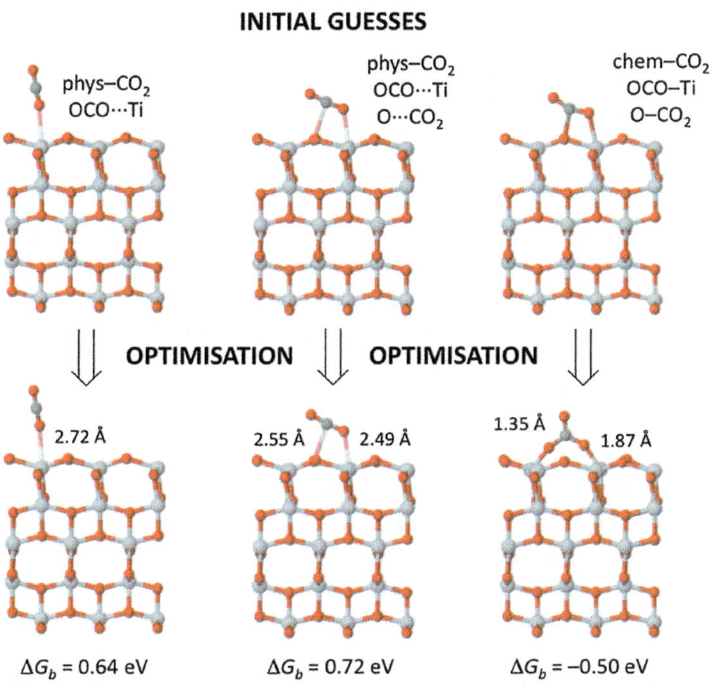

Figure 6.3 Initial structures and their corresponding optimised minima for the CO_2 fixation (physi- and chemisorbed) on $TiO_2(001)$.

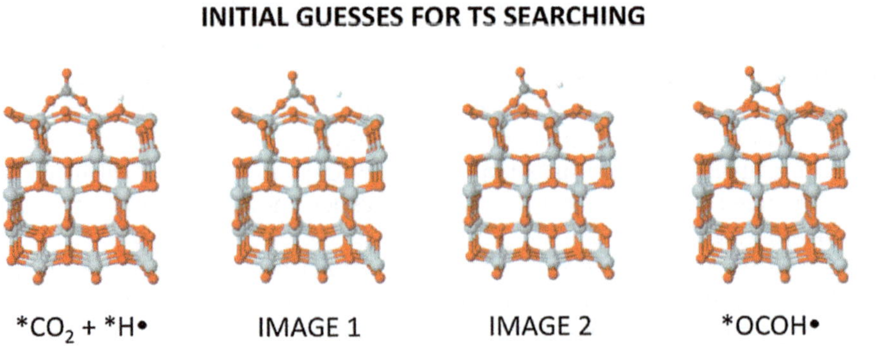

Figure 6.4 $^*CO_2 + {^*H^\bullet}$ and $^*OCOH^\bullet$ minima and the proposed two NEB images (1 and 2) for searching for the TS corresponding to the first H^+/e^- pair transfer on $TiO_2(001)$. This is an illustrative calculation; a larger number of images might be necessary for a more accurate description of the TS.

physisorbed minima exhibit non-spontaneous Gibbs free binding energies (0.64 and 0.72 eV), the chemisorption of CO_2 is profiled as being more stable with a spontaneous $\Delta G_b = -0.50$ eV. For this chemisorbed state we say that CO_2 is captured on the $TiO_2(001)$ surface. It deserves to be mentioned that, at this step, searching for the TS due to the change between physisorbed and chemisorbed CO_2 might be convenient, although the barrier of this process is expected to be lower than the ones that will be computed during the further conversion processes.

Once CO_2 is chemisorbed on the surface, the next step consists in the introduction of the first H^+/e^- pair. Between the two options of $OCHO^\bullet$ and $OCOH^\bullet$ radicals, we are going to focus on the second case. Thus, the optimised structure results in the one represented on the right of Figure 6.4. Applying eqn (6.27), the reaction Gibbs free energy at 298.15 K for this is hypothesised to be −0.32 eV *vs.* NHE, that is, the process is spontaneous. However, what happens to the activation barrier? According to eqn (6.31), a description of the $^*CO_2 + {^*H^\bullet}$ state is required, and therefore the TS between this and the $^*OCOH^\bullet$ radical. For this purpose, the nudge elastic band (NEB)[38] method was applied, despite there being other approaches for the location of the TS.

Finally, it is hypothesised (see Table 6.3) that the activation barrier for the electrochemical H^+/e^- pair injection from $^*CO_2$ to $^*OCOH^\bullet$ is 0.85 eV *vs.* NHE. This search should be done for all successive elementary steps in order to discriminate between the kinetic or thermodynamic controls of the reaction and locate the minimum energy path. Once the whole picture is gained, it is possible to conjecture the selectivity of the $TiO_2(001)$ surface towards the preference of one product against another and also hypothesise the maximum energy demanded for the whole process. If any larger barrier appears, once CO_2 is captured on the $TiO_2(001)$ surface, DFT advances that it

Table 6.3 Energy quantities (in eV, U_0 in V) for the calculation of the activation barrier for the first H^+/e^- pair injection from $*CO_2$ to $*OCOH^\bullet$.

Species	G	U_0	G^0_{act}	G_{act}
$*CO_2 + H^+ + e^-$	−1221.02			
$*CO_2 + *H^\bullet$	−1220.83			
TS	−1220.08	−0.19	0.75	0.85
$*OCOH^\bullet$	−1220.84			

is possible convert CO_2 into a certain hydrocarbon compound with a maximum energy input of 0.85 eV *vs.* NHE.

In summary, we have provided a set of computational tools to analyse the reaction mechanisms for the electrochemical CO_2 conversion into 'green' fuels through: (i) DFT characterisation of the material; (ii) finding structure/reactivity patterns; and (iii) estimation of the thermodynamics and kinetics of the reactions. In simpler words, providing a comprehensive guide in which we, as theoreticians, would be able to hypothesise promising candidate(s) for CO_2 conversion and deepening the understanding of the mechanistic events taking place at the atomic/electronic level. Our efforts are added to those already carried out by dozens of theoretical groups, not only on this specific topic,[39–44] but also in many other modelling studies such as hydrogen[45–50] and oxygen[51–56] evolution reactions (HER, OER) in water splitting or N_2 conversion into ammonia,[57,58] as some remarkable instances in which the foundations of this guide can be also applied. We are confident that the didactic point-of-view offered in this chapter will be of help to those researchers being introduced to the DFT modelling of heterogeneous catalysis materials chemistry and will serve as a basis to establishing a common methodology that takes into account both thermodynamic and kinetic evaluations with the aim of successful prediction and explanation.

Acknowledgements

The authors acknowledge the Australian Research Council (ARC) and the ARC Centre of Excellence for Electromaterials Science (ACES) for their support. Gratitude is also due to the National Computational Infrastructure (NCI) for providing the computational resources.

References

1. IUPAC Compendium of Chemical Terminology, Electronic version, http://goldbook.iupac.org/C00876.html.
2. S. A. Arrhenius, *Z. Phys. Chem.*, 1889, **4**, 96.
3. S. A. Arrhenius, *Z. Phys. Chem.*, 1889, **4**, 226.
4. P. Hohenberg and W. Kohn, *Phys. Rev.*, 1964, **136**, B864.
5. C. Møller and M. S. Plesset, *Phys. Rev.*, 1934, **46**, 618.
6. P. J. Knowles, C. Hampel and H. J. Werner, *J. Chem. Phys.*, 1993, **99**, 5219.

7. National Oceanic and Atmospheric Administration (NOAA) Earth System Research Laboratory, http://www.noaa.gov.
8. E. J. Maginn, *J. Phys. Chem. Lett.*, 2010, **1**, 3478.
9. T. R. Karl and K. E. Trenberth, *Science*, 2003, **302**, 1719.
10. J. Meyer, *Nature*, 2008, **455**, 733.
11. D. T. Whipple and P. J. A. Kenis, *J. Phys. Chem. Lett.*, 2010, **1**, 3451.
12. L. M. Azofra, *Chem. Phys.*, 2015, **453–454**, 1.
13. J. S. Murray and P. Politzer, *WIREs Comput. Mol. Sci.*, 2011, **1**, 153.
14. F. A. Bulat, A. Toro-Labbé, T. Brinck, J. S. Murray and P. Politzer, *J. Mol. Model.*, 2010, **16**, 1679.
15. Q. Sun, Z. Li, D. J. Searles, Y. Chen, G. Lu and A. Du, *J. Am. Chem. Soc.*, 2013, **135**, 8246.
16. X. Meng, S. Ouyang, T. Kako, P. Li, Q. Yu, T. Wang and J. Ye, *Chem. Commun.*, 2014, **50**, 11517.
17. H. Kiuchi, R. Shibuya, T. Kondo, J. Nakamura, H. Niwa, J. Miyawaki, M. Kawai, M. Oshima and Y. Harada, *Nanoscale Res. Lett.*, 2016, **11**, 127.
18. X. Zhang, X. Zhang, H. Dong, Z. Zhao, S. Zhang and Y. Huang, *Energy Environ. Sci.*, 2012, **5**, 6668.
19. M. Vogt, J. E. Bennett, Y. Huang, C. Wu, W. F. Schneider, J. F. Brennecke and B. L. Ashfeld, *Chem. – Eur. J.*, 2013, **19**, 11134.
20. C. Kunkel, F. Vines and F. Illas, *Energy Environ. Sci.*, 2016, **9**, 141.
21. M. J. Frisch, G. W. Trucks, H. B. Schlegel, G. E. Scuseria, M. A. Robb, J. R. Cheeseman, G. Scalmani, V. Barone, B. Mennucci, G. A. Petersson, H. Nakatsuji, M. Caricato, X. Li, H. P. Hratchian, A. F. Izmaylov, J. Bloino, G. Zheng, J. L. Sonnenberg, M. Hada, M. Ehara, K. Toyota, R. Fukuda, J. Hasegawa, M. Ishida, T. Nakajima, Y. Honda, O. Kitao, H. Nakai, T. Vreven, J. A. Montgomery Jr., J. E. Peralta, F. Ogliaro, M. Bearpark, J. J. Heyd, E. Brothers, K. N. Kudin, V. N. Staroverov, R. Kobayashi, J. Normand, K. Raghavachari, A. Rendell, J. C. Burant, S. S. Iyengar, J. Tomasi, M. Cossi, N. Rega, N. J. Millam, M. Klene, J. E. Knox, J. B. Cross, V. Bakken, C. Adamo, J. Jaramillo, R. Gomperts, R. E. Stratmann, O. Yazyev, A. J. Austin, R. Cammi, C. Pomelli, J. W. Ochterski, R. L. Martin, K. Morokuma, V. G. Zakrzewski, G. A. Voth, P. Salvador, J. J. Dannenberg, S. Dapprich, A. D. Daniels, Ö. Farkas, J. B. Foresman, J. V. Ortiz, J. Cioslowski and D. J. Fox, *Gaussian 09 (revision D.01)*, Gaussian, Inc., Wallingford, CT, 2009.
22. W. H. Koppenol and J. D. Rush, *J. Phys. Chem.*, 1987, **91**, 4429.
23. X. Nie, M. R. Esopi, M. J. Janik and A. Asthagiri, *Angew. Chem., Int. Ed.*, 2013, **52**, 2459.
24. X. Nie, W. Luo, M. J. Janik and A. Asthagiri, *J. Catal.*, 2014, **312**, 108.
25. J. P. Perdew, K. Burke and M. Ernzerhof, *Phys. Rev. Lett.*, 1996, 77, 3865.
26. S. Grimme, J. Antony, S. Ehrlich and H. Krieg, *J. Chem. Phys.*, 2010, **132**, 154104.
27. S. Grimme, S. Ehrlich and L. Goerigk, *J. Comput. Chem.*, 2011, **32**, 1456.
28. G. Kresse and J. Hafner, *Phys. Rev. B*, 1993, **47**, 558.
29. G. Kresse and J. Hafner, *Phys. Rev. B*, 1994, **49**, 14251.

30. G. Kresse and J. Furthmüller, *Phys. Rev. B*, 1996, **54**, 11169.
31. G. Kresse and J. Furthmüller, *Comput. Mater. Sci.*, 1996, **6**, 15.
32. VASP Manual, https://www.vasp.at.
33. P. E. Blöchl, *Phys. Rev. B*, 1994, **50**, 17953.
34. G. Kresse and D. Joubert, *Phys. Rev. B*, 1999, **59**, 1758.
35. J. Heyd, G. E. Scuseria and M. Ernzerhof, *J. Chem. Phys.*, 2003, **118**, 8207.
36. J. Heyd and G. E. Scuseria, *J. Chem. Phys.*, 2004, **121**, 1187.
37. J. Heyd, G. E. Scuseria and M. Ernzerhof, *J. Chem. Phys.*, 2006, **124**, 219906.
38. G. Mills, H. Jónsson and G. K. Schenter, *Surf. Sci.*, 1995, **324**, 305.
39. A. A. Peterson, F. Abild-Pedersen, F. Studt, J. Rossmeisl and J. K. Nørskov, *Energy Environ. Sci.*, 2010, **3**, 1311.
40. S. N. Habisreutinger, L. Schmidt-Mende and J. K. Stolarczyk, *Angew. Chem., Int. Ed.*, 2013, **52**, 7372.
41. H. C. Hsu, I. Shown, H.-Y. Wei, Y.-C. Chang, H.-Y. Du, Y.-G. Lin, C.-A. Tseng, C.-H. Wang, L.-C. Chen, Y.-C. Lin and K.-H. Chen, *Nanoscale*, 2013, **5**, 262.
42. J. Graciani, K. Mudiyanselage, F. Xu, A. E. Baber, J. Evans, S. D. Senanayake, D. J. Stacchiola, P. Liu, J. Hrbek, J. F. Sanz and J. A. Rodriguez, *Science*, 2014, **345**, 546.
43. L. M. Azofra, D. R. MacFarlane and C. Sun, *Phys. Chem. Chem. Phys.*, 2016, **18**, 18507.
44. L. M. Azofra, D. R. MacFarlane and C. Sun, *Chem. Commun.*, 2016, **52**, 3548.
45. E. Skulason, G. S. Karlberg, J. Rossmeisl, T. Bligaard, J. Greeley, H. Jonsson and J. K. Nørskov, *Phys. Chem. Chem. Phys.*, 2007, **9**, 3241.
46. H. A. Hansen, J. Rossmeisl and J. K. Nørskov, *Phys. Chem. Chem. Phys.*, 2008, **10**, 3722.
47. N. Holmberg and K. Laasonen, *J. Phys. Chem. Lett.*, 2015, **6**, 3956.
48. J. Kibsgaard, C. Tsai, K. Chan, J. D. Benck, J. K. Nørskov, F. Abild-Pedersen and T. F. Jaramillo, *Energy Environ. Sci.*, 2015, **8**, 3022.
49. Q. Tang and D.-E. Jiang, *ACS Catal.*, 2016, **6**, 4953.
50. G. Gao, A. P. O'Mullane and A. Du, *ACS Catal.*, 2017, **7**, 494.
51. M. G. Mavros, T. Tsuchimochi, T. Kowalczyk, A. McIsaac, L.-P. Wang and T. V. Voorhis, *Inorg. Chem.*, 2014, **53**, 6386.
52. Y. Zheng, Y. Jiao, Y. Zhu, L. H. Li, Y. Han, Y. Chen, A. Du, M. Jaroniec and S. Z. Qiao, *Nat. Commun.*, 2014, **5**, 3783.
53. S. Piccinin, A. Sartorel, G. Aquilanti, A. Goldoni, M. Bonchio and S. Fabris, *Proc. Natl. Acad. Sci. U. S. A.*, 2013, **110**, 4917.
54. C. Ma, S. Piccinin and S. Fabris, *ACS Catal.*, 2012, **2**, 1500.
55. H.-Z. Wu, L.-M. Liu and S.-J. Zhao, *Phys. Chem. Chem. Phys.*, 2014, **16**, 3299.
56. J. K. Nørskov, J. Rossmeisl, A. Logadottir, L. Lindqvist, J. R. Kitchin, T. Bligaard and H. Jónsson, *J. Phys. Chem. B*, 2004, **108**, 17886.
57. J. H. Montoya, C. Tsai, A. Vojvodic and J. K. Nørskov, *ChemSusChem*, 2015, **8**, 2180.
58. L. M. Azofra, N. Li, D. R. MacFarlane and C. Sun, *Energy Environ. Sci.*, 2016, **9**, 2545.

CHAPTER 7

Electrocarboxylation in Ionic Liquids

SHU-FENG ZHAO,[a,b] MICHAEL D. HORNE,[b] ALAN M. BOND[a] AND JIE ZHANG*[a]

[a] School of Chemistry and ARC Centre of Excellence for Electromaterials Science, Monash University, Clayton, VIC 3800, Australia; [b] CSIRO Mineral Resources, Private Bag 10, Clayton South, VIC 3169, Australia
*Email: Jie.Zhang@monash.edu

7.1 Introduction

The impact of human activity on the atmospheric carbon dioxide (CO_2) concentration is well documented. In 2012, emissions from fossil fuel combustion and land clearing exceeded 35 billion tonnes.[1] As a result, the atmospheric concentration of CO_2 has risen from 270 ppm in the pre-industrial era to *ca.* 404 ppm in October 2017.[2] If this rate of increase continues unabated, an average global temperature increase of 6 °C is predicted at the end of the current century by many climate simulations, which according to the study by Hansen *et al.*[3] will be sufficient to create an ice-free Antarctica.

CO_2 is perceived commonly to be a useless, ubiquitous and intractable waste product that is environmentally damaging and costly to remediate. However, despite its high kinetic stability, it can be activated by reactions with anions or radical anions and inserted as a carbon unit (C1) into a range of compounds, thereby transforming them into a surprisingly large variety of products (Figure 7.1).

Energy and Environment Series No. 21
Electrochemical Reduction of Carbon Dioxide: Overcoming the Limitations of Photosynthesis
Edited by Frank Marken and David Fermin

Published by the Royal Society of Chemistry, www.rsc.org

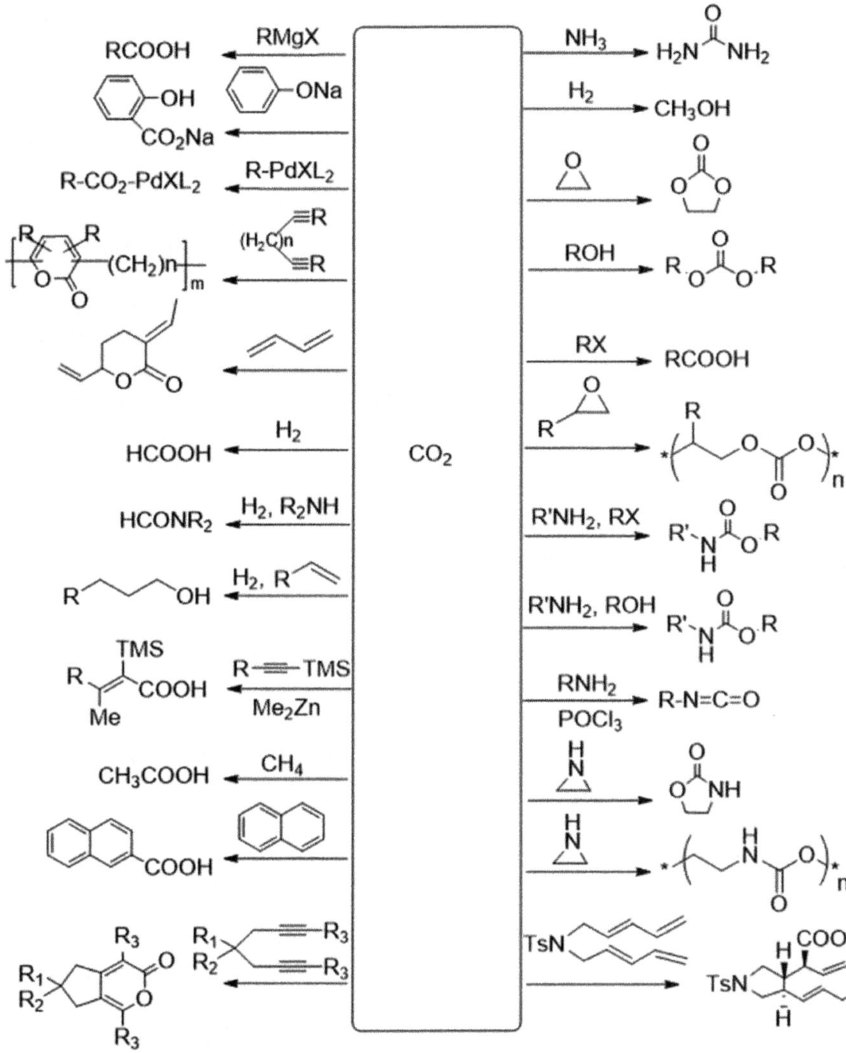

Figure 7.1 The utility of CO_2 as a C1 building block.
Adapted with permission from Sakakura *et al.*, *Chem. Rev.* 2007, **107**, 2365–2387.[4] Copyright 2015 American Chemical Society.

The combination of wide availability and potential chemical utility of CO_2 has stimulated a number of efforts to transform this compound into useful industrial commodities. Examples include the synthesis of salicylic acid,[4] inorganic and organic carbonates,[5] urea[6] and methanol.[7] Strategies[8] that have been employed to activate CO_2, include radio-chemical, thermo-chemical, photo-chemical, electrochemical, bio-chemical, bio-photochemical, photo-electrochemical, bio-electrochemical and biophoto-electrochemical. Among these methods, the electrochemical transformation of carbon dioxide is recognized as one of the most promising because in principle electrons in electrodes can be provided continuously, and their reducing power can be controllably varied by modifying the electrode potential. In addition, the electrochemical transformation of CO_2 to useful organic compounds has recently been recognized as a powerful technology for storing electricity from intermittent renewable resources, such as sunlight and wind. This use of CO_2 is a tantalizing prospect because it offers the possibility of closing the carbon cycle and mimicking photosynthesis.

Electrochemical studies involving carbon dioxide date back to the 19th century.[9] The electrolysis of a French wine to produce carbon dioxide was reported in 1801. Seventy years later, the first reproducible experimental results for electrochemical reduction of CO_2 in aqueous solution were reported.[10] Since then, electrochemical conversion of CO_2 has been undertaken in numerous solvent (electrolyte) media, with the aim of facilitating this transformation using a wide variety of scientific and technological advances.

Generally, the electrochemical conversion of carbon dioxide has been undertaken *via* one of two pathways:

(1) reduction to form fuel or fuel precursors[8,11] such as CO, ethylene, methanol or formic acid using either bulk materials, nanoparticles or molecular species as catalysts;
(2) utilization of carbon dioxide as a C1 building block for the formation of more complex organic molecules,[4] such as carboxylic acids, esters and lactones.

This chapter focuses on the second pathway, the electrochemical conversion of CO_2 to carboxylic acids, often called electrocarboxylation. In electrocarboxylation, a reactive organic radical species of either CO_2 or organic substrate is electrochemically generated which then reacts with its counterpart to form a carboxylate.

Carboxylic acids, generated from carboxylation reactions, are an important class of intermediates in the synthesis of pharmaceutical materials.[12,13] They are conventionally synthesized through multi-step reactions[4] involving the oxidation of aliphatic or aromatic moieties, or the hydration of cyanides, or by reacting CO_2 with strong nucleophiles such as Grignard reagents, alkyl lithiums, active methylene compounds, metal enolates or transition metal complexes. However, the use of these generally highly toxic or unstable reagents makes these methods less attractive for pharmaceutical applications. By contrast, the electrochemical transformation of CO_2 in aprotic organic solvents, such as acetonitrile (MeCN), *N*,*N*-dimethylformamide (DMF) and *N*-methyl-2-pyrrolidone (NMP),[14–17] can lead to the synthesis of carboxylic acids efficiently in high yields even at room temperature and atmospheric pressure.[15,16,18–27]

In recent years, room-temperature ionic liquids (RTILs) have received considerable attention in applied chemistry because they have broad yet tuneable solvent properties and are generally non-flammable and non-volatile. Moreover, selected ionic liquids have high thermal and chemical stability, and low toxicity.[28–33] They are particularly well suited as environmentally benign replacements for volatile organic solvents. Furthermore, their intrinsic conductivity and resistance to reductive and oxidative degradation make RTILs potentially useful media for electrochemistry[29,30,34–39] without the need for added supporting electrolytes.[30] Perhaps their greatest potential utility derives from their design flexibility. As their constituent ions generally include one and often two charged organic species, the freedom to modify the properties is immense. For electrocarboxylation and many other organic transformations, this freedom allows suitable ionic liquids to be designed that have added ability to control reaction pathways.[40,41]

An advantage of ionic liquids over conventional molecular solvents for electrocarboxylation is their ability to dissolve high concentrations of CO_2. Carbon dioxide is remarkably soluble in imidazolium based ionic liquids,[42–44] with the best indicator of solubility being the strength of the association of CO_2 with the ionic liquid anion. The cation plays a secondary role in CO_2 solubility.[43]

Several articles[22,45–47] have reviewed the preparative and mechanistic details of the electrocarboxylation of various compounds in molecular solvents. This review is focused instead on new progress made available by electrocarboxylation in ionic liquids. Initially, a survey of studies undertaken on electrocarboxylation in ionic liquids is provided. This material provides the background needed for a comparison with electrochemical carboxylation in conventional organic molecular solvents. The review concludes with a survey on the role that ionic liquids play in electrocarboxylation, and highlights the influence of proton availability from the ionic liquid itself or from water deliberately added, or present as an impurity.

7.2 Electrocarboxylation in Ionic Liquids

The electrocarboxylation of many different organic compounds including halides,[48–50] aromatic ketones,[51,52] epoxides,[53] alcohols,[54–59] amines[60] and activated olefins[61] has been attempted in ionic liquids. Of the experimental variables that have an impact on the reaction, the identity of electrode material is of particular significance. An ideal electrode material should facilitate the formation of the reactive intermediate but inhibit the direct electroreduction of CO_2. The nature of the ionic liquid has also received considerable attention, although most studies have been conducted using imidazolium based ionic liquids, especially 1-butyl-3-methylimidazolium tetrafluroborate ([Bmim][BF_4]), probably because this was one of the first commercially available ionic liquids.

7.2.1 Electrocarboxylation of Organic Halide Compounds in Ionic Liquids

Electrocarboxylation of organic halides provides a very convenient way of producing α-aryl-propionic acids, some of which have been widely used as non-steroidal anti-inflammatory drugs.[62] The mechanism and preparative aspects of electrocarboxylation of organic halides in aprotic solvent have been well established.[63] Typically, the mechanism is described by reactions (7.1)–(7.4).

$$RX + e^- \rightarrow RX^{\bullet -} \tag{7.1}$$

$$RX^{\bullet -} \rightarrow R^\bullet + X^- \tag{7.2}$$

$$R^\bullet + e^- \rightarrow R^- \tag{7.3}$$

$$R^- + CO_2 \rightarrow RCOO^- \tag{7.4}$$

In accordance with reactions (7.1)–(7.4), the carbon–halogen bond is reductively cleaved, generating a radical species that is further reduced to an anion. Carboxylation of the anion adds a carboxylic moiety to the reactant. The formation of $R^\bullet$ from RX may also occur *via* a single step concerted pathway instead of the consecutive two step pathway described above (reactions (7.1) and (7.2)). The electrocarboxylation of numerous organic halides has been investigated in undivided cells containing cathodes such as carbon,[64] Pt,[65,66] Ni,[67] Ag,[15,68,69] Hg,[64] Cu,[67] Ti,[67] Zn[70] and steel.[68,71] Of these materials, Ag is a powerful electrocatalyst for the formation of the reactive anion, which is the bottleneck step of the electrocarboxylation.[15] The electrocarboxylation of benzyl chloride in [Bmim][BF_4] was first reported at a Ag cathode.[50] Cyclic voltammograms in [Bmim][BF_4] (Figure 7.2) show the electroreduction of benzyl chloride occurring at a Ag electrode at potentials about 100 mV and 500 mV less negative than those at Cu and Ni electrodes, respectively. Similar results were obtained for the electrocarboxylation of 2-amino-5-bromopyridine in [Bmim][BF_4].[48]

The influence of temperature and pressure on the faradaic efficiency for electrocarboxylation of organic halides also has been investigated.[48–50] The increase in faradaic efficiency at higher temperature and pressure has been attributed to enhanced mass transport and increased CO_2 concentration. It is noteworthy that during electrocarboxylation of organic halides, ionic liquids have been reused up to four times without activity loss.[50]

The mechanism proposed for the electrocarboxylation of organic halides in ionic liquids is similar to that established in aprotic molecular solvents such as MeCN and DMF.[25,72] However, the side products are different in these two classes of solvents.[25,72] In organic molecular solvents, the main side products are dimers and alcohols resulting from the dimerization of $R^\bullet$ and a proton coupled electron transfer process involving $R^{\bullet -}$, respectively. However, in ionic liquids, dimers were the only side products reported in the

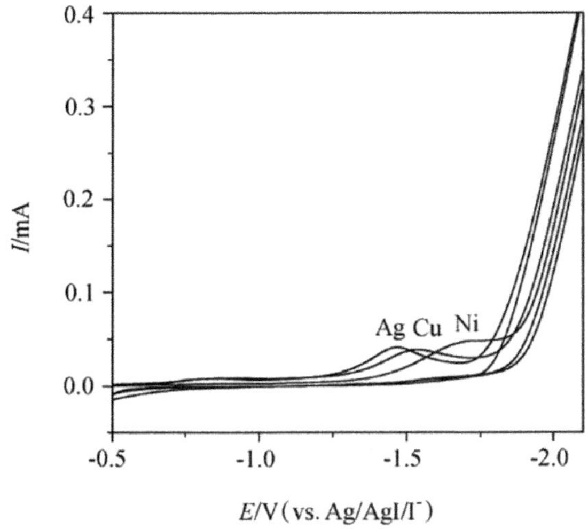

Figure 7.2 Cyclic voltammograms (presented in American convention) for the reduction of 15 mM benzyl chloride in [Bmim][BF_4] at Ag, Cu and Ni electrodes at a scan rate of 0.1 V s^{-1}, $T = 25$ °C.
Adapted with permission from Niu *et al.*, *Chin. J. Chem.* 2009, **27**, 1041–1044.[50] © 2009 SIOC, CAS, Shanghai & WILEY-VCH Verlag GmbH & Co. KGaA, Weinheim.

electrocarboxylation of benzyl chloride.[50] Clearly, the product distribution from the electrocarboxylation of organic halides is influenced by the properties of the reaction medium. The most probable reason for differences in products relates to the different stabilities of the radical and radical anion intermediates in the different solvent classes (essentially, charged *versus* uncharged environments). Stabilized radicals in dry ionic liquids will dimerize at a slower rate than in molecular solvents. Additionally, any adventitious proton source within the solvent, such as water, will rapidly react with a radical or radical anion to produce an alcohol. Thus, the water content in ionic liquids is expected to affect the product distribution as well, which will be discussed in detail later.

7.2.2 Electrocarboxylation of Aromatic Ketones in Ionic Liquids

α-aryl-propionic acids are important intermediates for anti-inflammatory drug synthesis and can be produced by the electrocarboxylation of aromatic ketones. Generally, electrocarboxylation of aromatic ketones involves the consumption of two electrons per molecule of substrate (reactions (7.5)–(7.8)) and the first step generates a resonance-stabilized radical anion. Common side products in the electrocarboxylation of aromatic ketones are pinacols arising from dimerization of the radical anion, and alcohols from a proton coupled electron transfer reaction in the presence of a proton source.

$$R_1R_2C{=}O + e^- \rightleftarrows R_1R_2C^{\bullet}\text{–}O^- \ (\mathbf{Ia}) \quad \text{or} \quad R_1R_2C^-O^{\bullet} \ (\mathbf{Ib}) \tag{7.5}$$

$$[R_1R_2C\text{–}O]^{\bullet -} + CO_2 \rightarrow [R_1R_2CO\text{–}COO]^{\bullet -} \tag{7.6}$$

$$[R_1R_2CO\text{–}COO]^{\bullet -} + e^- \rightarrow [R_1R_2CO\text{–}COO]^{2-} \tag{7.7}$$

$$[R_1R_2CO\text{–}COO]^{2-} + CO_2 \rightarrow [OOCO\text{–}(R_1R_2C)\text{–}COO]^{2-}. \tag{7.8}$$

The feasibility of using ionic liquids for the electrocarboxylation of aromatic ketones has been investigated in recent years.[51,52,73] The initial attempt was undertaken in [Bmim][BF_4] with acetophenone as the substrate for carboxylation.[51,52,73] Since the electroreduction of aromatic ketones occurs in a highly negative potential region where direct electroreduction of CO_2 may occur, an appropriate electrode should be chosen to avoid simultaneous electroreduction of both aromatic ketone and CO_2. Zhao *et al.* investigated the electroreduction of acetophenone and CO_2 in [Bmim][BF_4] on a range of electrode materials, including Ag, Cu, Ni, Pt, Ti and glassy carbon (GC) (Figure 7.3) and concluded that GC is the most suitable of these because it inhibits CO_2 reduction most effectively.

Zhao *et al.*[73] found that electrochemical reduction of acetophenone at a GC electrode in the presence of CO_2 is an overall chemically irreversible two-electron transfer process with the main product being 1-phenylethanol, which results from the proton coupled electron transfer reaction (Scheme 7.1) instead of the anticipated carboxylic acid. Protons required for this reaction were assumed to be derived from the C2–H position in $[Bmim]^+$, which are more readily available under a CO_2 atmosphere due to the formation of a strong complex between the *N*-heterocyclic carbene (NHC, the deprotonated form of $[Bmim]^+$) and CO_2. Electrochemical reduction of acetophenone in [Bmim][BF_4] was also investigated by Feng *et al.*[51] at Pt, Ni, standard steel and Cu electrodes even though these electrodes are not ideal for electrocarboxylation of acetophenone, as suggested above (Figure 7.3). Considerable quantities of carboxylic acid product were derived from the carboxylation reaction under all conditions. The apparent discrepancy between the results obtained from two research groups could be attributed to the difference in electrodes employed. Direct reduction of CO_2 could occur at Pt, Ni, steel and Cu electrodes to generate $CO_2^{\bullet -}$ (reaction (7.9)), which can then react with acetophenone to form a radical anion intermediate (reaction (7.10)) followed by the reactions described in reactions (7.7) and (7.8) to form a carboxylic acid as the final product.

$$CO_2 + e^- \rightarrow CO_2^{\bullet -} \tag{7.9}$$

$$CO_2^{\bullet -} + R_1R_2C{=}O \rightarrow R_1R_2CO\text{–}COO^{\bullet -}. \tag{7.10}$$

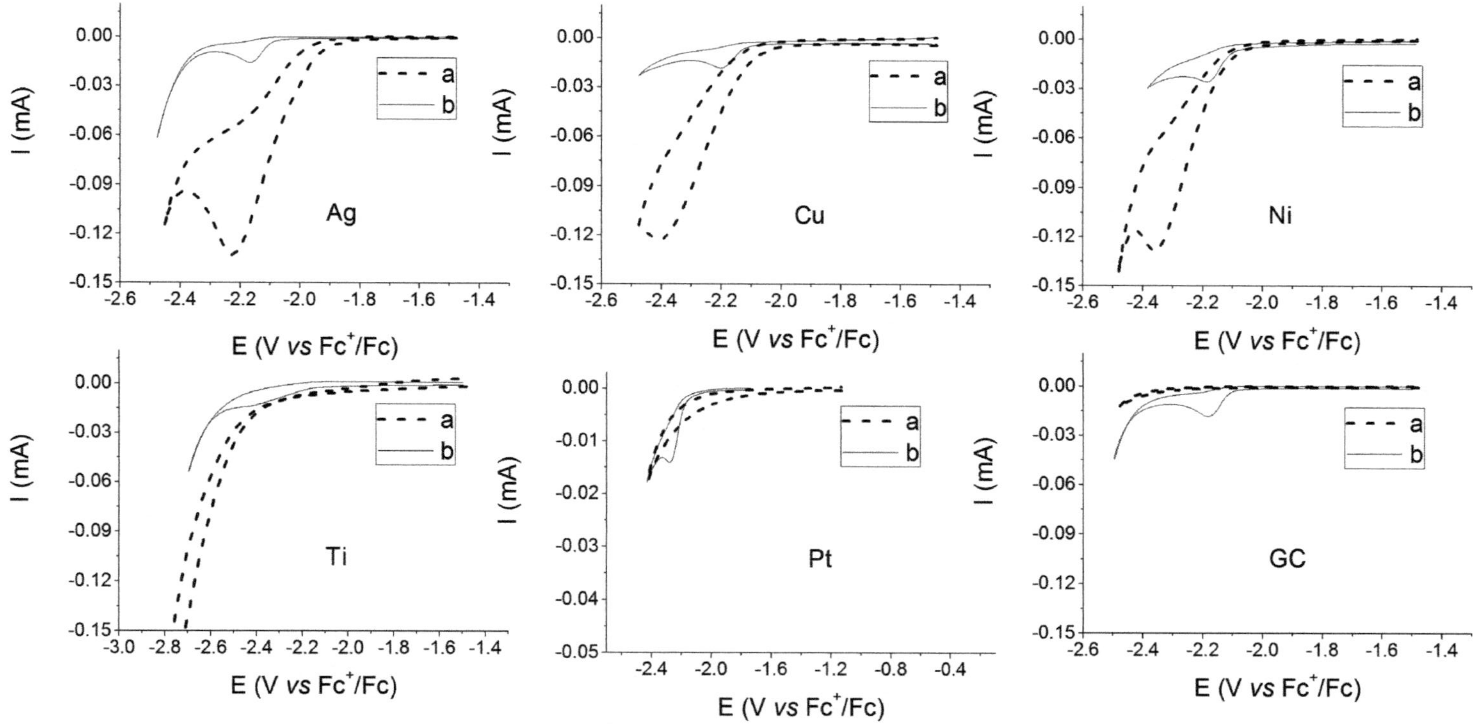

Figure 7.3 Cyclic voltammograms recorded at a scan rate of 0.05 V s^{-1} using designated working electrodes (Ag, Cu, Ni, Ti and Pt (2 mm diameter), and GC (3 mm diameter)) in [Bmim][BF_4] containing saturated CO_2 at 1 atm (dashed curves) and 10 mM acetophenone (solid curves). Fc = ferrocene.
Reproduced from ref. 73 with permission from The Royal Society of Chemistry.

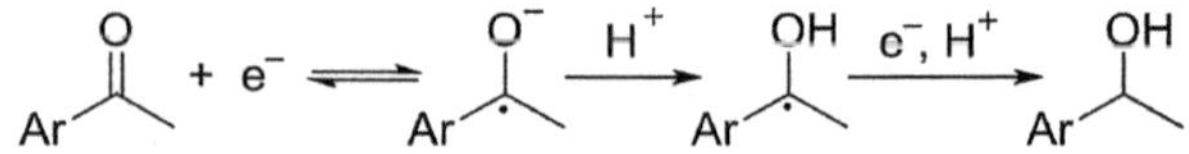

Scheme 7.1 A proton coupled electron transfer reaction for acetophenone reduction.

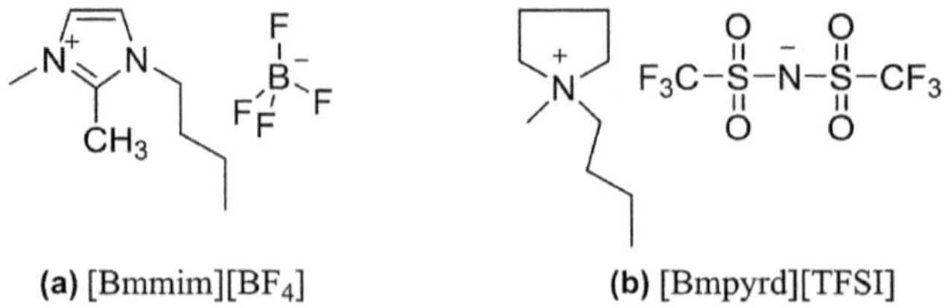

Scheme 7.2 Structures of (a) [Bmmim][BF_4] and (b) [Bmpyrd][TFSI].

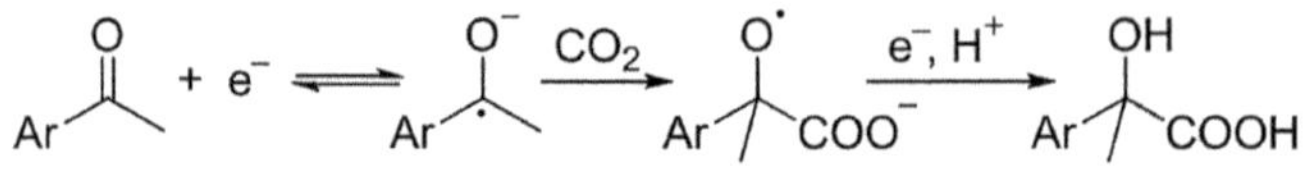

Scheme 7.3 Reaction scheme for electrocarboxylation of acetophenone.

Electrocarboxylation of acetophenone also has been investigated in other ionic liquids, including 1-butyl-2,3-dimethylimidazolium tetrafluroborate ([Bmmim][BF_4]) and 1-butyl-1-methylpyrrolidinium bis(trifluoromethyl-sulfonyl)imide ([Bmpyrd][TFSI]) (Scheme 7.2). The yield of carboxylic acid product was found to be strongly dependent on the proton availability in the ionic liquids.[52] When electroreduction of acetophenone was undertaken in the ionic liquid [Bmpyrd][TFSI], which has very low proton availability, a yield of carboxylic acid as high as 98% was obtained in the electrocarboxylation process (Scheme 7.3). Detailed discussions concerning the role of the ionic liquid in electrocarboxylation will be presented in the Section 7.3 of this chapter.

7.2.3 Electrocarboxylation of Other Substrates in Ionic Liquids

Electrocarboxylation of other substrates such as activated olefins,[61] alcohols,[54–59] epoxides[60] and amines[53] has been investigated in ionic liquids. The basic principles and experimental arrangements associated with these processes are similar to those described in Sections 7.2.1 and 7.2.2. Therefore, the relevant research outcomes are only briefly summarized below to demonstrate the wide applicability of the electrocarboxylation method.

Wang *et al.*[61] used [Bmim][BF_4] as the reaction medium for the electrocarboxylation of activated olefins. Moderate yields of monocarboxylic acids were obtained together with undesirable products resulting from hydrogenation reactions. By contrast, in MeCN, only monocarboxylic acids and dicarboxylic acids were obtained.[74] The influence of operational parameters such as temperature, electrode material and the applied current was examined. The highest yield (41%) of monocarboxylic acids was obtained at 50 °C when steel was used as the working electrode. It was also found that the ionic liquid could be reused at least five times without activity loss.

Carbonates and carbamates were also synthesized by carboxylation of substrates, such as epoxides, amines and alcohols, with electrogenerated $CO_2^{\bullet-}$.[53,60] Copper or other suitable metallic electrodes were often chosen to facilitate the formation of $CO_2^{\bullet-}$. High yields of cyclic carbonates were obtained by Yang *et al.*[60] in both imidazolium and pyridinium ionic liquids by reaction of epoxides with $CO_2^{\bullet-}$. Lu[59] and Cai[56,58] and their co-workers successfully synthesized linear carbonates with good yields in imidazolium based ionic liquids from alcohols, such as methanol, ethanol and aromatic alcohols. Similarly, high yields of carbamates were obtained in [Bmim][BF_4] from amines and carbon dioxide.[53] Cyclic carbonates were also synthesized from the reaction between diols and the NHC–CO_2 complex generated from the electroreduction of the imidazolium cation of the ionic liquid in the presence of CO_2; a very interesting concept but yet to be explored in detail.[55]

7.3 The Role of the Proton in Ionic Liquids in Determining the Reaction Pathway Accompanying Electroreduction of Aromatic Ketones under a CO_2 Atmosphere

As noted above, the use of ionic liquids as the reaction medium for electrocarboxylation has been explored for a wide range of substrates. However, despite the fact that the product distributions in ionic liquids and conventional organic solvents are often different, mechanistic investigations to identify the origin of this difference are scarce. Since aromatic ketones are the most widely investigated substrates and the medium and substrate effects have been systematically explored in a series of studies by Zhao *et al.*,[27,52,73,75–78] the discussion below is focused on this family of compounds.

Generally, the major competitive reactions in the electrocarboxylation of aromatic ketones are dimerization and proton coupled electron transfer reactions with the side products being dimers and alcohols, respectively. Thus, the three major reaction pathways during electroreduction of ketones under a CO_2 atmosphere are summarized in Scheme 7.4.

Formation of alcohols through the proton coupled electron transfer reactions requires a proton source. Therefore, the proton availability in ionic liquids, either from ionic liquid itself or from an adventitious impurity, such as H_2O, plays an important role in determining the reaction pathways that accompany aromatic ketones reduction under a CO_2 atmosphere.[52,73] Furthermore, it has been found that the presence of water in ionic liquids can also alter the kinetics of the dimerization reaction[52] and hence influence the reaction pathway.

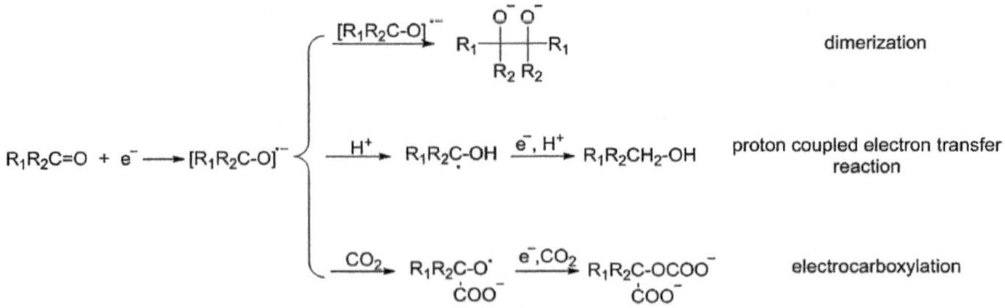

Scheme 7.4 Reaction pathways for electroreduction of aromatic ketones in ionic liquids under a CO_2 atmosphere.

7.3.1 The Influence of C2–H from Imidazolium

The C2–H in imidazolium is the most active with a pK_a value of 21–23.[79] In dry [Bmim][BF_4], reduced acetophenone undergoes dimerization under a N_2 atmosphere[73] and formation of alcohol is not favoured. However under a CO_2 atmosphere, acetophenone reduction occurs with a proton-coupled electron transfer pathway to form 1-phenylethanol as the main product with a yield of 97% (Table 7.1).[73] By contrast, in dry molecular solvents (MeCN, DMF and NMP),[21,80] carboxylation and dimerization are the major reaction pathways. The proton coupled electron transfer pathway leading to the formation of 1-phenylethanol requires the presence of a sufficiently strong proton donor, which is not available in neat [Bmim][BF_4] under a N_2 atmosphere. It was proposed by Zhao *et al.*[73] that the presence of CO_2 enhances the C2–H donating ability of $[Bmim]^+$ due to strong complex formation between NHC, the deprotonated form of $[Bmim]^+$, and CO_2, resulting in a thermodynamically favourable proton coupled electron transfer pathway (Scheme 7.5).

This mechanism proposed was supported by results derived from electroreduction of acetophenone under a CO_2 atmosphere in dry [Bmmim][BF_4] and [Bmpyrd][TFSI], which have lower proton availabilities (Table 7.1).[52] In [Bmmim][BF_4], the reactive C2–H proton is substituted by a methyl group and a significant amount of 2-hydroxy-2-phenylpropionic acid (yield of 15%) was obtained although 1-phenylethanol (yield of 60%) was still the major product. By contrast, in dry [Bmpyrd][TFSI] where any reactive proton source is essentially absent,[81] carboxylic acid with a yield as high as 98% was obtained. The clear implication is that the yield of carboxylate is strongly dependent on the proton availability in ionic liquids. Those with lower proton availability favour carboxylate formation.

Cremer *et al.*[82] showed by NMR measurements that the bonding strength and hence the acidity of the C2–H of imidazolium is strongly dependent on the anion of the ionic liquid. Ten imidazolium based ionic liquids with anions that include Cl^-, Br^-, I^-, $[NO_3]^-$, $[BF_4]^-$, $[TfO]^-$, $[PF_6]^-$, $[TFSI]^-$, $[Pf_2N]^-$ and $[FAP]^-$ ($[TfO]^-$ = trifluoromethanesulfonate ($[CF_3SO_3]^-$), $[Pf_2N]^-$ = bis(pentafluoroethylsulfonyl)imide, $[FAP]^-$ = tris(pentafluoroethyl)trifluorophosphate)

Table 7.1 Products[a] obtained from preparative scale reductive electrolysis of acetophenone in dry ionic liquids.

Ionic liquid	$C_{acetophenone}$ (mM)	Atmosphere ($p=1$ atm)	Charge consumption (F mol^{-1})	Product yield/%[b]		
				Dimer	Acid	Alcohol
$[Bmim][BF_4]$[73]	50	N_2	1.0	100	0	0
	50	CO_2	2.0	Trace	1	97
$[Bmmim][BF_4]$[52]	25	N_2	1.0	100	0	0
	25	CO_2	2.0	25	15	60
[Bmpyrd][TFSI][52]	25	N_2	1.0	100	0	0
	25	CO_2	2.0	2	98	0
NMP[80]	100	CO_2	1.7	35–40[c]	36–44[c]	4–6[c]

[a]Dimer = 1-[4-(1-hydroxy-ethylidene)-cyclohexa-2,5-dienyl]-1-phenyl-enthanol and 2,3-diphenyl-butane-2,3-diol; acid = 2-hydroxy-2-phenylpropionic acid; alcohol = 1-phenylethanol.
[b]Chromatography yields based on the reactant consumed.
[c]Yield based on HPLC analysis.

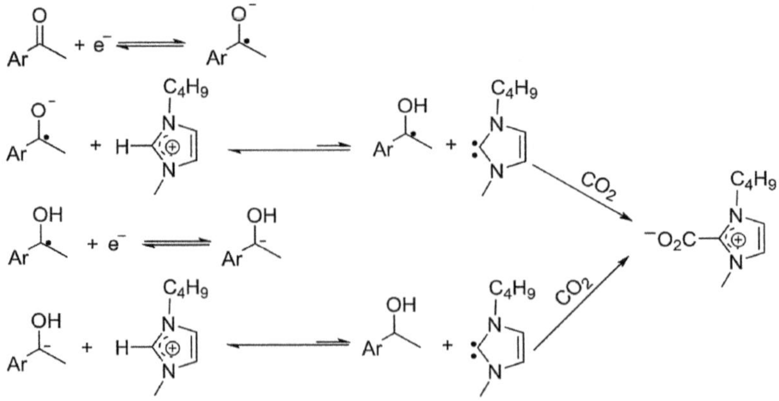

Scheme 7.5 Reaction scheme for electroreduction of acetophenone in [Bmim][BF_4] under a CO_2 atmosphere.
Reproduced from ref. 73 with permission from The Royal Society of Chemistry.

were assessed in their studies which showed that the C2–H donating ability increases as the size of the anion increases and the basicity and coordinating strength of the anions decreases. Of those investigated, ionic liquids containing the [FAP]$^-$ anion have the weakest C2–H donating ability. Since the C2–H hydrogen becomes more reactive in the presence of CO_2 due to the formation of a stable NHC–CO_2 complex, its activity should depend on the stability of NHC (deprotonated form of imidazolium) and CO_2, which is dependent on the anion of the ionic liquid.[83] Feroci *et al.* investigated the NHC reactivity towards CO_2 of a range of [Bmim]$^+$ based ionic liquids containing anions that included [BF_4]$^-$, [PF_6]$^-$, [CF_3CO_2]$^-$, [TfO]$^-$ and [TFSI]$^-$ using voltammetric, infrared and thermogravimetric methods of analysis. The results suggest that the rate of NHC–CO_2 complex formation is the slowest in [Bmim][TfO]. On this basis, [Bmim][TfO] should have the weakest C2–H donating ability, under a CO_2 atmosphere, of the ionic liquids. In principle, a higher yield of carboxylic acid product may be expected from imidazolium based ionic liquids with [TfO]$^-$ and [FAP]$^-$ anions. Zhao *et al.*[78] investigated the electroreduction of acetophenone, benzophenone and 4-phenylbenzophenone under a CO_2 atmosphere in [Bmim]$^+$ based ionic liquids having [BF_4]$^-$, [TfO]$^-$ and [FAP]$^-$ as the anions. The product distribution obtained in that study, based on HPLC analysis of the electrolysis products, is given in Table 7.2. In the case of acetophenone, 1-phenylethanol was the major product in all ionic liquids, implying that the difference in proton availability in these ionic liquids has no impact on the reaction pathway for acetophenone reduction. The authors attributed this observation to the fact that the acetophenone radical anion is a very strong base. Similar results were obtained with benzophenone, a derivative of acetophenone containing the electron withdrawing phenyl group, which should reduce the basicity of the radical anion. By contrast, when the basicity of the electrogenerated

Table 7.2 Products[a] derived from reduction of 50 mM aromatic ketone in imidazolium based ionic liquids under a CO_2 atmosphere. Reproduced from ref. 78 with permission from the PCCP Owner Societies.

Substrate	Ionic liquid	$E_{app}{}^{a}$ (V) *vs.* Fc/Fc^+	Product yield[b] (%)		
			Acid	Alcohol	Dimer
Acetophenone[c]	[Bmim][BF_4]	−2.25	0.0	90.6	9.4
	[Bmim][TfO]	−2.30	0.0	93.6	6.4
	[Bmim][FAP]	−2.32	0.0	90.6	9.4
Benzophenone[d]	[Bmim][BF_4]	−2.10	16.2	83.4	0.0
	[Bmim][TfO]	−2.12	14.5	85.3	0.0
	[Bmim][FAP]	−2.05	15.0	83.1	0.0
4-Phenylbenzophenone[e]	[Bmim][BF_4]	−2.05	32.8	66.5	0.0
	[Bmim][TfO]	−2.00	38.5	60.5	0.0
	[Bmim][FAP]	−2.07	40.7	59.0	0.0

[a]Applied potential for bulk electrolysis in V *vs.* Fc/Fc^+ (Fc = ferrocene).
[b]HPLC yields based on the consumed substrate.
[c]Acid = 2-hydroxy-2-phenylpropionic acid; alcohol = 1-phenylethanol; dimer = 1-[4-(1-hydroxyethylidene)-cyclohexa-2,5-dienyl]-1-phenyl-ethanol and 2,3-diphenyl-butane-2,3-diol.
[d]Acid = 2-hydroxy-2,2-diphenylacetic acid; alcohol = diphenylmethanol; dimer = 1,1,2,2,-tetraphenylethane-1,2-diol.
[e]Acid = 2-([1,1′-biphenyl]-4-yl)-2-hydroxy-2-phenylacetic acid; alcohol = [1,1′-biphenyl]-4-yl(phenyl)methanol; dimer = 1,2-di([1,1′-biphenyl]-4-yl)-1,2-diphenylethane-1,2-diol.

radical anion was further reduced when using 4-phenylbenzophenone, a strong effect of the ionic liquid anion on the product distribution was observed with the highest yield of acid product (40.7%) being obtained in [Bmim][FAP]. This study[78] reveals that the yield of electrocarboxylation of aromatic ketones in imidazolium based ionic liquids is influenced by the ketone substituent and the ionic liquid anion. Although the yields of carboxylic acids were improved when an optimal combination of aromatic ketone and ionic liquid anion was selected, the results obtained in the study[78] again confirmed that imidazolium based ionic liquids are not suitable reaction media for electrocarboxylation of aromatic ketones, as suggested in a previous report[73] from the same group.

7.3.2 The Influence of Adventitious Water

Despite having several attractive properties as outlined above, ionic liquids also suffer from major drawbacks, such as difficulty in purification, compared to volatile molecular solvents. The most commonly encountered impurity in ionic liquids is water, which may be reduced to around 10 ppm by prolonged drying at elevated temperatures and under high vacuum. However, water can be easily reintroduced into purified ionic liquids during the course of experiments that are performed under "bench-top" conditions or during storage.[84,85] Zhao *et al.*[75] investigated the influence of water on benzophenone reduction in the ionic liquids [Bmpyrd][TFSI] and 1-butyl-1-

methylpiperidinium bis(trifluoromethylsulfonyl)-imide ([Bmpipd][TFSI]) (Figure 7.4) and showed that the voltammetric characteristics of this process are highly sensitive to water. In dry ionic liquids, two chemically reversible one-electron reduction processes, corresponding to the formation of the radical anion and dianion, were observed. Upon addition of water, the first reduction process remains chemically reversible and the reversible potential relatively constant. By contrast, the second process becomes chemically irreversible and the peak potential shifts towards more positive values as the concentration of water increases. Eventually, when more than 0.2 M H_2O is present, the two resolved one-electron benzophenone reduction processes

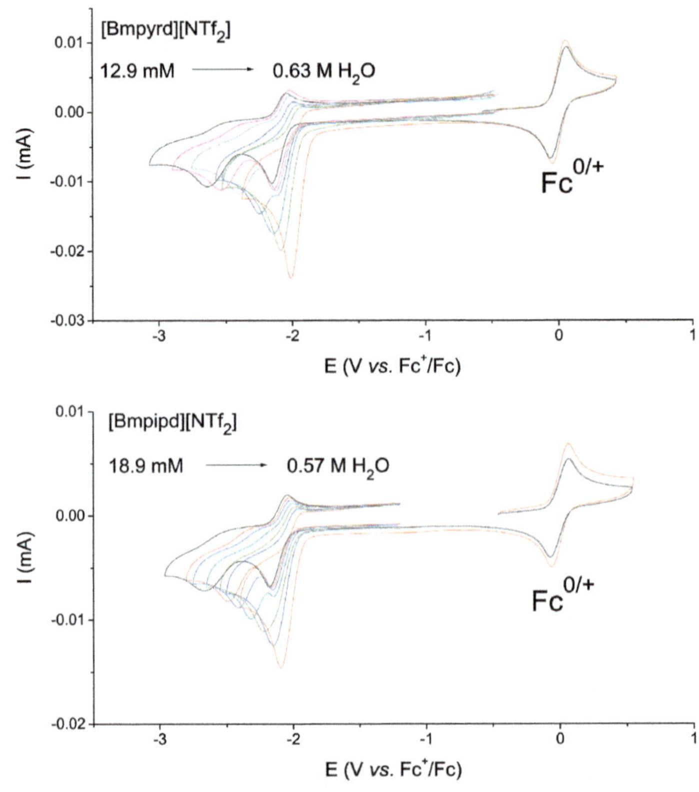

Figure 7.4 Comparison of the cyclic voltammograms obtained for the reduction of 10 mM benzophenone in [Bmpyrd][NTf_2] and [Bmpipd][NTf_2] under a N_2 atmosphere with a GC electrode (1.0 mm diameter) at a scan rate of 1.0 V s^{-1} as a function of water concentration. The voltammograms, from left to right, were obtained in "wet" ionic liquids that contained 0.0129, 0.0299, 0.0629, 0.1429, 0.2629, 0.4129 and 0.63 M water in [Bmpyrd][NTf_2], and 0.0189, 0.0449, 0.0689, 0.0989, 0.1789, 0.2989 and 0.57 M water in [Bmpipd][NTf_2]. The Fc/Fc^+ process ($E^{0'}=0$ V) with 10 mM Fc (Fc = ferrocene) is included for comparison.
Adapted with permission from Zhao *et al.*, *Chem. Eur. J.*, 2012, **18**, 5290–5301. © 2012 WILEYVCH Verlag GmbH & Co. KGaA, Weinheim.

merge into an overall two-electron reduction process with an enhanced peak current.

The reaction mechanism proposed for this process is described in Scheme 7.6, and involves protonation and hydrogen-bonding interactions between the benzophenone dianion. Based on the results obtained from a theory–experiment comparison exercise, as many as seven H_2O molecules were involved in the hydrogen-bonding interaction, in contrast with three found in the case of DMF.[86] Consequently, it is suggested that hydrogen-bonding interactions between the benzophenone dianion and water molecules are stronger in ionic liquids than in molecular solvents. On this basis, it may be expected that water would have a stronger influence on the reaction pathway of ketone reduction in ionic liquids under a CO_2 atmosphere. Indeed, it was found that only alcohol was obtained during the electrolysis of benzophenone under a CO_2 atmosphere in water saturated ionic liquids ($[H_2O] \approx 0.6$ M). By contrast, in relatively dry ionic liquids ($[H_2O] = 1.0$ mM), acid is the sole product.[52] Even when 0.4 M water is present in molecular

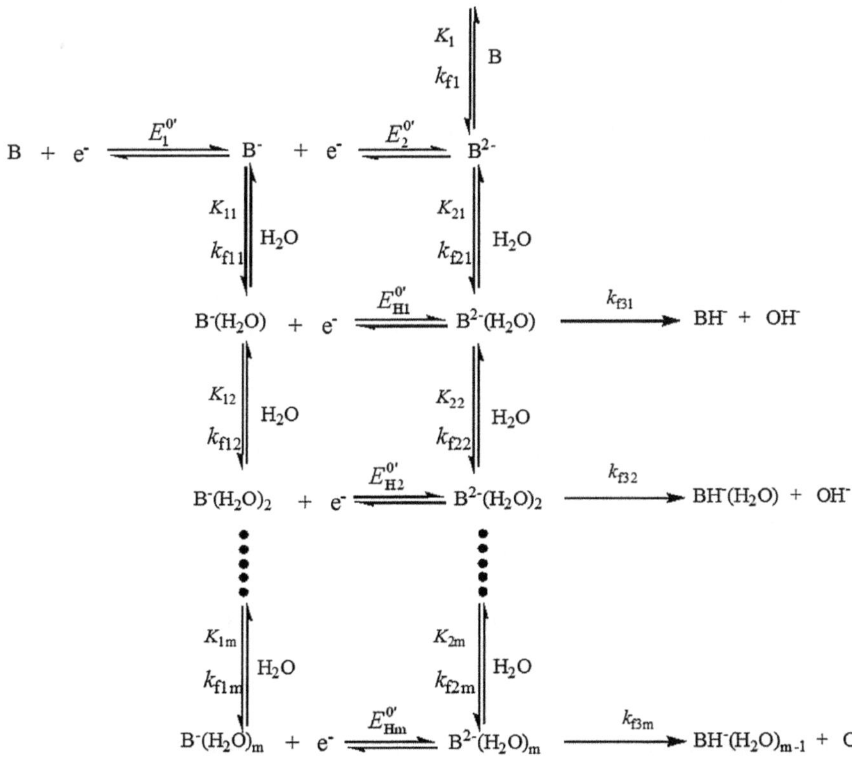

Scheme 7.6 Reaction scheme proposed for the reduction of benzophenone in "wet" [Bmpyrd][NTf$_2$] and [Bmpipd][NTf$_2$].
Adapted with permission from Zhao *et al.*, *Chem. Eur. J.*, 2012, **18**, 5290–5301. © 2012 WILEYVCH Verlag GmbH & Co. KGaA, Weinheim.

solvents (NMP, DMF and MeCN), acid is still the major product, although the amount of alcohol increases relative to that found in dry solvent.[14]

In [Bmpyrd][NTf$_2$], with low proton availability from the ionic liquid itself, it was found by Zhao *et al.*[52] that, under water-saturated conditions, the products obtained from the electrochemical reduction of acetophenone are exclusively dimers under both N_2 and CO_2 atmospheres. In a conventional organic solvent medium containing water, an increase in the number of dimers and amount of alcohol was observed and the yield of the carboxylic acid decreases significantly.[87] In water saturated [Bmpyrd][NTf$_2$], the dimerization rate constant of 1.0×10^7 M^{-1} s^{-1} under a N_2 atmosphere is more than three orders of magnitude higher than that obtained in dry [Bmpyrd][NTf$_2$]. This rate increase is attributed to strong interaction between the acetophenone radical anions and water through an extensive hydrogen-bonding network leading to larger degree of charge delocalization and thus favouring dimer formation.

7.4 Conclusions and Prospects

The use of ionic liquids as an alternative to conventional molecular solvents for the transformation of CO_2 into a variety of compounds through electrocarboxylation is receiving a growing level of attention. This can be attributed to their high conductivity and electrochemical stability and the ability to dissolve large amounts of CO_2 and organic substrates. Ionic liquids have demonstrated the potential for CO_2 capture and several reviews[88–90] summarizing recent advances in selective CO_2 capture using task-specific ionic liquids are available. However, it has also been noted that extensive energy is required for CO_2 desorption, making this step commercially unattractive. A potential alternative approach is to use an ionic liquid as the medium for CO_2 capture and subsequent electrochemical conversion of CO_2 through electrocarboxylation using electricity provided from renewable sources. However, despite the fact that considerable progress has been made in understanding how ionic environments and trace impurities influence reaction pathways, electrocarboxylation in ionic liquids is still in its infancy from the viewpoint of large-scale industrial applications. In particular, even though ionic liquids have green solvent properties in many aspects, their cost is much higher than conventional solvents. Thus, new synthetic routes are needed to reduce the cost of ionic liquids, and new strategies also are required to improve their recyclability in order to minimize the quantity of ionic liquids required.

In this review, we have only discussed the cathodic reaction that produces the target product in electrocarboxylation. However, in electrosynthesis, both the anode and cathode reactions occur to form a complete circuit. Most research has been conducted with either sacrificial anodes, such as magnesium, zinc or aluminium, or with a stable anode, like platinum or carbon. In order to make the entire electrolysis process commercially viable, the co-products generated at the anode should ideally be useful, or at least not harmful. Water oxidation to form O_2 would be an ideal anodic reaction to pair with electrocarboxylation at the cathode. Although some of the ionic liquids investigated to date show some promising properties for the synthesis of carboxylic acids with high yields, the results suggest that precautions need be taken to choose ionic liquids with unavailable protons sources. This implies use of dry ionic liquids with water carefully removes for electrocarboxylation applications. Thus, these ionic liquid environments are not ideal for oxygen evolution at the anode in an industrial application of electrocarboxylation. Suitable ionic liquids that can promote carboxylation pathways, while retarding the undesired dimerization and proton couple electron transfer reaction pathways and permitting a green anode reaction, still need to be identified.

Abbreviations

CO_2	Carbon dioxide
RTILs	Room temperature ionic liquids
[Bmim][BF_4]	1-butyl-3-methylimidazolium tetrafluroborate
[Bmmim][BF_4]	1-butyl-2,3-dimethylimidazolium tetrafluroborate
[Bmpyrd][TFSI]	1-butyl-1-methylpyrrolidiniumbis(trifluoromethylsulfonyl)-imide
[Bmpipd][TFSI]	1-butyl-1-methylpiperidinium bis(trifluoromethylsulfonyl)imide
[Bmim][TfO]	1-butyl-3-methylimidazolium trifluoromethylsulfonate
[Bmim][FAP]	1-butyl-3-methylimidazolium tris(pentafluoroethyl)trifluorophosphate
MeCN	Acetonitrile
DMF	*N*,*N*-dimethylformamide
NMP	*N*-methyl-2-pyrrolidone
Fc^+/Fc	Ferricenium/ferrocene

References

1. C. Le Quere, G. P. Peters, R. J. Andres, R. Andrew, T. A. Boden, P. Ciais, P. Friedlingstein, R. Houghton, G. Marland and R. Moriarty, *Earth Syst. Sci. Data*, 2014, **6**, 235.
2. P. Tans, NOAA/ESRL (www.esrl.noaa.gov/gmd/ccgg/trends/) and R. Keeling, Scripps Institution of Oceanography (scrippsco2.ucsd.edu/).
3. J. Hansen, M. Sato, P. Kharecha, D. Beerling, R. Berner, V. Masson-Delmotte, M. Pagani, M. Raymo, D. L. Royer and J. C. Zachos, *Open. Atmos. Sci. J.*, 2008, **2**, 217.
4. T. Sakakura, J. C. Choi and H. Yasuda, *Chem. Rev.*, 2007, **107**, 2365.
5. Xu and J. A. Moulijn, *Energy Fuels*, 1996, **10**, 305.
6. D. Fromm and D. Lützow, *Chem. Unserer Zeit*, 1979, **13**, 78.
7. G. A. Olah, A. Goeppert and G. K. S. Prakash, *J. Org. Chem.*, 2008, **74**, 487.

8. M. Mikkelsen, M. Jørgensen and F. C. Krebs, *Energy Environ. Sci.*, 2010, **3**, 43.
9. L. A. von Arnim, *Ann. Phys.*, 1801, **8**, 257.
10. N. N. Beketov, *Russ. Phys. Chem. Soc.*, 1869, **1**, 33.
11. E. V. Kondratenko, G. Mul, J. Baltrusaitis, G. Larrazabal and J. Perez-Ramirez, *Energy Environ. Sci.*, 2013, **6**, 3112.
12. E. J. Van Scott and R. J. Yu, *Arch. Dermatol.*, 1974, **110**, 586.
13. E. J. Van Scott and R. J. Yu, *Cutis*, 1989, **43**, 222.
14. A. A. Isse, A. Galia, C. Belfiore, G. Silvestri and A. Gennaro, *J. Electroanal. Chem.*, 2002, **526**, 41.
15. A. A. Isse and A. Gennaro, *Chem. Commun.*, 2002, 2798.
16. D. Ballivet-Tkatchenko, J. C. Folest and J. Tanji, *Appl. Organomet. Chem.*, 2000, **14**, 847.
17. H. Wang, K. Zhang, Y. Z. Liu, M. Y. Lin and J. X. Lu, *Tetrahedron*, 2008, **64**, 314.
18. S.-F. Zhao, M.-X. Zhu, K. Zhang, H. Wang and J.-X. Lu, *Tetrahedron Lett.*, 2011, **52**, 2702.
19. G. Filardo, S. Gambino, G. Silvestri, A. Gennaro and E. Vianello, *J. Electroanal. Chem. Interfacial Electrochem.*, 1984, **177**, 303.
20. H. Senboku, Y. Yamauchi, T. Fukuhara and S. Hara, *Electrochemistry*, 2006, **74**, 612.
21. O. Scialdone, C. Amatore, A. Galia and G. Filardo, *J. Electroanal. Chem.*, 2006, **592**, 163.
22. A. A. Isse and A. Gennaro, *Collect. Czech. Chem. Commun.*, 2003, **68**, 1379.
23. A. Gennaro, A. A. Isse, C. L. Bianchi, P. R. Mussini and M. Rossi, *Electrochem. Commun.*, 2009, **11**, 1932.
24. O. Scialdone, A. Galia, G. Silvestri, C. Amatore, L. Thouin and J. N. Verpeaux, *Chem. – Eur. J.*, 2006, **12**, 7433.
25. A. Gennaro, C. M. Sanchez-Sanchez, A. A. Isse and V. Montiel, *Electrochem. Commun.*, 2004, **6**, 627.
26. G. Silvestri, S. Gambino, G. Filardo and A. Gulotta, *Angew. Chem.*, 1984, **96**, 978.
27. S.-F. Zhao, H. Wang, Y.-C. Lan, X. Liu, J.-X. Lu and J. Zhang, *J. Electroanal. Chem.*, 2012, **664**, 105.
28. T. Welton, *Chem. Rev.*, 1999, **99**, 2071.
29. V. I. Pârvulescu and C. Hardacre, *Chem. Rev.*, 2007, **107**, 2615.
30. P. Hapiot and C. Lagrost, *Chem. Rev.*, 2008, **108**, 2238.
31. M. A. P. Martins, C. P. Frizzo, D. N. Moreira, N. Zanatta and H. G. Bonacorso, *Chem. Rev.*, 2008, **108**, 2015.
32. J.-C. Plaquevent, J. Levillain, F. D. R. Guillen, C. Malhiac and A.-C. Gaumont, *Chem. Rev.*, 2008, **108**, 5035.
33. J. P. Hallett and T. Welton, *Chem. Rev.*, 2011, **111**, 3508.
34. J. Zhang, A. I. Bhatt, A. M. Bond, A. G. Wedd, J. L. Scott and C. R. Strauss, *Electrochem. Commun.*, 2005, **7**, 1283.
35. J. Zhang and A. M. Bond, *Anal. Chem.*, 2003, **75**, 6938.
36. J. Zhang and A. M. Bond, *Anal. Chem.*, 2003, **75**, 2694.

37. J. Zhang and A. M. Bond, *Analyst*, 2005, **130**, 1132.
38. J. Zhang, A. M. Bond, D. R. MacFarlane, S. A. Forsyth, J. M. Pringle, A. W. A. Mariotti, A. F. Glowinski and A. G. Wedd, *Inorg. Chem.*, 2005, **44**, 5123.
39. J. Zhang, A. M. Bond, P. J. S. Richardt and A. G. Wedd, *Inorg. Chem.*, 2004, **43**, 8263.
40. M. Earle, P. McCormac and K. Seddon, *Green Chem.*, 1999, **1**, 23.
41. A. Rencurosi, L. Lay, G. Russo, E. Caneva and L. Poletti, *J. Org. Chem.*, 2005, **70**, 7765.
42. J. L. Anthony, E. J. Maginn and J. F. Brennecke, *J. Phys. Chem. B*, 2002, **106**, 7315.
43. C. Cadena, J. L. Anthony, J. K. Shah, T. I. Morrow, J. F. Brennecke and E. J. Maginn, *J. Am. Chem. Soc.*, 2004, **126**, 5300.
44. M. Hasib-ur-Rahman, M. Siaj and F. Larachi, *Chem. Eng. Process.*, 2010, **49**, 313.
45. H. Senboku, in *Transformation and Utilization of Carbon Dioxide*, ed. B. M. Bhanage and M. Arai, Springer, Berlin Heidelberg, 2014, ch. 10, pp. 245.
46. E. B. Cole and A. B. Bocarsly, in *Carbon Dioxide as Chemical Feedstock*, ed. M.Aresta, Wiley-VCH Verlag GmbH & Co. KGaA, Weinheim, 2010, ch. 11, pp. 297.
47. A.-H. Liu, J. Gao and L.-N. He, in *New and Future Developments in Catalysis*, ed. S.L. Suib, Elsevier, Amsterdam, 2013, ch. 4, pp. 82.
48. Q. Feng, K. Huang, S. Liu and X. Wang, *Electrochim. Acta*, 2010, **55**, 5741.
49. Y. Hiejima, M. Hayashi, A. Uda, S. Oya, H. Kondo, H. Senboku and K. Takahashi, *Phys. Chem. Chem. Phys.*, 2010, **12**, 1953.
50. D. F. Niu, J. B. Zhang, K. Zhang, T. Xue and J. X. Lu, *Chin. J. Chem.*, 2009, **27**, 1041.
51. Q. Feng, K. Huang, S. Liu, J. Yu and F. Liu, *Electrochim. Acta*, 2011, **56**, 5137.
52. S.-F. Zhao, M. Horne, A. M. Bond and J. Zhang, *Green Chem.*, 2014, **16**, 2242.
53. M. Feroci, M. Orsini, L. Rossi, G. Sotgiu and A. Inesi, *J. Org. Chem.*, 2007, **72**, 200.
54. F. Liu, S. Liu, Q. Feng, S. Zhuang, J. Zhang and P. Bu, *Int. J. Electrochem. Sci.*, 2012, 7, 4381.
55. L.-X. Wu, H. Wang, Z.-Y. Tu, B.-B. Ding, Y. Xiao and J.-X. Lu, *Int. J. Electrochem. Sci.*, 2012, **7**, 11540.
56. C. Yan, B. Lu, X. Wang, J. Zhao and Q. Cai, *J. Chem. Technol. Biotechnol.*, 2011, **86**, 1413.
57. D. Yuan, C. Yan, B. Lu, H. Wang, C. Zhong and Q. Cai, *Electrochim. Acta*, 2009, **54**, 2912.
58. X. Yuan, B. Lu, J. Liu, X. You, J. Zhao and Q. Cai, *J. Electrochem. Soc.*, 2012, **159**, E183.
59. L. Zhang, D. Niu, K. Zhang, G. Zhang, Y. Luo and J. Lu, *Green Chem.*, 2008, **10**, 202.

60. H. Yang, Y. Gu, Y. Deng and F. Shi, *Chem. Commun.*, 2002, 274.
61. H. Wang, G. R. Zhang, Y. Z. Liu, Y. W. Luo and J. X. Lu, *Electrochem. Commun.*, 2007, **9**, 2235.
62. O. Sock, M. Troupel and J. Perichon, *Tetrahedron Lett.*, 1985, **26**, 1509.
63. O. Scialdone, A. Galia, G. Errante, A. A. Isse, A. Gennaro and G. Filardo, *Electrochim. Acta*, 2008, **53**, 2514.
64. A. A. Isse and A. Gennaro, *Indian J. Chem., Sec. A*, 2003, **42**, 751.
65. R. R. Raju, S. K. Lateef, S. R. K. Mohan and S. R. J. Reddy, *Arkivoc*, 2006, 76.
66. A. J. Aishah, M. A. Hartini, S. Normala, A. M. Norhuda, H. H. N. Hanisi, H. M. Razif and T. Sugeng, *J. Nat. Gas Chem.*, 2007, **16**, 273.
67. D. F. Niu, L. P. Xiao, A. J. Zhang, G. R. Zhang, Q. Y. Tan and J. X. Lu, *Tetrahedron*, 2008, **64**, 10517.
68. C. L. Scortichini and S. J. Babinec, *J. Electroanal. Chem.*, 1994, **379**, 111.
69. J. L. Olloqui-Sariego, V. M. Molina, D. Gonzalez-Arjona, E. Roldan and M. Dominguez, *J. Electrochem. Soc.*, 2008, **155**, E157.
70. J. L. Olloqui-Sariego, V. M. Molina, D. Gonzalez-Arjona, E. Roldan and M. Dominguez, *J. Electrochem. Soc.*, 2010, **157**, E64.
71. S. Chanfreau, P. Cognet, S. Camy and J. S. Condoret, *J. Supercrit. Fluids*, 2008, **46**, 156.
72. A. A. Isse, M. G. Ferlin and A. Gennaro, *J. Electroanal. Chem.*, 2005, **581**, 38.
73. S.-F. Zhao, L.-X. Wu, H. Wang, J.-X. Lu, A. M. Bond and J. Zhang, *Green Chem.*, 2011, **13**, 3461.
74. H. Wang, Y. F. Du, M. Y. Lin, K. Zhang and J. X. Lu, *Chin. J. Chem.*, 2008, **26**, 1745.
75. S.-F. Zhao, J.-X. Lu, A. M. Bond and J. Zhang, *Chem. – Eur. J.*, 2012, **18**, 5290.
76. S. O'Toole, S. Pentlavalli and A. P. Doherty, *J. Phys. Chem. B*, 2007, **111**, 9281.
77. C. Lagrost, P. Hapiot and M. Vaultier, *Green Chem.*, 2005, 7, 468.
78. S.-F. Zhao, M. Horne, A. M. Bond and J. Zhang, *Phys. Chem. Chem. Phys.*, 2015, DOI: 10.1039/C5CP00095E.
79. T. L. Amyes, S. T. Diver, J. P. Richard, F. M. Rivas and K. Toth, *J. Am. Chem. Soc.*, 2004, **126**, 4366.
80. O. Scialdone, M. A. Sabatino, C. Belfiore, A. Galia, M. P. Paternostro and G. Filardo, *Electrochim. Acta*, 2006, **51**, 3500.
81. J. Ding, D. Zhou, G. Spinks, G. Wallace, S. Forsyth, M. Forsyth and D. MacFarlane, *Chem. Mater.*, 2003, **15**, 2392.
82. T. Cremer, C. Kolbeck, K. R. J. Lovelock, N. Paape, R. Wölfel, P. S. Schulz, P. Wasserscheid, H. Weber, J. Thar, B. Kirchner, F. Maier and H.-P. Steinrück, *Chem. – Eur. J.*, 2010, **16**, 9018.
83. M. Feroci, I. Chiarotto, G. Forte, S. Vecchio Ciprioti and A. Inesi, *ChemElectroChem*, 2014, **1**, 1407.
84. F. D. Francesco, N. Calisi, M. Creatini, B. Melai, P. Salvo and C. Chiappe, *Green Chem.*, 2011, **13**, 1712.

85. C. D. Tran, S. H. De paolilacerda and D. Oliveira, *Appl. Spectrosc.*, 2003, **57**, 152.
86. S. H. Wang, P. S. Singh and D. H. Evans, *J. Phys. Chem. C*, 2009, **113**, 16686.
87. O. Scialdone, A. Galia, A. A. Isse, A. Gennaro, M. A. Sabatino, R. Leone and G. Filardo, *J. Electroanal. Chem.*, 2007, **609**, 8.
88. X. Zhang, X. Zhang, H. Dong, Z. Zhao, S. Zhang and Y. Huang, *Energy Environ. Sci.*, 2012, **5**, 6668.
89. M. Ramdin, T. W. de Loos and T. J. H. Vlugt, *Ind. Eng. Chem. Res.*, 2012, **51**, 8149.
90. C. Wang, H. Luo, D.-E. Jiang, H. Li and S. Dai, *Angew. Chem.*, 2010, **122**, 6114.

CHAPTER 8

IR Spectro-electrochemistry and Group-6 α-diimine Catalysts of CO_2 Reduction

DAVID E. RYAN AND FRANTIŠEK HARTL*

Department of Chemistry, Whiteknights, Reading RG6 6AD, UK
*Email: f.hartl@reading.ac.uk

8.1 Introduction

8.1.1 Carbon Dioxide Reduction and Homogeneous Catalysis

The environmental impact of anthropogenic carbon dioxide is one of the most pertinent facets of mainstream science. Accordingly, the conversion of CO_2 into more useful carbon-based products represents a compelling, yet highly challenging pursuit of contemporary chemistry. Such an endeavour hence offers a value to society that is two-fold: reduced carbon-based products, converted conceivably from CO_2 that would otherwise be emitted into the atmosphere, may be used as fuels as a mean to satiate the incessant human demand for energy. In addition, materials procured from the reduction of CO_2 may serve as feedstocks for further synthetic transformations.[1] For example, carbon monoxide as a component of syngas (2 : 1 H_2/CO), can generate methanol, *via* Cu/ZnO/Al_2O_3-mediated catalysis,[2] or hydrocarbons, through the Fischer–Tropsch process.[3–5] Additionally, CO is a crucial reagent in the Monsanto/Cativa processes, from which acetic acid

Energy and Environment Series No. 21
Electrochemical Reduction of Carbon Dioxide: Overcoming the Limitations of Photosynthesis
Edited by Frank Marken and David Fermin

Published by the Royal Society of Chemistry, www.rsc.org

is produced.[6,7] Formic acid, another commonly encountered two-electron reduced product from CO_2, also finds use in subsequent organic synthesis applications.[8] Since CO_2 is formed as the ultimate thermodynamic product following the complete combustion of hydrocarbon fuels, its conversion into various reduced forms (delineated by either lower C/H, or higher C/O ratios) necessarily requires an input of energy.

A notable feature is the highly unfavourable one-electron reduction of CO_2 to the corresponding radical anion $CO_2^{\bullet-}$, as delineated by its decidedly negative formal reduction potential of −1.90 V *vs.* SHE.[9] Moreover, large overpotentials, associated with demanding kinetics, accompany the unfavourable thermodynamic parameters to this electron transfer, are due at least in part to the topological contortion from linear CO_2 to bent $CO_2^{\bullet-}$.[10,11] A unique electrocatalytic reduction of CO_2 to oxalate by a copper complex, offering a low-energy pathway for the formation of $CO_2^{\bullet-}$, presents an exception.[12] Conversely, proton-coupled multi-electron transfer pathways offer a more thermodynamically amenable route to reduced products, as demonstrated by their lower formal reduction potentials.[13] Subsequently, catalyst systems that are capable of selectively and efficiently mediating these reduction processes are highly sought, and the investigation thereof is an active and pertinent area of research.[14] Such strategies span the full ambit of heterogeneous and homogeneous systems, materials, and molecular chemistry, and frequently draw inspiration and insight from enzymatic CO_2 reduction processes utilised by organisms.[15–20]

Amongst molecular catalyst systems based on organometallic and coordination compounds, both photochemical and electrochemical approaches have received appreciable interest.[15–17] Here we focus on the latter strategy, that is, exploration and characterisation of electroactive transition metal-based catalyst systems, which permit efficient and selective $2e^-$ reduction of CO_2.[21] Molecular electrocatalysts based on transition metal compounds offer a unique advantage in this context, borne out of the aptitude to which their performance may be attuned by appropriate modification of their steric and electronic architecture. The cathodic behaviour of such compounds, which generates the catalytically active reduced species, is contingent on the energetic and topological nature of their frontier molecular orbitals. Hence, a key strategy to optimising the catalytic process involves suitable choice of ligands within the coordination sphere, and appropriate substitution of the organometallic scaffold. However, a pragmatic approach to this optimisation requires sufficient insight into the mechanistic features of the catalyst system.

8.2 Electroanalytical Techniques and Electrochemical Mechanisms

Investigating the mechanistic pathways of the catalyst system mandates the employment of unique analytical techniques; crucial information regarding the redox behaviour of electroactive catalyst precursors, reactivity of reduced or oxidised products, and the nature of intermediates may be ascertained. This section presents a brief treatment of the relevant start-up electroanalytical methods.

8.2.1 Cyclic Voltammetry

A skeletal outline of the redox-coupled reactions pertaining to an electroactive catalyst system can be initially provided by cyclic voltammetry, in which the potential at a working electrode is swept to populate virtual orbitals, or deplete populated orbitals of redox-active compounds, at more negative or positive electrode potentials, respectively. In this context, such methods allow a cursory insight into the ease with which an electroactive species may be oxidised or reduced, the nature of the redox products, and their subsequent reactivity. The first reduction potential, associated with transfer of an electron to an electroactive compound, reflects the capacity of the neutral species to accept an electron into its LUMO; a more negative reduction potential is therefore associated with a higher energy LUMO. The corresponding electrochemical oxidation of the complex, in which electrons in the HOMO of the neutral species are abstracted, can then allow approximation of the HOMO–LUMO gap, an important initial parameter when considering the reactivity of a species. Since this chapter is primarily concerned with the electrocatalytic *reduction* of CO_2, we resume our discussion of redox behaviour referring primarily to reduction processes.

Should the reduced species remain stable upon electron transfer (on the selected electrochemical timescale, as dictated by the scan rate), a reverse sweep towards more positive potentials at the working electrode produces an anodic counterwave as the $1e^-$ reduced species is reoxidised (eqn (8.1)). Deviation from the characteristic motif in the cyclic voltammogram (CV) delineating a reversible redox couple, is manifested by either chemical or electrochemical irreversibility. The latter is associated with slow electron transfer kinetics, most pertinently engendered by a significant distortion of molecular geometry accompanying the transfer of an electron. Chemical irreversibility concerns the aptitude of the reduced products to undergo some chemical transformation following the electron transfer process (eqn (8.2)), and the manifestation thereof in the CV is contingent on the timescale of the experiment. Less stable reduced products, which undergo more rapid chemical transformations, require faster scan rates to reacquire the anodic counterwave. In extreme cases, ultramicroelectrodes, which permit the use of very high scan rates, may be used in order to retain chemical reversibility of highly fissile redox products.[22] Such a mechanism, where an electron transfer prompts a chemical reaction (such as loss of a ligand coordinated to the metal centre), is denoted EC. If the chemical transformation yields a species with a more stabilised unoccupied or partially occupied orbital, then under the instantaneous potential conditions a further electron transfer occurs, which delineates an ECE mechanism (eqn (8.3)). Accordingly, further ECEC mechanisms may be envisaged. Qualitative information regarding the above may be obtained from cyclic voltammetric experiments, providing an initial insight into the reactivity and stability of reduced species.

$$\text{E} \qquad ML_n + e^- \rightleftharpoons ML_n^{\bullet -} \tag{8.1}$$

$$\begin{array}{llll} \text{E} & ML_n^{\bullet -} + e^- \longrightarrow ML_n^{2-} & E_{p,c-1} \\ \text{C} & ML_n^{2-} \longrightarrow ML_{n-1}^{2-} + L & \end{array} \tag{8.2}$$

$$\begin{array}{llll} \text{E} & ML_nX + e^- \longrightarrow ML_nX^{\bullet -} & E_{p,c-2} \\ \text{C} & ML_nX^{\bullet -} \longrightarrow ML_n^{\bullet} + X^- & \\ \text{E} & ML_n^{\bullet} + e^- \longrightarrow ML_n^- & E_{p,c-3} \end{array} \tag{8.3}$$

$$E_{p,c-3} > E_{p,c-2}$$

Eqn (8.4) presents a simplified example to illustrate the case where electrocatalytic reactivity is established; electron transfer to a neutral catalyst precursor prompts the generation of an active catalytic species (here, *via* the EC mechanism in eqn (8.2)), which upon completing a catalytic cycle reverts to a more oxidised entity. Under the instantaneous potential conditions, electrons are delivered to again regenerate the reduced catalyst species and fulfil the electrocatalytic cycle. In the case where the substrate is in relative excess, and sufficient catalyst turnover numbers are attained, large currents are seen in the CV as a diagnostic for electrocatalytic activity. Such electrocatalyst systems can considerably reduce the kinetic barrier (and accordingly the overpotential) associated with the overall reaction (eqn (8.5)), and work in this field is levelled at designing a catalyst system that accomplishes the reduction of carbon dioxide with electrochemical potentials close to thermodynamic values. This circumvention of the unfavourable kinetics associated with the uncatalysed reaction is also the provenance of the remarkable selectivity that may be achieved by an appropriately designed catalyst system; the catalysed reduction of CO_2 may proceed in favour of the now less kinetically favourable proton reduction to dihydrogen.[23,24]

$$ML_n + e^- \rightleftharpoons ML_n^{\bullet -}$$

$$ML_n^{\bullet -} + e^- \longrightarrow ML_n^{2-} \qquad E_{p,c-1}$$

$$ML_n^{2-} \longrightarrow ML_{n-1}^{2-} + L$$

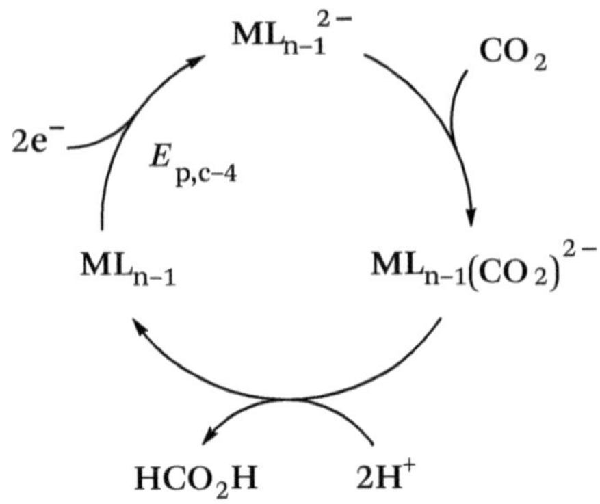

(8.4)

$$E_{p,c-4} > E_{p,c-1}$$

$$CO_2 + 2H^+ + 2e^- \longrightarrow HCO_2H \quad (8.5)$$

8.2.2 IR Spectro-electrochemistry

Beyond this superficial mechanistic insight, information regarding the structural nature of species in solution and the explication of detailed mechanistic features remain outside the scope of the routine cyclic voltammetric experiment. Instead, experimental methodology that allows for the combination of voltammetric and spectroscopic techniques may permit the *in situ* elucidation of involved species, which form and react under the prescribed potential conditions. Furthermore, the stability and chemical reversibility of electrogenerated species can be confirmed. Such experiments, dubbed spectro-electrochemical (SEC) techniques, incorporate a range of spectroscopic techniques, including, most commonly IR, UV–vis–NIR and EPR/ESR spectroscopies. Herein, we continue our discussion with regards to the IR SEC experiment, since it is most frequently this technique that has played, so far, the key role in mechanistic research on the electrocatalytic reduction of CO since the pioneering studies in the mid-1990s.[25,26]

The design of the IR SEC experiment warrants the use of a purpose-built 'optically transparent thin-layer electrochemical (OTTLE)' cell; the most important components of a versatile air-tight OTTLE cell are presented in

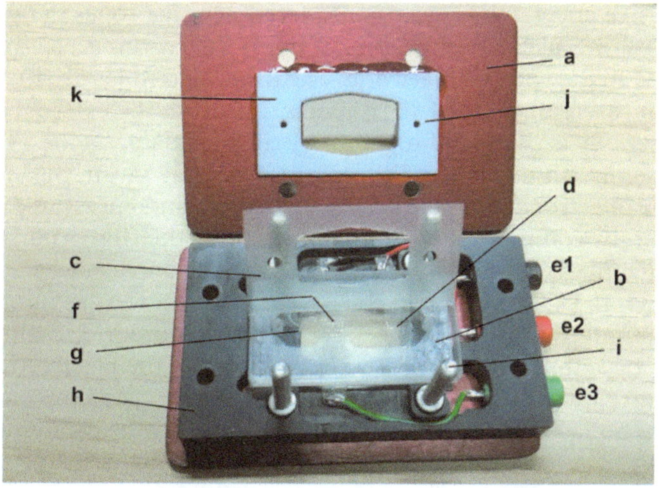

Figure 8.1 An air-tight optically transparent thin-layer electrochemical (OTTLE) cell. (a) Faceplate; (b) upper CaF_2 window; (c) silicon rubber spacer protecting the filling ports; (d) Pt minigrid counter electrode; (e) external potentiostat ports: 1, CE; 2, pRefE; 3, WE; (f) Pt minigrid working electrode (also Au, Cu, or perforated 0.1 mm graphite sheet); (g) Ag (or Pt) wire pseudoreference electrode; (h) modular electrode frame holding the optical windows; (i) mounting screws; (j) filling ports; (k) smooth PTFE spacer glued to the faceplate.

Figure 8.1. The thin-layer design of the electrochemical cell is necessary to ensure complete conversion of species in the optical path upon electrochemical reduction, and eliminate the phenomenon of diffusion in the thin solution layer from the experiment. The 0.2 mm optical path is constructed by placing a smooth polyethylene (PE) or ethylene–tetrafluoroethylene copolymer spacer with melt-sealed electrodes between the upper (drilled) and bottom optical windows (usually CaF_2, but also NaCl or high-density PE for the IR region below 1100 or 500 cm^{-1}). The path length here is also minimised to avoid strong solvent absorption. The working and counter electrodes typically take the form of a minigrid (Pt, Au, Cu; or a perforated thin graphite sheet) that permits the transmission of the infrared beam. For special applications in the realms of vibrational spectroscopy (VCD, 2D-IR) the minigrid working electrode bears a 3 mm circular aperture[27] and may feature a longer optical path (up to 1.2 mm), not compromising the thin-layer electrochemical behaviour. In the thin-layer cell design, problems associated with solvent resistivity (potential zonation; IR drop) are not inconsequential, and must be taken into account during the experiment, especially in the frequent case where resistive organic solvents are employed. Accordingly, higher concentrations of supporting electrolyte (up to 0.3 M) are typically used in comparison to conventional cyclic voltammetric experiments.[28] Stable reference electrodes (saturated calomel electrode (SCE), Ag/Ag^+) are typically not used for practical reasons (solvent bridge and associated edge-effects); instead pseudo-reference electrodes (a twinned tiny Ag or Pt wire) are used in close proximity to the working electrode at the opposite side to the counter electrode. The three electrodes are then wired to a convenient (preferably small-sized) potentiostat *via* the external ports. The air-tight assembly is then held together by the faceplate, secured by the four mounting screws; air-free conditions are essential given the anticipated sensitivity of electrogenerated species to oxygen and moisture, as well as often the parent compound, which is injected *via* the filling ports.

A problem may arise in the course of an efficient electrocatalytic conversion of CO_2 to CO, resulting in the development of CO gas bubbles in the optically monitored area of the working electrode minigrid, casing drifts of the spectral base line and ultimately replacement of the electrolyte. In some cases of $[Mn(CO)_3(\alpha\text{-diimine})]^-$ catalysts (*cf.* Chapter 5) electrogenerated CO was found to replace the α-diimine ligand and convert the catalyst to inactive $[Mn(CO)_5]^-$.[29,30]

Installed into an IR interferometer, intermittently recorded IR spectra show the advent and disappearance of various species in solution as the potential is varied. Chemical reactivity of electrogenerated species may be tracked and the resultant spectral motifs assigned to products. Naturally, the limitations of IR spectroscopy with regards to structural assignment apply in this case, and the relatively long spectrum acquisition times (*ca.* 1 min) in the conventional IR SEC experiment prohibit the elucidation of short-lived species. However, low-temperature spectro-electrochemical experiments with a cryostatted version of the OTTLE cell,[31] or rapid-scan FTIR combined with stopped flow mixing,[32–34] can reveal such transient species, which may provide valuable mechanistic insight. For a more thorough treatment of spectro-electrochemical techniques, the reader is directed to more comprehensive texts.[35,36]

In a typical IR SEC experiment, loading of the OTTLE cell with electrolyte does not require surrounding inert atmosphere. However, in a special case it may be convenient, or even necessary, to place the cell in an argon-pressurised glove box that eventually also accommodates the sample compartment of an FT-IR spectrometer (Figure 8.2). The selected example represents $2e^-$ reduction of the parent complex *trans(Cl)*-$[Os(CO)_2(bpy)Cl_2]$ in THF (ν(CO) at 1967 and 1892 cm^{-1}) producing the insoluble polymer $[Os(CO)_2(bpy)]_n$ (ν(CO) at 1958 and 1870 cm^{-1}) deposited as a strongly air-sensitive blue film in the area of the minigrid working electrode. The subsequent bpy-localised reduction of $[Os(CO)_2(bpy)]_n$ to its catalytically active anionic form (ν(CO) at 1930 and 1859 cm^{-1}), converting CO_2 to CO and formate, required reloading of the OTTLE cell under the inert atmosphere with a fresh electrolyte solution prepared directly in the glove box,[37] to remove the liberated chloride ions.

Systematic *in situ* mechanistic IR SEC studies of the reduction paths of the complexes $[M(CO)_3(\text{R-bpy})X]$ (M = Mn, Re) presented in Chapter 5 were conducted by Kubiak and co-workers[38] who employed the air-tight thin-layer specular (external) reflectance cell schematically depicted in Figure 8.3.

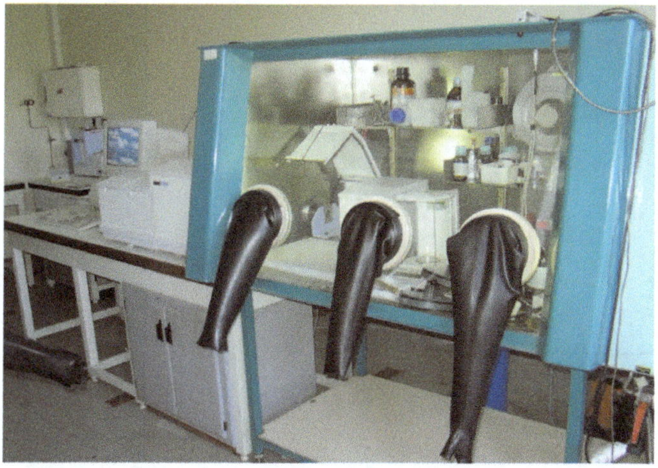

Figure 8.2 A glove box (University of Grenoble Alpes, France) with a built-in sample compartment of a Nicolet FT-IR spectrometer suited for IR spectro-electrochemistry of extremely air- and moisture-sensitive parent compounds and intermediates generated within an OTTLE cell (as shown in Figure 8.1).

The cell can be used in the temperature range from −55 to +80 °C, fitting to a holder fixed in an external sample compartment of an FT-IR spectrometer. It utilizes diverse polished, optically reflecting working electrode materials (Pt, Au, GC).

IR reflectance SEC cells are generally suited better than OTTLE cells for investigation of surface/interface phenomena and immobilised catalyst systems. On the other hand, OTTLE cells offer easy construction, handling, maintenance, and facile adaptation to specific experimental conditions (*e.g.*, wide spectral range down to the far-IR region, and luminescence and Raman studies using epifluorescence and Raman microscopes). Emerging new SEC techniques in this field are introduced in Chapter 9.

8.3 Group-6 Carbonyl Complexes Bearing Redox Non-innocent Ligands

Following on from the growing interest in CO_2 electrocatalysis mediated by Earth-abundant metals bearing non-innocent, mainly polypyridine-based ligands,[39] which has provided much of the motivation towards investigation of the catalysts [Mn(CO)$_3$(bpy)X], and electro-generated Fe(0) porphyrin catalysts,[21] there has been some recent attention devoted to analogous group-6 metal systems. Stable 18-valence electron congeners to the group-7 counterparts are delineated by the formula [M(CO)$_3$(N$^\frown$N)L] (M = Cr, Mo, W; N$^\frown$N = α-diimine, L = neutral two-electron donor ligand). Perhaps the most quintessential examples thereof are where L = CO; hence, the tetracarbonyl complexes [M(CO)$_4$(N$^\frown$N)] (**1**) have served as the prototype in this field, at

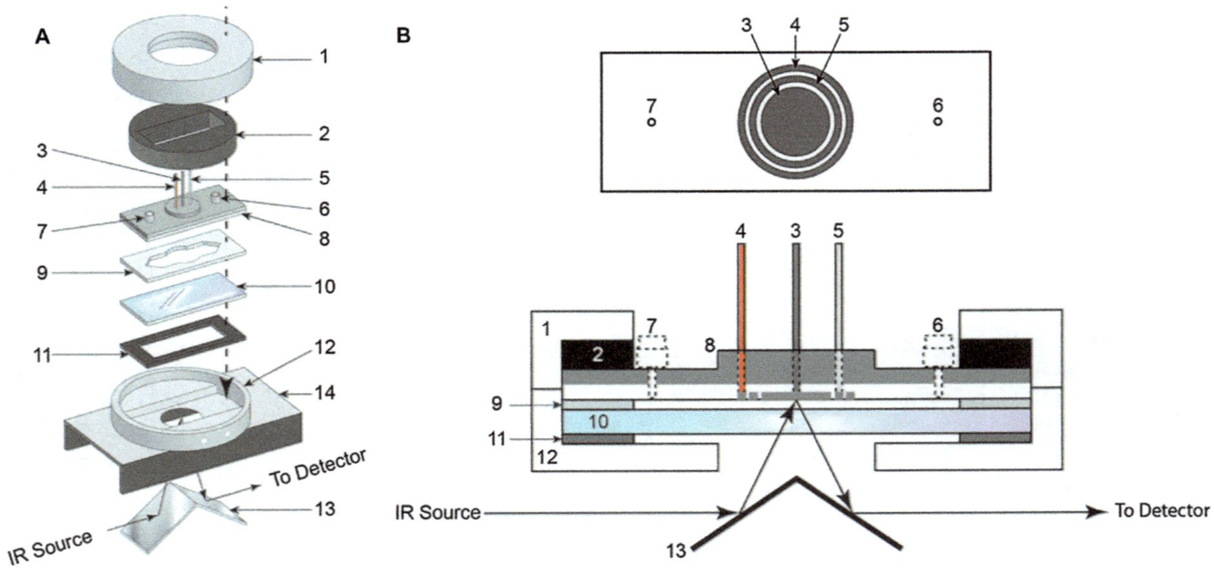

Figure 8.3 Disassembled view of the specular reflectance IR spectro-electrochemical cell: (1) tightening brass cap (threaded inside); (2) brass ring required to tighten the cell; (3) working electrode; (4) auxiliary/counter electrode; (5) pseudo-reference electrode; (6), (7) injection ports; (8) cell body, top part aluminium, lower part Teflon (all three electrodes and both filling ports are press-fitted into the cell body, so that they can be replaced if needed); (9) Teflon spacer; (10) CaF_2 window; (11) rubber gasket; (12) hollow brass cell body with threaded inlet and outlet ports (Swagelock) for connection to circulating thermostatting bath; (13) mirrors; (14) two-mirror reflectance accessory.
Reprinted with permission from C. W. Machan, M. D. Sampson, S. A. Chabolla, T. Dang and C. F. Kubiak, *Organometallics*, 2014, **33**, 4550–4559.[38] Copyright 2015 American Chemical Society.

least in part due to their ease of preparation from the readily available hexacarbonyl precursor, $[M(CO)_6]$. In addition, the tetracarbonyl group-6 compounds may be compared to the Re/Mn tetracarbonyl intermediates in the CO_2 electrocatalytic cycle, which infer that a CO ligand can be labilised in much the same way as the X-type ligand of $[M(CO)_3(N^\frown N)X]$ (**2**; M = Mn, Re), when sufficiently negative potentials are applied to group-6 metal complexes **1**; the resultant five-coordinate species can participate in an analogous catalytic cycle that effects the $2e^-$ reduction of CO_2.

8.3.1 Cathodic Behaviour and CO_2 Catalysis Pertaining to 2,2′-bipyridine Complexes

The cathodic behaviour of the group-6 2,2′-bipyridine (bpy) complexes $[M(CO)_4(bpy)]$ (**3**) is apparently somewhat more straightforward than that of the group-7 analogues (Figure 8.4, inset). Upon the $1e^-$ reduction of **3**, a stable radical anion $\mathbf{3}^{\bullet-}$ is generated,[40–45] and the remarkable stability of this $1e^-$ reduced species also extends to a number of other tricarbonyl complexes of

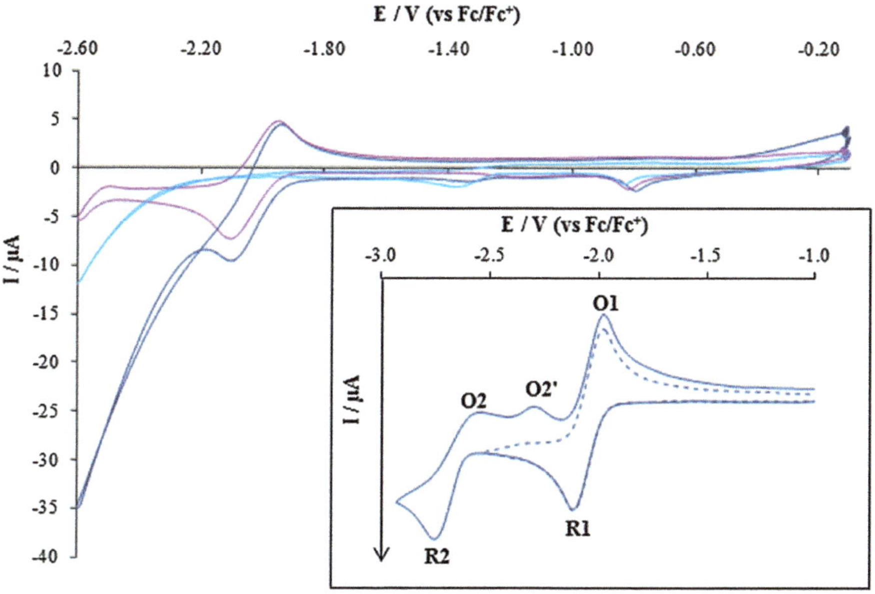

Figure 8.4 Cyclic voltammograms showing the $1e^-$ reduction of $[Mo(CO)_4(bpy)]$ (**3-Mo**) in THF at a gold microdisc electrode, both with (blue curve) and without (pink curve) CO_2 dissolved in the electrolyte. The turquoise curve represents the background cathodic scan of the electrolyte saturated with CO_2. Inset: cyclic voltammogram of **3-Mo** showing the reversible first reduction and irreversible second reduction at a platinum microdisc electrode. The anodic wave O2′ belongs to oxidation of $\mathbf{4\text{-}Mo}^{2-}$.

the type [M(CO)$_3$(N$^\frown$N)(CNR)] (M = Cr, Mo, W; CNR = isocyanide).[46,47] The subsequent irreversible 1e$^-$ reduction of these radical anions induces the dissociation of an axial carbonyl ligand to afford the five-coordinate dianions [M(CO)$_3$(bpy)]$^{2-}$ (**4**$^{2-}$).[48,49] The latter species, which might be considered formally isoelectronic with [M(CO)$_3$(bpy)]$^-$ (M = Mn, Re), apparently constitute the active catalyst in the 2e$^-$ reduction of CO_2; this is elucidated by the observation of a current enhancement at the second reduction wave, as shown by CVs recorded in CO_2 saturated solutions. A noteworthy feature in the CV of complex **3** is the oxidation of **4**$^{2-}$ to **4**$^{\bullet -}$, which is observed at more positive potentials than the second reduction wave of the tetracarbonyl parent, **3**$^{\bullet -}$ to **3**$^{2-}$. Interestingly, it was observed[49] that when a gold working electrode (microdisc or minigrid) was used instead of Pt, the onset of catalytic current was shifted more positively to near the oxidation potential pertaining to **4**$^{2-}$ → **4**$^{\bullet -}$ (Figure 8.4). This finding suggests that by some dissociative mechanism related to non-innocence of the electrode surface in this system, the catalytically active five-coordinate species **4**$^{2-}$ can be generated at potentials significantly less negative than the 1e$^-$ reduction of **3**$^{\bullet -}$ (Scheme 8.1). Another interesting finding uncovered by IR SEC studies was the appearance of a set of ν(CO) signals, therein assigned to a protonated reduced tetracarbonyl species, which are more prevalent when *N*-methyl-2-pyrrolidone (NMP) is used as a solvent instead of THF.[49] Given that NMP is aprotic, but also water-miscible, the proton source is likely adventitious water present in solution. Due to the infancy of the research into this facet of CO_2 electrocatalysis, only limited mechanistic insight or unambiguous characterisation of various implicated species has thus far been gleaned. Nevertheless, a strong interaction of the radical anion [Mo(CO)$_4$(bpy)]$^{\bullet -}$ (**3-Mo**$^{\bullet -}$) with the cathodic gold surface resulting in the dissociative formation of the 2e$^-$ reduced CO_2 catalyst

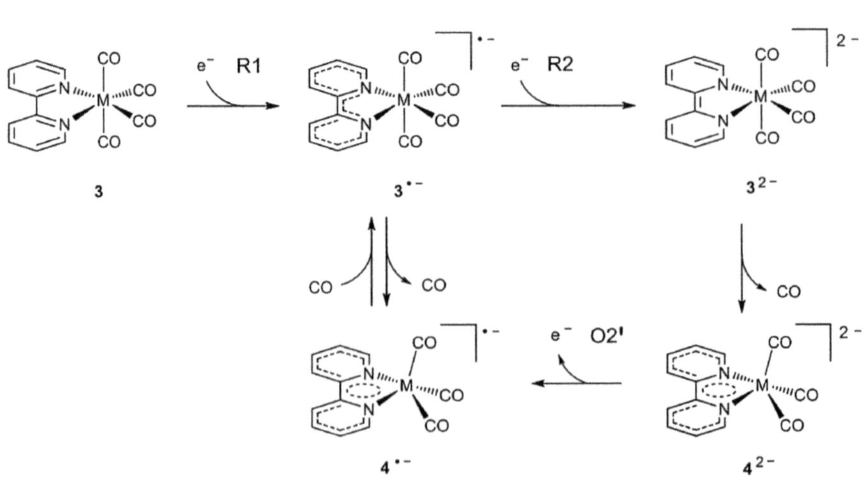

Scheme 8.1 Cathodic pathway for group-6 complexes [M(CO)$_4$(bpy)].

$[Mo(bpy)(CO)_3]^{2-}$ (**4-Mo^{2-}**) at a *ca.* 600 mV less negative potential than is the reduction of **3-Mo$^{\bullet-}$** (Scheme 8.1) has recently been proven by IR–vis sum frequency generation spectroscopy representing one of the modern analytical methods introduced to this field. A more detailed presentation of this interface survey study is given in Chapter 9.

8.3.2 Electronic Structure

Our discussion to this point has largely been confined to complexes containing 2,2′-bipyridine as the redox non-innocent ligand. However, beyond the subtle substitution strategies that attune the ligand donor properties, substantial modification of the α-diimine ligand backbone affords markedly different redox and electrocatalytic behaviour of the corresponding complexes. A vast array of chelating pnictogen donor ligands, bearing low-lying LUMOs constructed from conjugated π-networks, have been investigated in various facets of organometallic chemistry relevant to that of the topic in this work. Perhaps most predominantly explored are α-diimine ligands, which are typified by the three examples shown schematically in Chart 8.1: 2,2′-bipyridine, pyridyl-2-carbaldimine (pyim), and non-aromatic 1,4-diazabutadiene (dab); note that in the literature, the abbreviation ‘pyca’ is often used for the pyridyl-2-carbaldimine ligand; however, it is also used for pyridyl-2-carbaldehyde and thus we adopt the abbreviation pyim, derived from ‘pyridyl-monoimine’, to avoid confusion.

These ligands differ in the aromatic/non-aromatic nature of the imine moieties, and offer a spectrum of electronic structure and chemical properties when bound to, pertinently, organometallic fragments of $\{M(CO)_3X\}$ and $\{M(CO)_3L\}^+$ (M = Mn, Re), as well as $\{M(CO)_4\}$ (M = Cr, Mo, W). A selection of rhenium(I) tricarbonyl complexes with non-innocent α-diimine ligands, **5-12** (Chart 8.2), along with their carbonyl infrared stretching frequencies and reduction potentials, are collated in Table 8.1. IR data for the corresponding 1e$^-$ reduced radical anions are listed in Table 8.2, with the average shift in ν(CO) stretching frequencies ($\Delta\nu_{av}$) that accompanies the first electrochemical reduction. Generally (subject to the various substitution motifs possible on the α-diimine ligand backbone), the energies of the ligand-based LUMOs decrease in the order bpy > pyim > dab (Chart 8.1), which is accompanied by more positive 1e$^-$ reduction potentials of the neutral parent compound in the same order. This can be rationalised by the consideration that the aromaticity borne by heteroaromatic moieties in bpy and pyim is lifted on population of the LUMO, which is associated with an

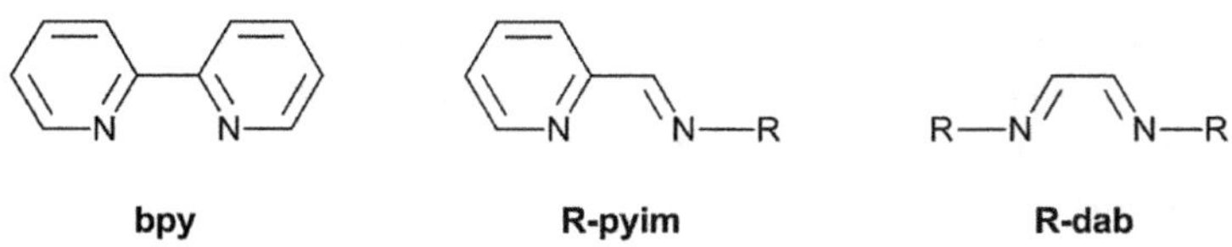

Chart 8.1 Molecular structures of the discussed series of α-diimine ligands.

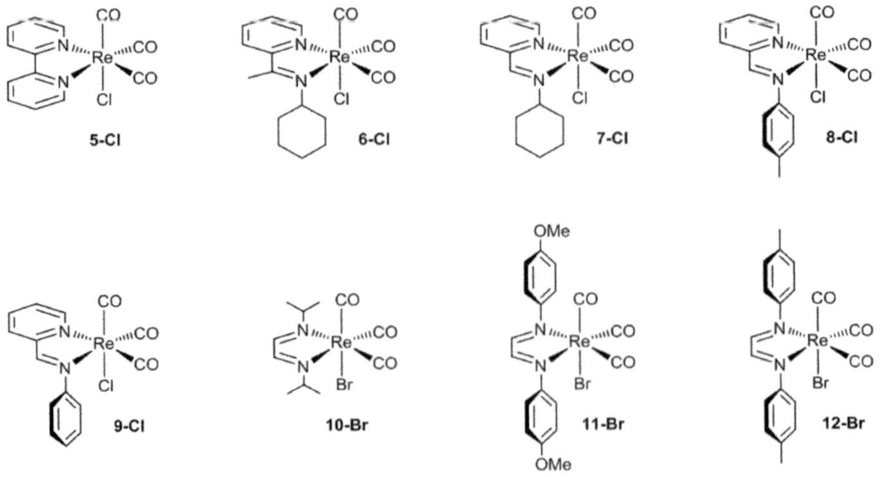

Chart 8.2 Molecular structures of the discussed series of rhenium(I) tricarbonyl α-diimine complexes.

Table 8.1 Cyclic voltammetric data and infrared ν(CO) frequencies[a] for various complexes $[Re(CO)_3(\alpha\text{-diimine})X]$ (X = Cl, Br), as delineated in Chart 8.2.

			IR ν(CO)/cm^{-1}			
Complex	$E_{1/2}$[b] (V)	$E_{p,c}$[c] (V)	A′	A″	A′	Ref.
5-Cl	−1.75	−2.14	2025	1918	1902	26, 50
6-Cl	−1.70	−2.18	2021	1911	1899	51
7-Cl	−1.62	—[d]	2023	1921	1900	52
8-Cl	−1.35	—[d]	2022	1918	1894	53
9-Cl	−1.35	—[d]	2027	1927	1902	53
10-Br	−1.40[e]	−2.41	2020	1921	1905	54
11-Br	−1.10	−1.99	2024	1929	1909	54
12-Br	−1.05	−1.92	2024	1931	1909	54

[a]Experimental parameters vary amongst different sources.
[b]Half-wave potential for the first reversible 1e$^-$ reduction *vs.* Fc/Fc$^+$ (converted from other quoted reference electrodes in required cases).[55,56]
[c]Peak potential of the cathodic wave, corresponding to the second irreversible reduction.
[d]Not reported.
[e]Other sources report an irreversible first reduction for **10-Br**, though **10-Cl** maintains reversibility in this regime ($E_{1/2} = -1.43$ V).[57]

energy cost over the relatively facile 1e$^-$ reduction of the non-aromatic dab ligand. The carbonyl IR stretching frequencies concur with this assessment; the higher ν(CO) wavenumbers reported for the dab complexes **10-Br–12-Br** (Chart 8.2) are diagnostic of augmented π-back donation (*vide infra*) into the lower energy π* orbitals associated with the dab ligands, which competes with that of the carbonyl ligands. Conversely, the successive reduction to access the doubly reduced state is liable to the effects of Pauli repulsion, whereby the aptitude of pyim and bpy ligands to delocalise the accumulated charge over a larger number of atoms throughout their extended π-system

Table 8.2 IR ν(CO) wavenumbers[a] for the 1e$^-$ reduced species of some complexes in Table 8.1 (Chart 8.2), along with the average shift in ν(CO) frequencies from their corresponding neutral precursors.

	IR ν(CO) (cm^{-1})			
Complex	A′	A″	A′	$\Delta\nu_{av}$ (cm^{-1})
5-Cl$^{\bullet-}$	1998	1885	1867	32
6-Cl$^{\bullet-}$	2012	1897	1882	13
10-Br$^{\bullet-}$	1990	1872	1846	46
11-Br$^{\bullet-}$	1997	1884	1864	39
12-Br$^{\bullet-}$	1998	1886	1864	39

[a]Experimental parameters vary amongst different sources. For references see Table 8.1.

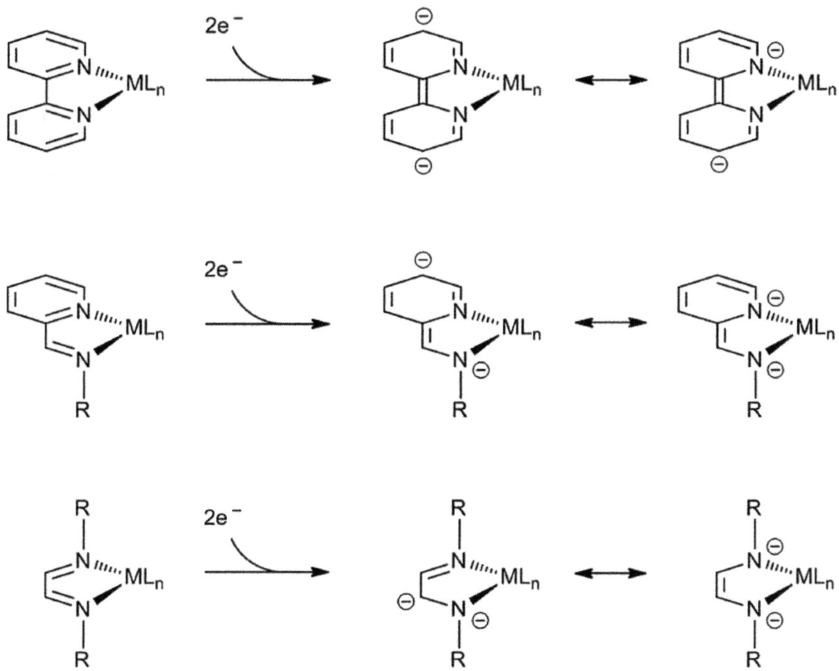

Figure 8.5 Contributing resonance structures of metal complexes with bpy, pyim and dab in their doubly reduced states, illustrating the extent of delocalisation, and distribution of electron density in the now fully occupied LUMO of the complex.

tends to permit milder cathodic potentials. This is depicted qualitatively by the principal contributing resonance forms shown in Figure 8.5; by virtue of the heteroaromatic moieties present, bpy and pyim ligands can separate the electrochemically delivered dianionic character such that adjacent charges can be avoided. Likewise, analysis of the ν(CO) wavenumbers associated with the ground and 1e$^-$ reduced states (Table 8.2) supports this conjecture; a larger shift in carbonyl stretching frequencies upon the 1e$^-$ reduction of the dab complexes is consistent with the augmented exertion of accrued electron density onto the metal centre, by virtue of the smaller conjugated π-framework of the dab ligand, which is in turn compensated by larger π-back donation to the carbonyl ligands. However, the overpotential associated with the electrochemical and chemical irreversibility of the second reduction also alters the corresponding cathodic wave peak potential ($E_{p,c}$), such that this value likely varies somewhat with the leaving ligand X. Furthermore, it is imperative to note that IR stretching frequencies and cathodic potentials do often vary with the solvent/electrolyte used.

The above discussion applies most authoritatively when one considers a largely ligand-based LUMO with little contribution from metal-centred orbitals. As mentioned before, this is substantiated by experimental data for these complexes,[26,54,58–60] but the nature of the metal based orbitals of appropriate symmetry, and their overlap with the α-diimine ligand LUMO, will have varying degrees of influence on the energy and topology of the latter. The principle bonding interaction between metal based orbital (d_{xz} in our chosen axes) and the ligand LUMO in the ground state is illustrated qualitatively in Figure 8.6 with the simple case of the dab ligand, which is formulated as π back-bonding in competition with the other π-acid ligands (most pertinently ancillary carbonyls) within the coordination sphere. Naturally, the energy of the α-diimine π*, and its coefficient centred on the coordinating nitrogen atoms, will have a marked effect on the overlap with

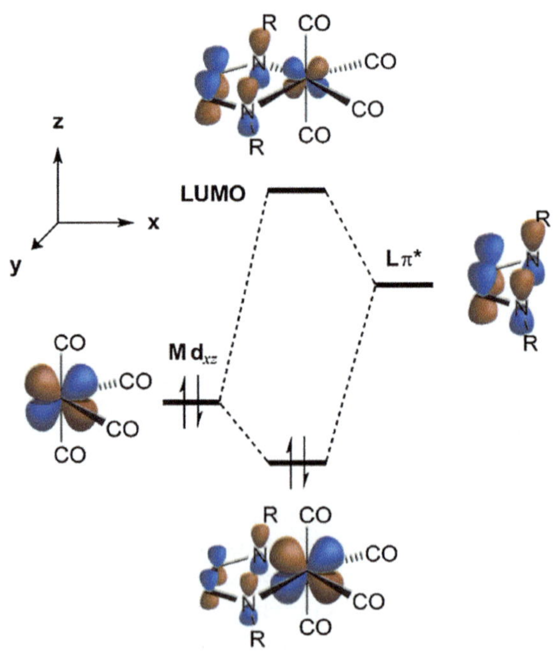

Figure 8.6 Qualitative diagram showing heuristically the interaction between metal d_{xz} and α-diimine π* orbitals in the model complex $[M(CO)_4(R\text{-dab})]$.

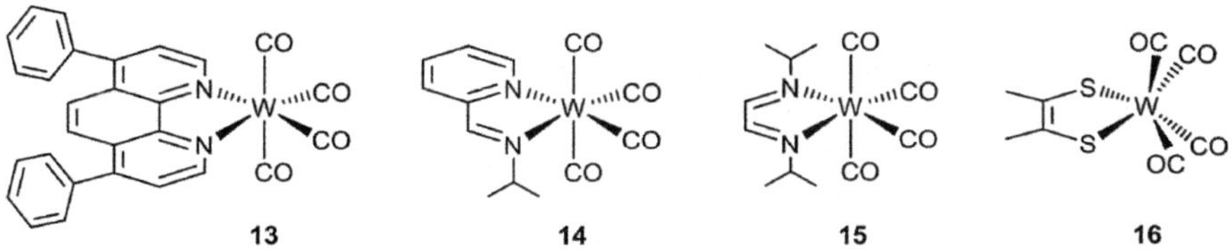

13 14 15 16

Chart 8.3 Molecular structures of the discussed series of tungsten tetracarbonyl complexes.

the metal d_{xz} orbital; more efficient overlap will result in a destabilisation of the LUMO, since this attains an α-diimine π*–metal d_π antibonding character (Figure 8.6). The latter is realised to an increasing extent from bpy <pyim <dab; complexes containing the bpy ligand possess largely ligand-based LUMOs, whereas comparable dab complexes manifest an appreciable dab π*–metal d_π antibonding character to their LUMO as a result of the larger coefficient placed on the diazabutadiene nitrogen atoms.[61] Accordingly, the MLCT character of the strong *x*-polarised electronic transition characteristic of these complexes,[62–66] becomes in fact more metal–ligand bonding to anti-bonding ($d_\pi + \pi^* \rightarrow \pi^* - d_\pi$) in its nature. To this end, resonance Raman (RR) excitation profiles measured for tungsten tetracarbonyl complexes [W(CO)$_4$(α-diimine)], **13–15** (Chart 8.3), bearing the series of ligands Bphen (4,7-diphenyl-1,10-phenanthroline), ^{i}Pr-pyim and ^{i}Pr-dab, respectively, have provided salient evidence.[67] The strong RR effect observed for symmetrical ligand stretching vibrations in [W(CO)$_4$(Bphen)] (**13**) confirms the MLCT nature of the electronic transition, which stipulates the reduction of the ligand in the excited state. These vibrations are markedly attenuated in the excitation profiles collected for [W(CO)$_4$(^{i}Pr-pyim)] (**14**), but nonetheless still present. Conversely, these disappear completely in the case of [W(CO)$_4$(^{i}Pr-dab)] (**15**), and instead the RR effect for ν(W–N) bands is hereby testament to the $d_\pi + \pi^* \rightarrow \pi^* - d_\pi$ character of the electronic transition. In consummation, the authors advocate that **15** loses entirely its MLCT character, suggesting comparable energies and equal contributions from W $5d_\pi$ and ^{i}Pr-dab π* orbitals. This is somewhat contested by more recent computational work done on **15**, which reports contributions of 41% (W $5d_{xz}$), 31% (^{i}Pr-dab π*); and 8% (W $5d_{xz}$), 68% (^{i}Pr-dab π*) to the HOMO-3 and LUMO respectively.[68] It is noteworthy, however, that transitions from the close-lying HOMO-2 and HOMO-1 (though the order of the aforementioned three highest-lying occupied MOs is subject to change), which bear predominantly metal d/carbonyl π* character, still maintain MLCT character that is largely independent of the α-diimine ligand.[61] An extreme example of M→L π-back donation to chelate ligands analogous to the herein discussed α-diimines is the curious case of the dithiolene complex [W(CO)$_4$(mdt)] (**16**; mdt = 1,2-dimethyl-1,2-dithiolate; Chart 8.3).[69] Presumably, the double-bond rule plays a large role in stabilising the π* orbital of the mdt ligand, which engenders C=C double and S–C single bond orders. Accordingly, structural data advocates a prevailing W^{II}(mdt^{2-})ene-1,2-dithiolate character, owing the

extent of back donation in **16** that populates the especially low-lying mdt π* orbital. Perhaps the most conspicuous result of this electronic phenomenon is the trigonal prismatic structure of the ground state complex in lieu of the more conventionally adopted octahedral geometry. However, upon the 1e⁻ reduction of **16** to **16•⁻**, an electronic rearrangement, yielding a W^0(mdt•⁻) metallacycle, accompanies a structural rearrangement to octahedral geometry.

The opportunity for N-substitution in pyim and dab ligands allows one to tune the nitrogen-centred coefficient of the LUMO more directly, as can be observed by comparing the reduction potentials belonging to rhenium dab complexes **10-Br–12-Br** (Table 8.3). More strongly donating N-substituents, as is the case of ^{i}Pr-dab complex **10-Br**, result in a markedly more negative reduction potential reflecting the larger N-coefficient in the ^{i}Pr-dab π* orbital that serves to destabilise the Re d_π–^{i}Pr-dab π* antibonding LUMO. A likewise trend is observed amongst collated pyim complexes **6-Cl–9-Cl**; the latter two bearing an N-aryl moiety, wherein the aromatic N-substituent engenders the opposite effect, is more easily reduced by *ca.* 300 mV. Such a conjecture provides a reasonable explanation for the large discrepancy in reduction potentials between complexes bearing aliphatic and aromatic N-substituted α-diimine ligands of a given type. Since, due to steric constraints of the bound chelate, aromatic N-substituents are orientated orthogonal to the metallacycle plane and are unable to align their π-framework orbitals with the LUMO of the α-diimine moiety; mesomeric extension of the latter over such aryl substituents is hence not the provenance of the milder reduction potentials belonging to **8-Cl, 9-Cl, 11-Br** and **12-Br.** An additional consideration is prompted by an orbital interaction reported to be present in iridium–NHC complexes [IrCl(COD)(NHC)] (COD = 1,4-cyclooctadiene,

Table 8.3 Cyclic voltammetric data[a] and average CO force constants for $[M(CO)_{4-n}L_n(DBSQ)]$ (M = Mn, Re; L = phosphine/phosphite; DBSQ = 3,5-di-*tert*-butyl-1,2-benzosemiquinone; $n = 0, 1, 2$).[79]

Complex	k_{av} [b] (N m^{-1})	$E_{1/2}$ [c] (V)
$[Re(CO)_4(DBSQ)]$	1636	−0.51[e,f]
$[Re(CO)_3(P(OPh)_3)(DBSQ)]$	1569	−0.67[e,f]
$[Re(CO)_3(PPh_3)(DBSQ)]$	1540	−0.87
$[Re(CO)_2(P(OPh)_3)_2(DBSQ)]$	1489[d]	−0.92[g]
$[Re(CO)_2(PPh_3)_2(DBSQ)]$	1426[d]	−1.14
$[Mn(CO)_4(DBSQ)]$	1656[d]	−0.80[f,g]
$[Mn(CO)_3(PPh_3)(DBSQ)]$	1553[d]	−0.92
$[Mn(CO)_2(PPh_3)_2(DBSQ)]$	1441[d]	−1.26[f,g]

[a]In CH_2Cl_2; WE = Pt, RefE = Ag wire (pseudo-reference); supporting electrolyte = $[N^nBu_4][PF_6]$; ν = 100 mV s^{-1}. Voltammetric experiments were in some cases conducted in the presence of excess ligand, to combat chemical irreversibility.
[b]Force constants derived from IR ν(CO) frequencies, measured in C_5H_6.
[c]First 1e⁻ reduction potential *vs.* Fc/Fc⁺.
[d]Measured in CH_2Cl_2.
[e]Measured at 203 K.
[f]Electrochemically irreversible.
[g]Chemically irreversible.

NHC = *N*,*N*-diaryl *N*-heterocyclic carbene). In this example, it was noted that substitution of the C_2 and C_3 positions of the NHC backbone had a lesser effect on the metal-centred electrochemical oxidation potentials of the corresponding complexes than did variation of the *para*-substituent on the N-aryl moieties.[70–75] For the same reasons discussed above, this was an unusual result, since transmission of the mesomeric influence stipulated by the aryl *para*-substituents to the metal centre is unlikely to occur through π bonds, owing to the orthogonality of π frameworks belonging to the NHC ligand backbone and the N-aryl substituents. Instead, computational methods demonstrated that a 'through space' interaction established between metal centred orbitals and the *ipso*-carbon of the N-aryl substituents was responsible for this phenomenon.[76,77] However, a link between the aforementioned and the dab complexes listed in Table 8.1 is somewhat tenuous given the inherently different geometry involved, and it is instructive to note that the redox potentials and carbonyl IR stretching frequencies do not vastly vary between the *p*-tolyl-dab and the more donating *p*-anisyl-dab complexes **12-Br** and **11-Br**, respectively, amongst other examples in the literature.[78] One might lastly also consider the synergistic nature between σ-donicity and π-acidity of the chelating ligand; N-aryl substituents, in their observed orientation, might well reduce the basicity of the nitrogen lone pairs. Since weaker field donors serve to stabilise occupied metal-based orbitals, π-back donation is in turn attenuated, leading to milder reduction potentials as described above.

With regards to the metal fragment, the energy of the metal d_{π} orbitals, and their spatial orientation (considering hybridisation and diffuseness) is altered by the nature of the ancillary ligands that constitute the remainder of the coordination sphere. This is aptly demonstrated by representatives of $[M(CO)_{4-n}L_n(DBSQ)]$ (M = Mn, Re; L = phosphine/phosphite; DBSQ = 3,5-di-*tert*-butyl-1,2-benzosemiquinone; $n = 0,\ 1,\ 2$).[79] Selected examples are presented in Table 8.3, along with their average CO stretching force constant values (k_{av}) derived from IR ν(CO) bands, and first reduction potentials. Substitution of carbonyl ligands with weaker π-acid phosphine ligands raises the energy of the dπ orbitals, which is compensated by augmented π-back donation to the remaining carbonyls. For the complexes in their ground states, this is manifested in the smaller k_{av} values upon iterative substitution of carbonyl ligands with phosphines. Likewise, the increased π-acidity of phosphite ligands over phosphine representatives is demonstrated by lower k_{av} values concerning the latter. The higher-lying dπ orbitals of phosphine-substituted complexes in turn interact with the semiquinone LUMO in a destabilising manner (in an entirely analogous manner to that described for α-diimines; see Figure 8.6). Accordingly, greater incorporation of phosphine ligands within the coordination sphere results in higher reduction potentials. A curious feature in the listed properties in Table 8.3 is the difference in k_{av} and redox potentials between complexes of Re and Mn. The 3d orbitals of Mn are inherently lower in energy, as expected for a first-row transition metal in relation to its third-row counterpart. This will naturally lead to curtailed back donation to π^* orbitals of the ligated carbonyl ligands, aptly substantiated by the higher k_{av} values for the listed Mn complexes in comparison to the corresponding Re congeners. The opposite effect exerted on the respective reduction potentials is at variance with this consideration, but perhaps this provides a salient testimony for the importance of orbital overlap between the metal-based d_π orbital and the nitrogen-centred coefficients of the α-diimine π^* orbital, in addition to the compatibility with regards to their relative energies. The more localised and less diffuse nature of the Mn $3d_{xz}$ orbital perhaps achieves a more efficient overlap with the α-diimine π^* orbital, giving rise to a 'resonance' that destabilises the LUMO of the complex, despite being more energetically mismatched in comparison to the corresponding Re $5d_{xz}$ orbital. The same rationale may in part explain the behaviour exhibited by complexes [Mn(CO)$_3$(α-diimine)X] (see Chapter 5) under reducing potentials, *viz.* the observed lability of the anionic ligand and more marked spin density borne by the Mn centre. It is worth noting, however, that many of the complexes listed in Table 8.3 displayed electrochemically and/or chemically irreversible cathodic behaviour at the reported potentials, which presumably perturbs the reduction potentials to some extent.[79]

Returning our attention to more specifically the group-6 tetracarbonyl complexes of the type [M(CO)$_4$(α-diimine)], the foregoing discussion point applies also down this triad; more positive reduction potentials go as Cr<Mo<W (−2.16, −2.07 and −1.99 V, respectively).[49] In this case the ν(CO) bands belonging to each complex are less congruent, for which higher wavenumbers are manifested in the order W<Cr<Mo. The latter sequence (albeit inverted in terms of magnitude) is upheld in M–CO bond dissociation energies[80,81] and M–C force constants of the respective hexacarbonyl complexes.[82] However, interestingly for group-6 tetracarbonyl complexes [M(CO)$_4$(pbo)] (**17**) and [M(CO)$_4$(pbt)] (**18**) (M = Cr, Mo, W; pbo = 2-(2-pyridyl)benzoxazole, pbt = 2-(2-pyridyl)benzothiazole; Chart 8.4), the order of the cathodic peak potentials belonging to the 1e$^-$ quasireversible reductions follows as W<Mo<Cr (becoming more negative down the triad; Table 8.4).[83] In this example, the highly delocalised aromatic nature of the ligand, and the presence of electronegative heteroatoms within its backbone, presumably engenders a very low-lying LUMO, which affords a predominantly ligand-based reduction. This is supported by the 94% contribution from the pbo ligand π^* orbitals to the LUMO, as gleaned from

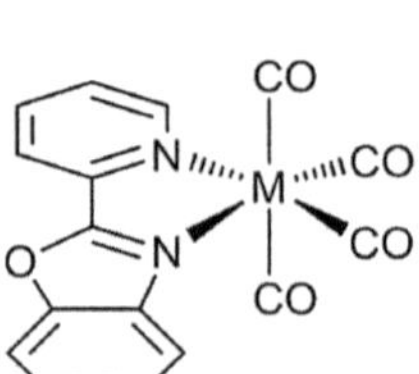

17

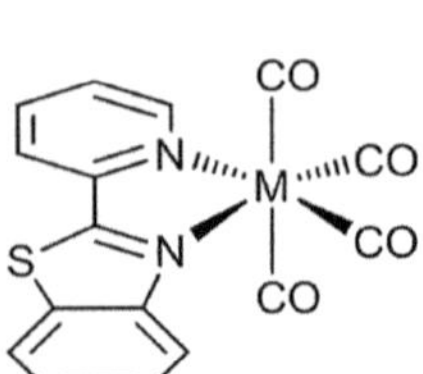

18

Chart 8.4 Molecular structures of the group-6 tetracarbonyl pbo (left) and pbt (right) complexes.

Table 8.4 Cyclic voltammetric data[a] for complexes **17** and **18**.[83]

Complex[b]		$E_{p,c}$[c] (V)	$E_{p,a}$[d] (V)
17	Cr	−1.69	0.22
	Mo	−1.76	0.40
	W	−1.80	0.31
18	Cr	−1.80	0.33
	Mo	−1.86	0.45
	W	−1.89	0.39

[a]In MeCN; WE = Pt, RefE = SCE (converted to potentials *vs.* Fc/Fc^+)[55,56] CE = Pt; supporting electrolyte = $[N^nBu_4][ClO_4]$; $\nu = 50\ mV\,s^{-1}$.
[b]Refer to Chart 8.4.
[c]Cathodic peak potentials of the ligand-based $1e^-$ reduction.
[d]Anodic peak potentials of the metal-based $1e^-$ oxidation.

DFT calculations on **17** (*cf.* 68% from ^{i}Pr-dab in **15**, *vide supra*). Also listed in Table 8.4 are the predominantly metal-based oxidation potentials for **17** and **18**, which instead follow the order Cr<W<Mo, at variance with the order of increasing ν(CO) IR frequencies W<Cr<Mo.[83]

8.3.3 Ligand-based Reactivity of Pyridyl-2-carbaldimine Complexes

Intuitively successive to the tetracarbonyl bpy complexes of group-6 metals (Section 8.3.1), pyim complexes $[Mo(CO)_4(Ar\text{-}pyim)]$ (Ar = Ph, 2,6-diisopropylphenyl; **19** and **20**, respectively; Chart 8.5) have also been investigated.[84] The cathodic behaviour of such complexes is very similar to the aforementioned bpy congeners, albeit, as expected, with milder $1e^-$ reduction potentials associated with the first cathodic wave. The presence of analogous species in solution during the iterative $2e^-$ reduction of **19** and **20** was evidenced by IR SEC studies. However, whilst current enhancement is indeed observed under reducing potentials in the CO_2-saturated MeCN solution, this is observed only at the very negative second reduction potential; no advanced catalysis, closely following the first reduction wave was reported for **19** and **20**, at variance with the bpy complexes. Furthermore, a successive scan in cyclic voltammetry experiments revealed markedly diminished catalytic current, indicating that catalyst turnover is compromised by competing unproductive side reactions. It is noteworthy also that no significant current enhancement was observed in CO_2-saturated THF solution, suggesting an important role played by the solvent. However, this enabled the elucidation of a valuable mechanistic insight upon the chemical reduction of **20** with a potassium-graphite intercalation compound in THF-d_8 (Scheme 8.2).

Reaction of **21** with excess KC_8 generated five-coordinate dianion $\mathbf{22}^{2-}$, was elucidated by interpretation of ^1H and ^{13}C NMR spectra, as well as gas chromatography headspace analysis, which showed loss of ligated CO. Subsequent reaction of $\mathbf{22}^{2-}$ with CO_2 afforded high yields of complex **23**, which, supported by X-ray crystallographic analysis, shows the remarkable

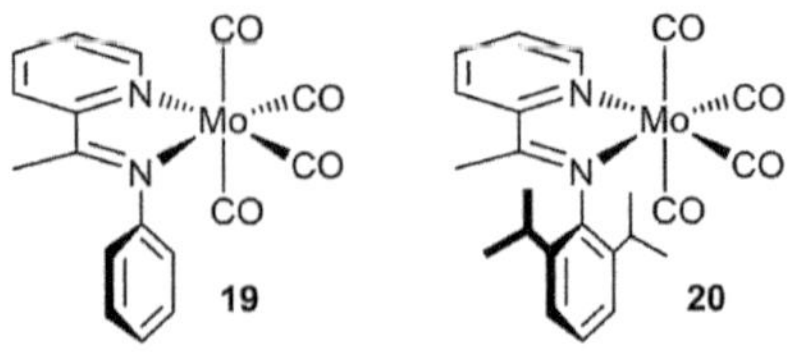

Chart 8.5 Molecular structures of the discussed molybdenum tetracarbonyl pyim complexes.

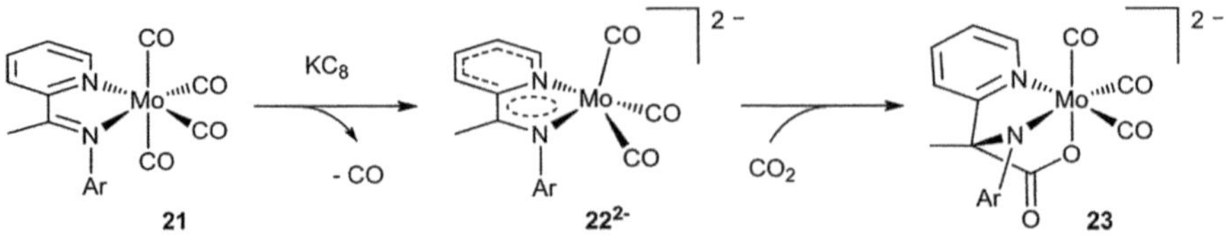

Scheme 8.2 Carbon dioxide binding at the reduced Mo-pyim metala-ring.

attack on CO_2 by the imine carbon of the pyim ligand. Such a reaction demonstrates the reactivity of the non-aromatic imine moieties; our previous discussion (Section 8.3.2) surrounding the nature of the α-diimine π* based LUMO of these complexes sheds some light on this. In an analogous manner to the distribution of π* orbital coefficient in the quintessential C=O carbonyl functional group, we should anticipate the larger coefficient to reside on the imine carbon atom. This is presumably exacerbated by the N-aryl substituents in **19** and **20**, which reduce the coefficient borne by the coordinating nitrogen atoms (Section 8.3.2). Combined with presumably poor Mo–Ar-pyim^{2-} overlap in the now fully occupied ligand based LUMO of **22**$^{2-}$, it is perhaps not surprising that CO_2 binds to the imine carbon of the pyim ligand. Further testimony regarding the nature of the LUMO in analogous pyim complexes stems from the intramolecular reaction which follows the reaction of [Mo(CO)$_3$(η^3-allyl)(Me-pyim)(R-Im)][OTf] (R-Im = N-alkyl substituted imidazole; OTf = trifluoromethanesulfonate; **24**) when treated with a strong amide base (Scheme 8.3).[85] The deprotonation of the imidazole moiety leads to a C–C coupling reaction targeting the imine carbon to yield neutral complex **25**. Here, the electrophilic behaviour of the unoccupied π* orbital of **24**, based most pertinently on the imine carbon, demonstrates the complementary nucleophilic behaviour of the corresponding doubly occupied orbital in **22**$^{2-}$. Conversely, the analogous reaction with bpy congener **26** leads, *via* a mechanism assisted by the ostensibly ancillary carbonyl ligands,[85] to C-metalated imidazolate complex **27**, which undergoes no C–C coupling reaction.[86]

The application of pyim ligands in the Re(I) mediated CO_2 catalysis provide some additional insight in this context.[51] The complex [Re(CO)$_3$(Cy-pyim)Cl] (**6-Cl**; Chart 8.2) has been investigated under reducing potentials, which ultimately generates the doubly reduced five-coordinate species [Re(CO)$_3$(Cy-pyim)]$^-$ (**28**$^-$; Scheme 8.4). Anion **28**$^-$ then purportedly reacts with CO_2 in two

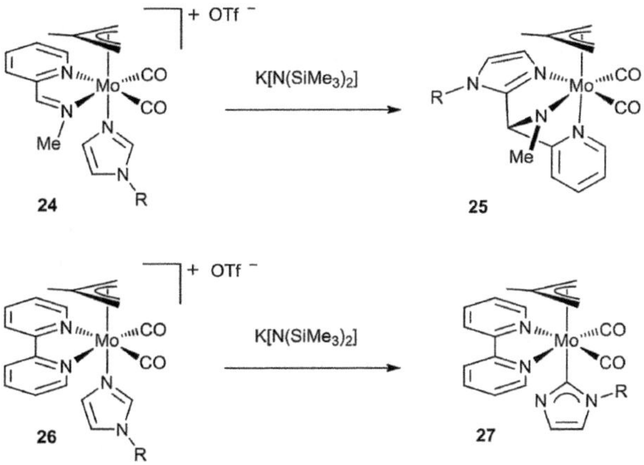

Scheme 8.3 Difference in the reactivity of Mo imidazole bpy and pyim complexes induced by a strong reductant amide base.

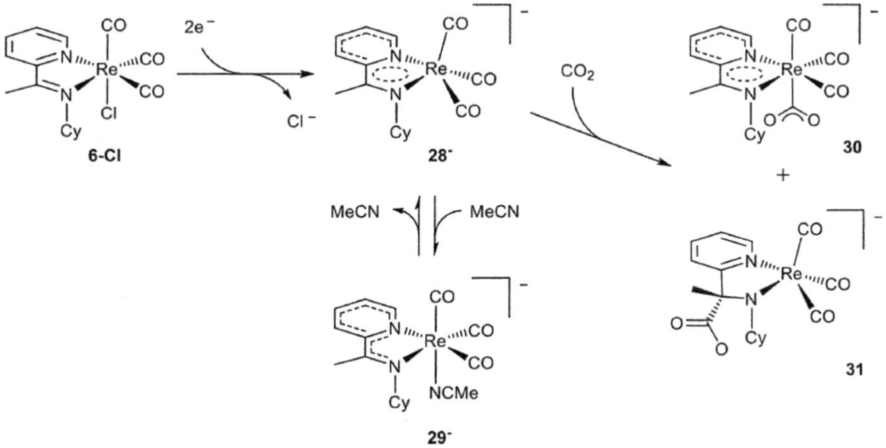

Scheme 8.4 Reported reactivity of the doubly reduced five-coordinate Re-pyim species, **28**$^-$, with CO_2 and MeCN.

different ways, which is representative of those we have previously discussed; direct metal attack of CO_2 yields **30**$^-$, whilst ligand-based attack affords **31**$^-$; the latter is assumed to be deleterious to catalytic turnover. Additionally, the two observed products of CO_2 electrocatalysis are CO and CO_3^{2-}.

In acetonitrile solution, **28**$^-$ is ostensibly in equilibrium with the six-coordinate MeCN adduct $[Re(CO)_3(Cy\text{-}pyim)(MeCN)]^-$ (**29**$^-$), for which markedly higher ν(CO) wavenumbers are reported (1943, 1839 cm^{-1} *vs.* 1986, 1852 cm^{-1}, for **28**$^-$ and **29**$^-$, respectively).[51] It was suggested that the two complexes differ in their oxidation state at the rhenium centre; **28**$^-$ bears the expected Re^0(Cy-pyim$^-$) metallacycle character, whereas **29**$^-$ is instead Re^I(Cy-pyim^{2-}) in nature. However, it is instructive to point out that this conjecture is at variance with the above discussion regarding (a) the instability of doubly reduced states of pyim and dab in comparison to more extensive π-networks of bpy and phen, and (b) the far from inconsequential π^*-d_π overlap between the cyclohexyl-substituted imine nitrogen and Re based orbitals. Upon the first $1e^-$ reduction of **6-Cl**, two species are formed, which comprise **6-Cl**$^{\bullet-}$ and **33**$^\bullet$, the latter generated *via* loss of Cl^- (Scheme 8.5). Here instead lies the opportunity for coordination of MeCN; by virtue of being easier to reduce, the pyim ligand engenders a greater contribution from resonance tautomer **D** (Scheme 8.5) in the complex **33**$^\bullet$, which naturally encourages the coordination of MeCN to form **29**$^\bullet$. Indeed,

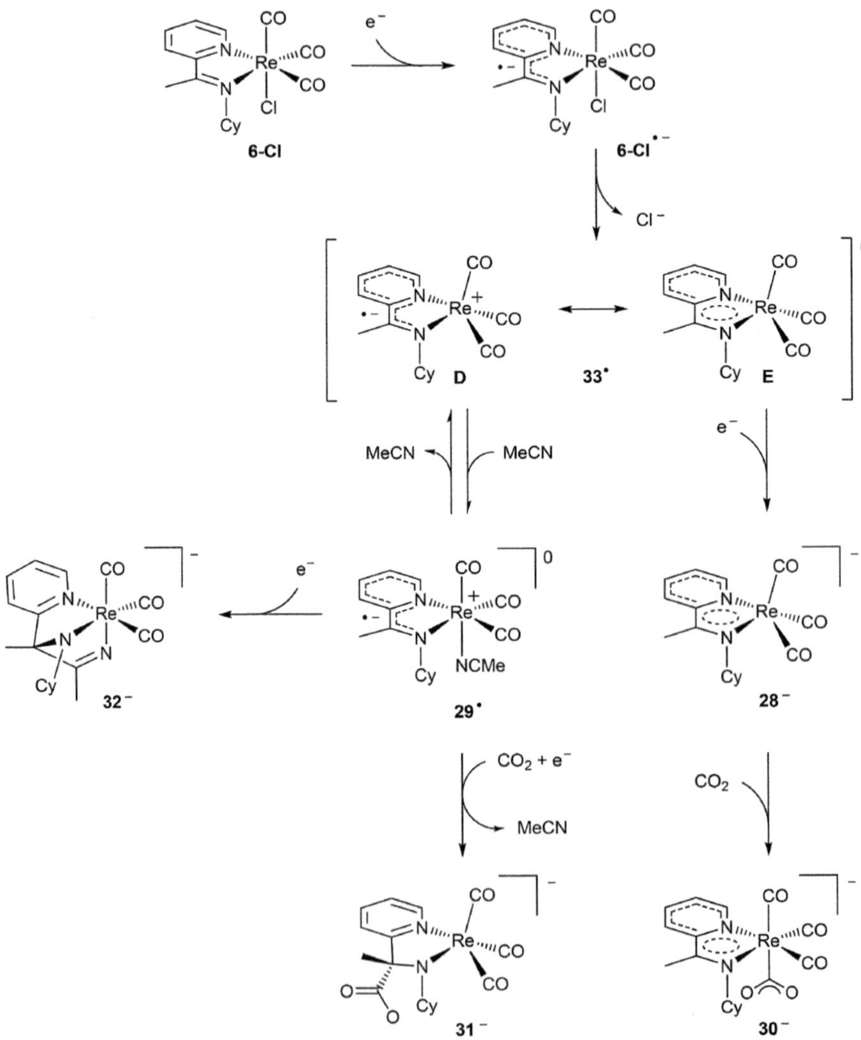

Scheme 8.5 Tentative reaction scheme for reduced rhenium–pyim complex **6-Cl** in the presence of carbon dioxide.

the advent of an additional set of ν(CO) bands in the IR SEC spectra is observed at 1984 and 1844 cm^{-1}, consistent with this assignment. These stretching frequencies corresponding to **29**• persist and continue to grow upon the second reduction, suggesting that **29**• is reduced at more negative potentials than the remaining **33**•. The latter is converted here, as expected, into **28**⁻, but it is worthwhile to consider an alternative assignment for the adventitious ν(CO) bands ascribed originally to **29**⁻ (*vide supra*). It might be suggested that the nucleophilic reactivity of the imine carbon on the reduced pyim ligand is once again demonstrated; the C≡N functional group belonging to the coordinated acetonitrile ligand presents a suitably electrophilic moiety to form an adduct in an entirely analogous manner to ligand-based attack on CO_2. In Scheme 8.5 the resultant complex is tentatively drawn as **32**⁻, which bears a κ^3-*N,N,N* ligand, comprising of two amido and one amino (of the pyridine) coordinating termini. The overall negative charge of the complex then naturally leads to the oxidation state +1 at the Re centre, and explains the hypsochromic shift of the respective ν(CO) bands. Perhaps more likely is that this reaction does not require the delivery of an additional electron, or **32**⁻ undergoes further reaction to form a neutral complex (*via* protonation at the imine nitrogen, or loss of hydride at the tertiary cyclohexyl carbon).[87]

To frame the ligand-based reactivity of these pyim complexes differently, one might consider the reaction to be a [3+2] cycloaddition between a M–N=C metala-1,3-dipole and a C≡N or CO_2 dipolarophile. This type of reaction is readily found in the literature, indeed particularly pertaining to organometallic complexes bearing pyim and dab ligands.[87] Whereas, to the best of our knowledge, no literature on the applicability of this reaction to nitriles exists, numerous other dipolarophiles, including alkynes,[88–94] ketones[95] and heteroallenes[96–98] (particularly analogous to CO_2) form metallabicyclo[2.2.1] adducts with the M(α-diimine) moiety. To aid this description, the isolobal relationship between the aforementioned metalla-1,3-dipole and an azomethine ylide may be considered (Scheme 8.6). It is constructed from the well-established isolobal relationship between a carbene (:CR_2) and a d^8 ML_4 fragment.[99–101] The exemplar organometallic complexes for this isolobal relationship are the iron dab carbonyls, [$Fe(CO)_3$(R-dab)] (**34**; Scheme 8.6), which meet the requirements of the d^8 ML_4 fragment; an imine group of the dab ligand then furnishes the isolobal analogy with azomethine ylides. Consequently, complexes of this type react readily in the aforementioned [3+2] cycloaddition manner with various dipolarophiles from their ground state.[87] Application of this isolobal relationship to similar complexes of manganese to test its robustness, provides an invaluable insight to this discussion. The radicals [$Mn(CO)_3$(R-dab)]•, generated *via* photolysis, were expected to resist the [3+2] cycloaddition reaction, on the grounds that the formally d^6 ML_4 fragment (borne out of the Mn^I(R-dab•⁻) character of the metallacycle) does not meet the requirements for the isolobal analogy. Indeed, this was the case, but the [3+2] cycloaddition reaction would proceed readily in the coordinating solvent THF. It was

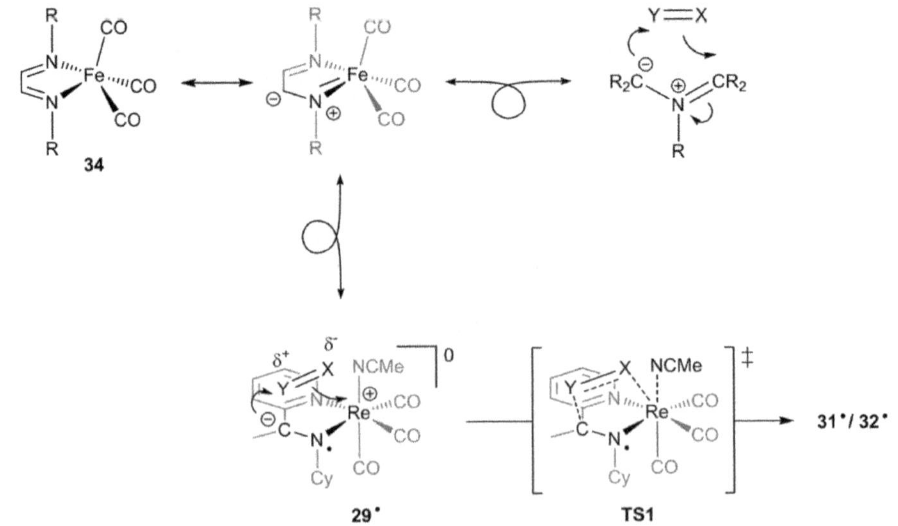

Scheme 8.6 Isolobal relationships used to interpret the apparent [3 + 2] cycloaddition reactivity of certain α-diimine complexes.

thereby suggested that coordination of THF would afford a formally d^6 ML_5 fragment, apparently suitably isolobal with the prerequisite d^8 ML_4 moiety. This does not hold for complexes of $[M(CO)_4(dab)]$ or $[M(CO)_4(pyim)]$, however, which likewise bear a d^6 ML_5 moiety; a fact that highlights the need for (a) population of the dab/pyim π* orbital, to render the 1,3-dipole suitably reactive, and (b) a sufficiently labile ligand or free coordination site (as in **34**). The former consideration is achieved for **34**, presumably on account of the highly efficient overlap between the dab π* and the Fe d_π orbitals;[102] the resonance phenomenon shown for **34** in Scheme 8.6 also accounts for the regioselectivity of the [3 + 2] cycloaddition. The forgoing discussion provides salient evidence for the suggested reactivity of **29•** (*via* **TS1**; Scheme 8.6), on the basis that it appears to fulfil the requirements set out above. In fact, pyim ligands apparently represent the worst-case scenario with regards to the assumedly non-productive ligand-based attack on CO_2; the nucleophilicity pertaining to the C-terminus of the 1,3-dipole is exacerbated by both the presence of a methyl substituent, and the vehement aromaticity of the neighbouring pyridyl moiety, which discourages delocalisation of charge towards the pyridyl nitrogen. In the case of molybdenum complex $\mathbf{22}^{2-}$ (Scheme 8.2), which is more directly isolobal with **34** by virtue of being d^8 ML_4 in its doubly reduced state, the reactivity of the 1,3-dipole is perhaps further augmented by the aromatic substituent on the imine nitrogen, rendering the latter more electron deficient.

The endeavours in the field of electrocatalytic CO_2 activation should pursue further the behaviour of group-6 tetracarbonyls, despite the somewhat discouraging results shown by the aforementioned pyim complexes. Still yet unexplored have been the corresponding R-dab complexes.

8.4 Summary and Outlook

When drawing comparisons between infrared spectroscopic techniques,[103] and those which are generally considered as gold standard analytical methods for structure determination (*viz.* NMR and X-ray crystallography), a little is left to be desired regarding the ambiguity of spectral data. However, it is the marriage of IR spectroscopy to the field of electrochemistry where the true power of the somewhat niche technique of spectro-electrochemistry is fully realised. Indeed, the relative incompatibility of the other, more powerful analytical techniques, with *in situ* monitoring of electrochemical phenomena is perhaps the reason for the rising interest in IR spectro-electrochemistry. In the particular context of the electrocatalytic reduction of carbon dioxide mediated by metal carbonyl complexes, IR SEC studies have elucidated the vast diversity in reactivity induced by often seemingly small changes in various parameters, including the solvent, nature and quantity of added Lewis or Brønsted acids, structure and substitution motifs of α-diimine ligands, and the presence of other (organometallic) species that can work in tandem to promote catalysis. Indeed, it seems that every time a slightly different catalyst structure is trialled, remarkable and unexpected behaviour is often manifested. Given also the infancy of this field of molecular catalysis in general, there is still much work needed before catalyst systems may be pragmatically optimised. Specifically, group-6 tetracarbonyls of topic herein have been somewhat side-lined in pursuit of their better-established rhenium and manganese carbonyl congeners. The experiments[49] outlined in this chapter should surely spur on more research into this Cr, Mo or W mediated catalysis[104] and confirm or contest the insight grafted herein. Several opportunities have been emerging in electrocatalytic CO_2 reduction, targeting increased stability, activity and selectivity of the electrocatalysts and their low-cost practical, industrial and commercial exploitation. Increasing attention is paid to development and modelling of heterogeneous electrocatalysts,[105] and immobilisation of well-explored efficient but non-recyclable transition-metal-based homogeneous catalysts and their incorporation onto various electrode surfaces.[106] Identification and selective monitoring of the catalytically active species and reaction intermediates with *in situ* spectro-electrochemical techniques has further potential for development.

References

1. M. Aresta, in *Activation of Small Molecules*, ed. W. B. Tolman, Wiley-VCH Verlag GmbH & Co. KGaA, Weinheim, Germany, 2006, pp. 1–41.
2. K. C. Waugh, *Catal. Today*, 1992, **15**, 51–75.
3. T. Riedel, M. Claeys, H. Schulz, G. Schaub, S.-S. Nam, K.-W. Jun, M.-J. Choi, G. Kishan and K.-W. Lee, *Appl. Catal., A*, 1999, **186**, 201–213.
4. D. A. Hickman and L. D. Schmidt, *Science*, 1993, **259**, 343–346.
5. G. Centi and S. Perathoner, *Catal. Today*, 2009, **148**, 191–205.

6. P. M. Maitlis, A. Haynes, G. J. Sunley and M. J. Howard, *J. Chem. Soc., Dalton Trans.*, 1996, 2187–2196.
7. G. J. Sunley and D. J. Watson, *Catal. Today*, 2000, **58**, 293–307.
8. R. Reithmeier, C. Bruckmeier and B. Rieger, *Catalysts*, 2012, **2**, 544–571.
9. H. A. Schwarz and R. W. Dodson, *J. Phys. Chem.*, 1989, **93**, 409–414.
10. E. Fujita, *Coord. Chem. Rev.*, 1999, **185–186**, 373–384.
11. E. Fujita and B. S. Brunschwig, in *Electron Transfer in Chemistry*, ed. Z. Balzani, Wiley-VCH Verlag GmbH, Weinheim, Germany, 2001, pp. 88–126.
12. R. Angamuthu, P. Byers, M. Lutz, A. L. Spek and E. Bouwman, *Science*, 2010, **327**, 313–315.
13. J.-M. Lehn and R. Ziessel, *Proc. Natl. Acad. Sci. U. S. A.*, 1982, **79**, 701–704.
14. K. Tanaka and D. Oyama, *Coord. Chem. Rev.*, 2002, **226**, 2011–2018.
15. J. Qiao, Y. Liu, F. Hong and J. Zhang, *Chem. Soc. Rev.*, 2014, **43**, 631–675.
16. C. Finn, S. Schnittger, L. J. Yellowlees and J. B. Love, *Chem. Commun.*, 2012, **48**, 1392–1399.
17. A. J. Morris, G. J. Meyer and E. Fujita, *Acc. Chem. Res.*, 2009, **42**, 1983–1994.
18. V. P. Indrakanti, J. D. Kubicki and H. H. Schobert, *Energy Environ. Sci.*, 2009, **2**, 745.
19. S. N. Habisreutinger, L. Schmidt-Mende and J. K. Stolarczyk, *Angew. Chem., Int. Ed.*, 2013, **52**, 7372–7408.
20. B. Kumar, M. Llorente, J. Froehlich, T. Dang, A. Sathrum and C. P. Kubiak, *Annu. Rev. Phys. Chem.*, 2012, **63**, 541–569.
21. C. Constentin, G. Passard, M. Robert and J.-M. Saveant, *Proc. Natl. Acad. Sci. U. S. A.*, 2014, **111**, 14990–14994.
22. J. Heinze, *Angew. Chem. Int. Ed.*, 1993, **32**, 1268–1288.
23. R. Kortlever, J. Shen, K. J. P. Schouten, F. Calle-Vallejo and M. T. M. Koper, *J. Phys. Chem. Lett.*, 2015, **6**, 4073–4082.
24. D. T. Whipple and P. J. A. Kenis, *J. Phys. Chem. Lett.*, 2010, **1**, 3451–3458.
25. P. Christensen, A. Hamnett, A. V. G. Muir and J. A. Timney, *J. Chem. Soc., Dalton Trans.*, 1992, 1455–1463.
26. F. P. A. Johnson, M. W. George, F. Hartl and J. J. Turner, *Organometallics*, 1996, **15**, 3374–3387.
27. S. R. Domingos, H. Luyten, F. C. J. van Anrooij, H. J. Sanders, B. H. Bakker, W. J. Buma, F. Hartl and S. Woutersen, *Rev. Sci. Instrum.*, 2013, **84**, 033103-1–033103-4.
28. L. K. Morris, E. A. Abu, C. Bowman, C. E. Estridge, S. E. Andria, C. J. Seliskar and W. R. Heineman, *Electroanalysis*, 2011, **23**, 939–946.
29. Q. Zeng, J. Tory and F. Hartl, *Organometallics*, 2014, **33**, 5002–5008.
30. S. J. Spall, T. Keane, J. Tory, D. C. Cocker, H. Adams, H. Fowler, A. J. H. M. Meijer, F. Hartl and J. Weinstein, *Inorg. Chem.*, 2016, **55**, 12568–12582.

31. T. Mahabiersing, H. Luyten, R. Nieuwendam and F. Hartl, *Collect. Czech. Chem. Commun.*, 2003, **68**, 1687–1709.
32. B. Jin, F. Tao and P. Liu, *J. Electroanal. Chem.*, 2008, **624**, 179–185.
33. M. Razavet, S. J. Borg, S. J. George, S. P. Best, S. A. Fairhurst and C. J. Pickett, *Chem. Commun.*, 2002, 700–701.
34. M. D. Sampson, J. D. Froehlich, J. M. Smieja, E. E. Benson, I. D. Sharp and C. P. Kubiak, *Energy Environ. Sci.*, 2013, **6**, 3748.
35. S. P. Best, S. J. Borg and K. A. Vincent, in *Spectroelectrochemistry*, ed. W. Kaim and A. Klein, Royal Society of Chemistry, Cambridge, 2008, pp. 1–30.
36. R. J. Gale, *Spectroelectrochemistry: Theory and Practice*, Plenum Press, New York, 1988.
37. F. Hartl, T. Mahabiersing, S. Chardon-Noblat, P. Da Costa and A. Deronzier, *Inorg. Chem.*, 2004, **43**, 7250–7258.
38. C. W. Machan, M. D. Sampson, S. A. Chabolla, T. Dang and C. F. Kubiak, *Organometallics*, 2014, **33**, 4550–4559.
39. N. Elgrishi, M. B. Chambers, X. Wang and M. Fontecave, *Chem. Soc. Rev.*, 2017, **46**, 761–796.
40. Y. Kaizu and H. Kobayashi, *Bull. Chem. Soc. Jpn.*, 1972, **45**, 470–477.
41. D. Miholová and A. A. Vlček, *J. Organomet. Chem.*, 1985, **279**, 317–326.
42. D. Miholová, B. Gaš, S. Záliš, J. Klíma and A. A. Vlček, *J. Organomet. Chem.*, 1987, **330**, 75–84.
43. A. Vlček, Jr, F. Baumann, W. Kaim, F.-W. Grevels and F. Hartl, *J. Chem. Soc., Dalton Trans.*, 1998, 215–220.
44. K. D. Franz, H. tom Dieck, K. A. Ostoja Starzewski and F. Hohmann, *Tetrahedron*, 1975, **31**, 1465–1469.
45. K. D. Franz, H. tom Dieck, U. Krynitz and I. W. Renk, *J. Organomet. Chem.*, 1974, **64**, 361–366.
46. A. Bell and R. A. Walton, *Polyhedron*, 1986, **5**, 951–958.
47. L. D. Detter and R. A. Walton, *Polyhedron*, 1986, **5**, 1321–1328.
48. M. L. Clark, K. A. Grice, C. E. Moore, A. L. Rheingold and C. P. Kubiak, *Chem. Sci.*, 2014, **5**, 1894.
49. J. Tory, B. Setterfield-Price, R. A. W. Dryfe and F. Hartl, *ChemElectroChem*, 2015, **2**, 213–217.
50. J. M. Smieja and C. P. Kubiak, *Inorg. Chem.*, 2010, **49**, 9283–9289.
51. C. W. Machan, S. A. Chabolla and C. P. Kubiak, *Organometallics*, 2015, **34**, 4678–4683.
52. R. N. Dominey, B. Hauser, J. Hubbard and J. Dunham, *Inorg. Chem.*, 1991, **30**, 4754–4758.
53. P. J. Ball, T. R. Shtoyko, J. A. Krause Bauer, W. J. Oldham and W. B. Connick, *Inorg. Chem.*, 2004, **43**, 622–632.
54. B. D. Rossenaar, F. Hartl and D. J. Stufkens, *Inorg. Chem.*, 1996, **35**, 6194–6203.
55. A. M. Stolzenberg and M. T. Stershic, *J. Am. Chem. Soc.*, 1988, **110**, 6391–6402.

56. M. C. Elvington and K. J. Brewer, in *Applications of Physical Methods to Inorganic and Bioinorganic Chemistry*, ed. R. A. Scott and C. M. Lukehart, Wiley, 2007, p. 23.
57. A. Klein, C. Vogler and W. Kaim, *Organometallics*, 1996, **15**, 236–244.
58. W. Kaim and S. Kohlmann, *Inorg. Chem.*, 1990, **29**, 2909–2914.
59. W. Kaim and S. Kohlmann, *Chem. Phys. Lett.*, 1987, **139**, 365–369.
60. B. P. Sullivan, C. M. Bolinger, D. Conrad, W. J. Vining and T. J. Meyer, *J. Chem. Soc., Chem. Commun.*, 1985, 1414.
61. D. J. Stufkens, *Coord. Chem. Rev.*, 1990, **104**, 39–112.
62. R. W. Balk, D. J. Stufkens and A. Oskam, *J. Chem. Soc., Dalton Trans.*, 1982, 275–282.
63. R. W. Balk, T. Snoeck, D. J. Stufkens and A. Oskam, *Inorg. Chem.*, 1980, **19**, 3015–3021.
64. R. W. Balk, D. J. Stufkens and A. Oskam, *Inorg. Chim. Acta*, 1979, **34**, 267–274.
65. H. tom Dieck and I. W. Renk, *Chem. Ber.*, 1972, **105**, 1419–1430.
66. H. tom Dieck and I. W. Renk, *Chem. Ber.*, 1971, **104**, 110–130.
67. P. C. Servaas, H. K. Van Dijk, T. L. Snoeck, D. J. Stufkens and A. Oskam, *Inorg. Chem.*, 1985, **24**, 4494–4498.
68. S. Záliš, I. R. Farrell and A. Vlček, *J. Am. Chem. Soc.*, 2003, **125**, 4580–4592.
69. Y. Yan, P. Chandrasekaran, J. T. Mague, S. DeBeer, S. Sproules and J. P. Donahue, *Inorg. Chem.*, 2012, **51**, 346–361.
70. M. Süssner and H. Plenio, *Chem. Commun.*, 2005, 5417–5419.
71. M. Süssner and H. Plenio, *Angew. Chem., Int. Ed.*, 2005, **44**, 6885–6888.
72. S. Leuthäusser, D. Schwarz and H. Plenio, *Chem. – Eur. J.*, 2007, **13**, 7195–7203.
73. S. Leuthäuber, V. Schmidts, C. M. Thiele and H. Plenio, *Chem. – Eur. J.*, 2008, **14**, 5465–5481.
74. S. Wolf and H. Plenio, *J. Organomet. Chem.*, 2009, **694**, 1487–1492.
75. X. Hu, I. Castro-Rodriguez, K. Olsen and K. Meyer, *Organometallics*, 2004, **23**, 755–764.
76. I. Fernández, N. Lugan and G. Lavigne, *Organometallics*, 2012, **31**, 1155–1160.
77. R. Credendino, L. Falivene and L. Cavallo, *J. Am. Chem. Soc.*, 2012, **134**, 8127–8135.
78. M. V. Vollmer, C. W. Machan, M. L. Clark, W. E. Antholine, J. Agarwal, H. F. Schaefer, C. P. Kubiak and J. R. Walensky, *Organometallics*, 2015, **34**, 3–12.
79. F. Hartl and A. Vlček, *Inorg. Chem.*, 1996, **35**, 1257–1265.
80. Y. Ishikawa and K. Kawakami, *J. Phys. Chem. A*, 2007, **111**, 9940–9944.
81. M. J. Schadt and A. J. Lees, *Inorg. Chem.*, 1986, **25**, 672–677.
82. L. H. Jones, R. S. McDowell and M. Goldblatt, *Inorg. Chem.*, 1969, **8**, 2349–2363.
83. P. Datta, D. Sardar, A. P. Mukhopadhyay, E. López-Torres, C. J. Pastor and C. Sinha, *J. Organomet. Chem.*, 2011, **696**, 488–495.

84. D. Sieh, D. C. Lacy, J. C. Peters and C. P. Kubiak, *Chem. – Eur. J.*, 2015, **21**, 8497–8503.
85. A. Cebollada, M. E. Viguri, J. Pérez, J. Díaz, R. López and L. Riera, *Inorg. Chem.*, 2015, **54**, 2580–2590.
86. M. Brill, J. Díaz, M. A. Huertos, R. López, J. Pérez and L. Riera, *Chem. – Eur. J.*, 2011, **17**, 8584–8595.
87. H.-W. Frühauf, *Coord. Chem. Rev.*, 2002, **230**, 79–96.
88. H.-W. Frühauf, F. Seils and C. H. Stam, *Organometallics*, 1989, **8**, 2338–2343.
89. M. van Wijnkoop, P. P. M. de Lange, H.-W. Frühauf, K. Vrieze, W. J. J. Smeets and A. L. Spek, *Organometallics*, 1995, **14**, 4781–4791.
90. M. Van Wijnkoop, P. P. M. De Lange, H.-W. Frühauf, K. Vrieze, Y. Wang, K. Goubitz and C. H. Stam, *Organometallics*, 1992, **11**, 3607–3617.
91. H.-W. Frühauf and F. Seils, *J. Organomet. Chem.*, 1986, **302**, 59–64.
92. R. Siebenlist, M. de Beurs, N. Feiken, H.-W. Frühauf, K. Vrieze, H. Kooijman, N. Veldman, M. T. Lakin and A. L. Spek, *Organometallics*, 2000, **19**, 3032–3053.
93. H.-W. Frühauf, F. Seils, R. J. Goddard and M. J. Romao, *Organometallics*, 1985, **4**, 948–949.
94. H.-W. Frühauf, F. Seils, M. J. Romao and R. J. Goddard, *Angew. Chem., Int. Ed.*, 1983, **22**, 992–993.
95. J. Scholz, H. Görls, H. Schumann and R. Weimann, *Organometallics*, 2001, **20**, 4394–4402.
96. P. P. M. de Lange, E. Alberts, M. van Wijnkoop, H.-W. Frühauf, K. Vrieze, H. Kooijman and A. L. Spek, *J. Organomet. Chem.*, 1994, **465**, 241–249.
97. N. Feiken, H.-W. Frühauf, K. Vrieze, N. Veldman and A. L. Spek, *J. Organomet. Chem.*, 1996, **511**, 281–291.
98. N. Feiken, H.-W. Frühauf, K. Vrieze, J. Fraanje and K. Goubitz, *Organometallics*, 1994, **13**, 2825–2832.
99. R. Hoffmann, *Angew. Chem.*, 2006, **94**, 725–739.
100. Y. Jean, *Molecular Orbitals of Transition Metal Complexes*, Oxford University Press, 2005.
101. F. G. A. Stone, *Angew. Chem., Int. Ed.*, 1984, **23**, 89–99.
102. M. W. Kokkes, D. J. Stufkens and A. Oskam, *J. Chem. Soc., Dalton Trans.*, 1983, 439–445.
103. F. Zera, *Chem. Soc. Rev.*, 2014, **43**, 7624–7663.
104. J. Tory, G. Gobaille-Shaw, A. M. Chippindale and F. Hartl, *J. Organomet. Chem.*, 2014, **760**, 30–41.
105. J. Qiao, Y. Liu, F. Hong and J. Zhang, *Chem. Soc. Rev.*, 2014, **43**, 631–675.
106. B. Reuillard, K. H. Ly, T. E. Rosser, M. F. Kuehnel, I. Zebger and E. Reisner, *J. Am. Chem. Soc.*, 2017, **139**, 14425–14435.

CHAPTER 9

Probing CO_2 Reduction Intermediates Employing in situ *Spectroscopy and Spectrometry*

S. PÉREZ-RODRÍGUEZ,[a] G. GARCÍA,[b] M. J. LÁZARO*[a] AND E. PASTOR*[b]

[a] Instituto de Carboquímica (CSIC), Miguel Luesma Castán 4, 50018 Zaragoza, Spain; [b] Instituto de Materiales y Nanotecnología, Departamento de Química, Universidad de La Laguna, POB 456, 38200, La Laguna, Santa Cruz de Tenerife, Spain

*Email: mlazaro@icb.csic.es; epastor@ull.es

9.1 Introduction

Despite the extended literature related to the electrochemical reduction of CO_2 on different substrates at different conditions, the detailed mechanism and the implied intermediates of the reaction are not completely clear. The last is in part associated with studies in which the electrochemical reduction of CO_2 was only carried out by conventional potentiostatic techniques such as cyclic voltammetry and chronoamperometry without determining the final compounds.[1–5] Although interesting information can be obtained by these methods, the activity and specificity toward the CO_2 electrochemical

Energy and Environment Series No. 21
Electrochemical Reduction of Carbon Dioxide: Overcoming the Limitations of Photosynthesis
Edited by Frank Marken and David Fermin

Published by the Royal Society of Chemistry, www.rsc.org

reaction cannot be fully studied, since the hydrogen evolution reaction (HER) also occurs as a competitive reaction in aqueous electrolytes. Therefore, a higher current density at the electrode, does not always lead to an increase of the conversion rate of CO_2, since the HER may be favoured. On the other hand, CO_2 may lead to adsorbed species at the electrode surface, such as CO_{ad} and formate species, decreasing the catalytic area and consequently the delivered current density.

On the other hand, there is a lack of work employing *in situ* techniques coupled to the electrochemical systems in order to detect and quantify the reduction products, and several works have reported gaseous and liquids compounds using *ex situ* techniques.[6–15] In general, gaseous products (such as hydrocarbons, H_2, carbon monoxide) were quantitatively determined by gas chromatography, while products dissolved in the electrolyte (*e.g.*, formic acid) were analyzed by high performance liquid chromatographic (HPLC). Steam chromatography was used for the determination of alcohols and aldehydes (methanol, ethanol, formaldehyde). Other techniques, such as ion chromatography, mass spectrometry or iodometric titration were also used in these studies. These *ex situ* methods are versatile, commercial, affordable and easy to use; however, the reaction products can change substantially during the experimental procedure. In this context, the reaction products are commonly trapped in gas collectors or in liquid solvents, in which a further reaction may occur and other by-product compounds may be produced. In addition to those described, reaction intermediates are difficult to follow by *ex situ* techniques, and therefore mechanistic and kinetic studies are tricky if not impossible.

Because of the overall aim of generating knowledge about the mechanism pathways for the CO_2 reduction reaction and helping to design catalysts with high performance toward this process, *in situ* techniques should be employed. These approaches allow the study of the electrochemical reduction of CO_2 occurring at the electrocatalyst by potential or current control, and simultaneously they may detect, identify and quantify the intermediates and products of the reaction, as well as monitor the overall parameters, such as the surface structure or the adsorption strength, changing along the reaction. Consequently, *in situ* techniques appear as adequate tools in order to elucidate the reaction mechanism of the electrochemical processes. Nevertheless, *in situ* systems are mostly not commercial, are expensive and are not easy to use.

Several *in situ* methodologies have been developed and used to study the CO_2 electroreduction reaction in a variety of materials and media, such as *in situ* Fourier transform infrared (FTIR), Raman and UV–vis spectroscopy and differential electrochemical mass spectrometry (DEMS). In the current chapter, a review of the most important and well-developed *in situ* techniques for a better understanding of CO_2 electroreduction reaction is presented. In addition, an explanation of the assembly and procedure of these novel techniques will be also displayed.

9.2 Differential Electrochemical Mass Spectrometry (DEMS)

DEMS is a valuable analytical tool for the on-line detection of volatile and gaseous intermediates and products during the electroreduction of CO_2. The DEMS technique consists mainly of an electrochemical system that is directly attached to the vacuum chamber of a mass spectrometer through a porous membrane. In this way, only gaseous and volatile species formed during the electrochemical reactions pass through the membrane pores, whereas liquid or solid species are prevented. In aqueous systems and in some organic electrolytes, such as propylene carbonate, the separation of the electrolyte from the vacuum chamber can be achieved by using porous hydrophobic membranes based on polytetrafluoroethylene (PTFE).

Thus, DEMS technique allows the *in situ* detection of volatile and gaseous products/intermediates generated in electrochemical reactions with excellent sensitivity. In this way, mechanistic and kinetic studies of particular electrochemical systems can be achieved by the simultaneous acquisition of conventional cyclic voltammograms (CVs), current transients (CTs) or other electrochemical methods with the corresponding mass spectrometric signal for selected m/z (mass to charge) ratios. Quantitative analysis of the substances produced or consumed on the electrode is also possible by means of the DEMS calibration for each component, which is based on relating the faradic current and its corresponding mass spectrometric signal through a constant, which is different for each component. More information about the main principles of DEMS can be found elsewhere.[16–19]

Despite the great benefits of DEMS, the literature is not very extensive on the utilization of this technique for studying CO_2 electrochemical reduction reaction.[20–36] The most commonly monitored products from the CO_2 reduction reaction are methane ($m/z=16$: $CH_4^{+\bullet}$ or $m/z=15$: CH_3^{+}), ethane ($m/z=27$: CH_2CH^{+} or $m/z=26$: $CHCH^{+}$), carbon monoxide ($m/z=28$: CO^{+}), formaldehyde ($m/z=30$: $H_2CO^{+\bullet}$), methanol ($m/z=32$: $CH_3OH^{+\bullet}$ or $m/z=31$: CH_3O^{+}) and formic acid ($m/z=45$: $HCOO^{+}$). It is remarkable that a specific m/z signal may be related to more than one species, *e.g.* ethane and formaldehyde may be detected by the $m/z=30$ signal and therefore a different ionic fragment (*e.g.* $m/z=27$ or 26 for ethane) should be studied to distinguish between species.

Furthermore, the mass signal $m/z=44$, indicative of carbon dioxide, and $m/z=2$, of hydrogen, are usually registered in order to analyze the CO_2 consumption or H_2 formation due to water reduction, which occurs as a competitive reaction. In this way, it is possible to study whether the electrode favours or inhibits the HER or to determine the faradic efficiency of the CO_2 reduction process, (E^{CO_2}).[20–30,37] The estimation of this last parameter requires a previous determination of the DEMS calibration constant for the formation of hydrogen (K^{H_2}). In the absence of CO_2, the main reaction taking place at cathodic potentials is the HER. Thus, the ratio between the ionic $m/z=2$ ($Q_i^{H_2}$) and the faradic charges ($Q_f^{H_2}$) can be obtained and related according to eqn (9.1), taking into account the number of electrons involved in the formation of a hydrogen molecule ($n=2$):

$$K^{H_2} = \left[2\frac{Q_i^{H_2}}{Q_f^{H_2}}\right]. \quad (9.1)$$

In the presence of CO_2, at cathodic enough potentials, the most important contributions to the faradic current are CO_2 reduction and the HER. Thus, the current efficiency for CO_2 reduction can be determined from the subsequent expression:

$$E^{CO_2} = 1 - \left[2\frac{Q_i^{H_2}}{K^{H_2}Q_f^{T}}\right] \quad (9.2)$$

where 2 is the number of electrons involved in the HER, (Q_f^T) is the charge associated to both faradic processes, the HER and CO_2 reduction (obtained from the faradic current in the presence of CO_2) and ($Q_i^{H_2}$) is the ionic charge associated to the $m/z=2$ signal recorded during the electrochemical reduction of CO_2.

Most of the DEMS measurements for the CO_2 reduction reaction have been carried out on thin film sputtered electrodes.[20–22,28,29] Metals with low overpotentials for the HER, such as Pt and Pd, have attracted special attention, since adsorbed hydrogen acts as a reactant for the reduction process.[4,20,21,23,26,30,38–42] Adsorbed hydrogen favours the formation of other adsorbed species derived from CO_2 reduction reaction at the metal surface, and can act as intermediates or be just a reaction product. Several works have indirectly demonstrated the formation of adsorbed carbon monoxide (CO_{ad}) on Pt and Pd electrodes utilizing the DEMS technique.[20–23,36] CO_{ad} may adsorb strongly on a metallic catalyst, thus further CO_2 reduction is usually hindered.[23,41–43]

Remarkable is the work of Brisard *et al.*,[21] in which the formation of formaldehyde, methanol and formic acid on polycrystalline Pt in 0.1 M $HClO_4$ was detected by DEMS. CTs and corresponding mass CTs (MCTs) were recorded after a potentiostatic pulse from 0.2 to −0.9 V *vs.* RHE in presence of dissolved CO_2 or in its absence, using a CO saturated solution in the last case. They found an increment of the mass signals $m/z=30$ and 32, associated to the production of formaldehyde and methanol that follows the same trend in CO_2 as in CO saturated solution. However, formic acid formation was only detected in the presence of dissolved CO_2. Therefore, they concluded that CO_{ad} is not a reaction intermediate during formic acid formation, while methanol and formaldehyde production occurs *via* CO_{ad}.

On the other hand, DEMS measurements performed by Kolbe *et al.*[20] showed the formation of CO_{ad} on Pd at −1.0 V *vs.* Ag/AgCl in a CO_2 saturated $KHCO_3$ solution, while volatile CO was obtained at potentials below −1.4 V *vs.* Ag/AgCl. This study is notable since it implies a partial reduction and posterior adsorption of CO_2 at the Pd surface, which can be further reduced to CO, which is an important reaction intermediate to produce value-added compounds such as methanol or formaldehyde.

Indeed, we recently confirmed the formation of different adsorbed species other than CO (*e.g.* $COOH_{ad}$ and COH_{ad}) from the CO_2 electroreduction on Pd nanoparticles supported on diverse nanostructured carbon materials (Vulcan XC-72R, carbon nanofibers, CNFs; carbon nanocoils, CNCs; and ordered mesoporous carbons, OMCs),[26] using a novel DEMS configuration, which allows the direct adaptation of any electrochemical system to a commercial mass spectrometer. Further details of the set-up can be found in ref. 44. Cyclic voltammetry studies in 0.1 M $NaHCO_3$ showed that CO_2 was effectively reduced to other species, which were adsorbed onto Pd/C electrocatalysts at −1.0 V *vs.* Ag/AgCl. Consistent with these results, molecular hydrogen production ($m/z = 2$) was significantly decreased by the presence of CO_2. Figure 9.1 displays the results achieved using a Pd/Vulcan catalyst.

By means of the comparison of the oxidation charges of a monolayer of CO or reduced CO_2 adsorbed on the catalyst surface at −0.5 and −1.0 V *vs.* Ag/AgCl, respectively; ($Q_{CO_{2,red}}/Q_{CO}$), it was demonstrated that in addition to carbon monoxide, other adsorbates were also generated on Pd/C electrocatalysts.[26] Furthermore, a different ratio ($Q_{CO_{2,red}}/Q_{CO}$) was obtained for the electrocatalysts indicating that the product distribution from CO_2 reduction was influenced by the nature of the carbon support, which could be associated to differences in the strength of Pd–H_{ad}.

Similarly, the adsorption of species derived from CO_2 reduction onto carbon substrates has been recently confirmed by DEMS.[25] The same nanostructured carbon materials were used as electrodes: CNFs, CNCs and OMCs. The commercial material Vulcan XC-72R was also used for comparison. A lower faradic current was obtained at −0.7 V *vs.* RHE in 0.1 M $NaHCO_3$ in the presence of dissolved CO_2 and the formation of hydrogen was inhibited at the carbon electrodes. indicating the presence of adsorbed species. In addition, the formation of electrolysis products was not detected by DEMS and, thus, volatile species were not formed at these potentials. At more negative potentials (−1.1 V *vs.* RHE) the adsorption of reduced species was more evident at all the carbon surfaces, although with a different affinity. CNFs, CNCs and Vulcan showed a higher inhibition of the faradic current in the presence of CO_2, and therefore, a higher formation of adsorbates from the CO_2 reduction reaction, in comparison to OMCs. This different behavior was explained by a poor adsorption of hydrogen on OMCs, which inhibits the interaction with reduced CO_2.

Interestingly, Kortlever *et al.*[36] recently studied the electrochemical CO_2 reduction on Pd–Pt electrode by DEMS. Products in the electrolyte were detected by on-line HPLC. They confirmed that these catalysts, which are active for the oxidation of formic acid, were also active for the electrochemical reduction of CO_2 to formic acid. In fact, the formation of formic

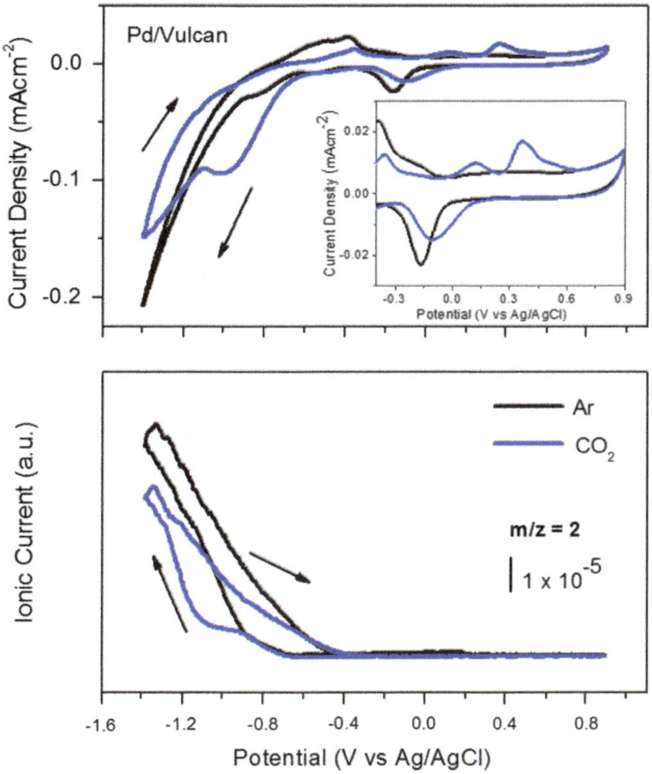

Figure 9.1 CVs (upper panel) and MSCVs for H_2, $m/z = 2$, (bottom panel) for a Pd/Vulcan catalyst in 0.1 M $NaHCO_3$ ($\nu = 10$ mV s^{-1}). Black curves: Ar saturated solution. Blue curves: CO_2 saturated solution. Inset shows a zoom of the oxidation region.
Reproduced from *Applied Catalysis B: Environmental*, **163**, S. Pérez-Rodríguez, N. Rillo, M. J. Lázaro and E. Pastor, Pd catalysts supported onto nanostructured carbon materials for CO_2 valorization by electrochemical reduction, 83–95,[26] Copyright 2015, with permission from Elsevier.

acid was detected at more positive potentials on the Pd–Pt catalyst, in comparison to bare palladium electrode. In addition, the catalyst was able to reduce bicarbonate as well as to reversibly convert CO_2 into formic acid and *vice versa*.

The electrochemical reduction of CO_2 on Au–Pd core–shell nanostructures has been also studied,[27] obtaining promising results. Surprisingly, the faradic efficiency for CO_2 reduction was strongly affected by the thickness of Pd nanoshells, increasing twofold (from 41 to 92%) when the thickness was decreased from 10 to 1 nm.

Copper-based materials have been also extensively used for CO_2 conversion due to their exceptional activity for the formation of hydrocarbons and alcohols.[5,11,12,28,29,32–34,45–52] Formation of methane, ethane, formaldehyde and methanol on Cu electrodes has been revealed by DEMS measurements.[28,29,32–34]

Remarkable is the work of Dubé *et al.*,[28] in which the generation of methane, alcohol, formaldehyde and ethene on bulk copper in H_2SO_4 and $HClO_4$ was detected by DEMS. CTs and CVs were recorded in the presence of dissolved CO_2 or CO in both working solutions. In general, a lower formation of the reduction products was obtained in CO saturated solution due to the strong adsorption of this species at the copper surface. In addition, it was shown that the electrolyte composition affected the formation of the reduction compounds, being the production rate higher in the perchloric acid medium, in most cases. However, it was observed that the formation of methane and formaldehyde were kinetically faster, compared with those of alcohol and ethene formation in both electrolytes.

On the basis of these results, the reaction mechanism of CO_2 electroreduction on Cu in acidic media was proposed. First, adsorbed hydrogen is formed during the cathodic sweep by eqn (9.3):

$$H^+ + e^- \leftrightarrows H_{(ad)}. \tag{9.3}$$

Or by water dissociation:

$$H_2O + e^- \leftrightarrows H_{(ad)} + OH^-. \tag{9.4}$$

Followed by hydrogen evolution:

$$2H_{(ad)} \rightarrow H_2. \tag{9.5}$$

Then, dissolved carbon dioxide may adsorb (eqn (9.6))

$$CO_2 \rightarrow CO_{2(ad)} \tag{9.6}$$

and react at the Cu surface following two main pathways:

(i) Formation of formic acid according to eqn (9.7):

$$CO_{2(ad)} + 2H_{(ad)} \rightarrow HCOOH. \tag{9.7}$$

(ii) Formation of CO as reaction intermediate (eqn (9.16)): methanol, formaldehyde and CH_2 radicals can be produced (reactions (9.9)–(9.11)). Then, CH_2 radicals may interact with adsorbed hydrogen, with another CH_2 radical or with CO_{ad} producing methane, ethene and ethanol, respectively (eqn (9.12)–(9.14)).

$$CO_{2(ad)} + 2H_{(ad)} \rightarrow CO_{(ad)} + H_2O \tag{9.8}$$

$$CO_{(ad)} + 4H_{(ad)} \rightarrow CH_3OH \tag{9.9}$$

$$CO_{(ad)} + 2H_{(ad)} \rightarrow HCHO \tag{9.10}$$

$$CO_{(ad)} + 4H_{(ad)} \rightarrow CH_{2(ad)} + H_2O \tag{9.11}$$

$$CH_{2(ad)} + 2H_{(ad)} \rightarrow CH_4 \tag{9.12}$$

$$CH_{2(ad)} + CH_{2(ad)} \rightarrow CH_2CH_2 \tag{9.13}$$

$$CH_{2(ad)} + CO_{(ad)} + 4H_{(ad)} \rightarrow CH_3CH_2OH. \tag{9.14}$$

Most of these studies involve thin film sputtered electrodes onto the PTFE membrane.[20–22,28] The main advantage of this set-up is that the response time is good enough for most electrochemical reactions due to the short distance between the electrode and the vacuum system. However, a quick depletion of the active phase, caused by the strong formation of bubbles due to water reduction at such negative potentials, can occur. In order to overcome this limitation, Vielstich and coworkers[29] developed a new type of DEMS set-up based on rotating electrodes. In addition, this configuration enhances the mass transport of the volatile species produced at the catalyst surface to the mass chamber.[29,53] A noticeable formation of methane and ethene was observed below −1.8 V *vs.* SCE in 0.5 M $KHCO_3$ at bulk and electrodeposited copper.[29]

Later, the group of Koper and co-workers[32–35] studied the electrochemical reduction of CO_2 on copper-based electrodes. They observed a reduction peak at around −0.6/−0.7 V *vs.* RHE during the CO_2 electroreduction on polycrystalline copper electrode at pH > 6.5.[34] By means of on-line HPLC measurements it was shown that this peak occurs due to the production of formate *via* direct bicarbonate reduction, and not to the formation of a CO ad-layer on the electrode surface, as was previously reported in the literature. In agreement with these results, the formation of volatile products, such as methane and ethylene, which require CO as an intermediate, were detected by DEMS at more negative potentials. In addition, the intensity of the cathodic peak was influenced by the electrode morphology and the identity of the cations and anions in solution. In fact, the formate peak was observed in the SO_4^{2-}, ClO_4^- and Cl^- containing electrolytes, while it did not appear within Br^- and I^-.

The formation of methane and ethylene as main products from CO_2 reduction was also observed on Cu overlayers of varying thickness supported on Pt in 0.1 M $KHCO_3$.[33] Interestingly, a higher relative faradic selectivity for both hydrocarbons was obtained with increasing the copper layer thickness to 15 nm, as can be seen in Figure 9.2. On the other hand, the formation of methane decreased much faster than that of ethylene as the Cu layer thickness decreased, sharply increasing the faradic efficiency to ethylene. Thus, these results show that the selectivity for CO_2 reduction to hydrocarbons can be tuneable with the copper layer thickness.

Similarly, the electrocatalytic activity toward CO_2 reduction has been recently studied on Au@Cu core@shell nanoparticles.[32] It was found that the activity and selectivity of the reaction can be tweaked as a function of the thickness of Cu shells. The H_2 and CH_4 formation increased with the number of Cu layers, while the formation of ethylene decreased. In addition, the formation of formate (monitored by on line HPLC) was also strongly

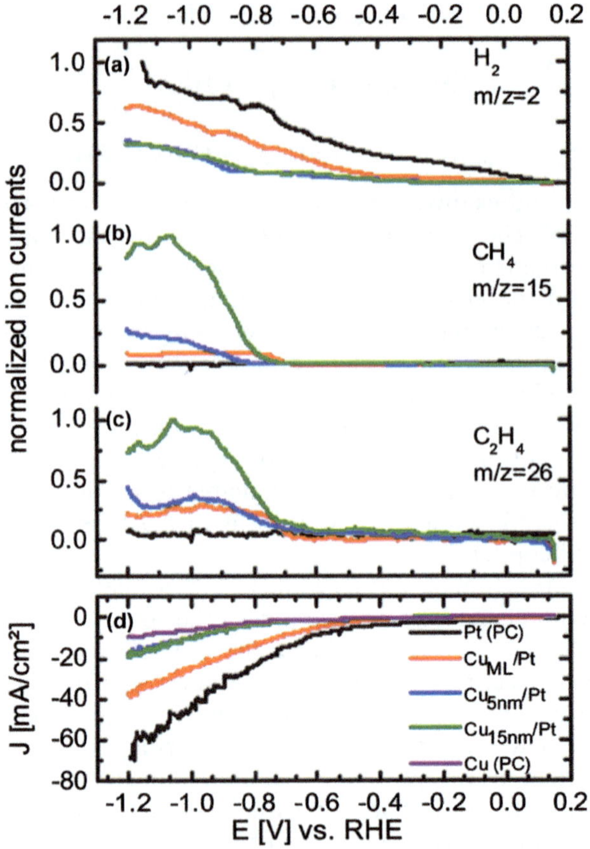

Figure 9.2 The normalized ions currents for hydrogen, methane and ethylene probed by DEMS in dependence of the applied potential (a)–(c) and polarization curves for CO_2 electroreduction on different nanometer-scaled layer thickness of Cu supported on pure Pt in 0.1 M $KHCO_3$ at room temperature.
Reprinted with permission from R. Reske, M. Duca, M. Oezaslan, K. J. P. Schouten, M. T. M. Koper and P. Strasser, *J. Phys. Chem. Lett.*, 2013, **4**, 2410–2413.[33] Copyright 2013 American Chemical Society.

dependent on the thickness of Cu shells, showing the largest conversion on Au nanoparticles with 7–8 layers of Cu.

The electrochemical reduction of CO_2 has also been studied on Co electrodes by DEMS.[31] An enhanced activity for CO_2 electroreduction to CO and CH_4 has been recently achieved by Koper and coworkers[31] at a cobalt protoporphyrin immobilized on a pyrolytic graphite electrode in aqueous acidic solution at relatively low overpotential (0.5 V). It was shown that the activity and selectivity of the reaction was influenced by the pH of the media. This pH-dependent efficiency was explained by a mechanism in which CO_2 is activated by the cobalt protoporphyrin through the stabilization of a catalyst-bound $CO_2^{\bullet-}$ radical anion. This intermediate acts as a Brønsted base, abstracting a proton from water, and, hence, leading to an overall reactivity for CO_2 reduction whose pH dependence is substantially different from the competing H_2 evolution. Thus, lowering the overpotential for the formation of this intermediate is the key to designing improved catalysts for CO_2 reduction. Furthermore, a suitable pH adjustment may contribute to increasing the faradic efficiency of the reaction.

However, all these studies were carried out in different conditions from those in real electrodes (*i.e.* bulk electrodes or sputtered onto a glassy carbon disk). In this context, several properties of the real electrodes are missed on these configurations, such as the use of gas diffusion layers (GDLs), the porous structure of both catalytic and gas diffusion layers, the metal loading into the catalyst layer, hydrophobicity degree of both layers, reactants and products diffusion rate inside the GDL, and so on. Because of this our research group has developed a novel DEMS configuration for the spectro-electrochemical characterization of gas diffusion electrodes (GDEs) for CO_2 reduction.[24,54] The configuration consists of a working electrode (a layer of the catalyst deposited onto a carbon cloth) fixed between a PTFE membrane and a carbon glassy rod, which is connected to a gold wire outside the cell to keep the electrical contact. In this way, the electrode is in direct contact with the membrane, allowing the *in situ* detection of the products despite of the huge formation of hydrogen by the HER, which takes places in parallel to the electrochemical reduction of CO_2. Furthermore, the spectro-electrochemical cell can work at different temperatures and allows electrolyte exchange under control of the working electrode potential. The novel DEMS set-up was previously tested in a well-known reaction, *i.e.* the electrochemical oxidation of methanol, which takes places in the anode side of PEMFCs (when methanol is used as fuel).[54] The design was demonstrated to be appropriate for simultaneous evaluation of the catalytic and the diffusion properties of GDEs, as well as the identification of products with high sensibility.

The activity of carbon-supported iron oxide electrodes for CO_2 reduction was studied by means of the novel DEMS configuration.[24] Carbon Vulcan XC-72R was utilized as a catalyst support and treated with different oxidation procedures by concentrated HNO_3 (Nc) or HNO_3–H_2SO_4 (NS) at room (Ta) or boiling (Tb) temperatures to create functional groups. In this way, the influence of the surface chemistry of the carbon support on the electrochemical properties of the GDEs for CO_2 reduction was scrutinized. DEMS measurements revealed the formation of $m/z = 29$ ($CH_3CH_2^+$ or CHO^+: ethane/ethanol or an aldehyde) and $m/z = 45$ ($HCOO^+$ or CH_3CHOH^+: formic acid or ethanol) on Fe/C electrodes at potentials below −0.4 and −0.6 V *vs.* Ag/AgCl, respectively, in the base electrolyte (0.5 M H_2SO_4) saturated with CO_2 (Figure 9.3), at room temperature and atmospheric pressure.

Additionally, it was found that the catalytic activity and the product distribution were strongly influenced by the catalyst support. In fact, all the electrodes presented a lower H_2 production in CO_2 dissolved solution than in Ar saturated media, with the exception of the Fe/Vulcan NSTa0.5 sample, on which H_2 evolution reaction was enhanced, as can be seen in Figure 9.4.

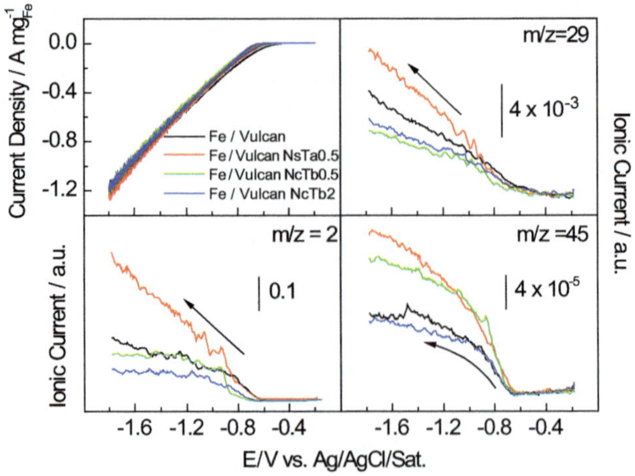

Figure 9.3 CVs and MSCVs on Fe/C catalysts in a CO_2 saturated solution 0.5 M H_2SO_4 (sweep rate 10 mV s^{-1}, $E_i = -0.2$ V *vs.* Ag/AgCl/Sat.).
Reproduced from S. Pérez-Rodriguez, G. Garcia, L. Calvillo, V. Celorrio, E. Pastor and M. J. Lázaro, *International Journal of Electrochemistry*, 2011, 249804,[24] http://dx.doi.org/10.4061/2011/249804.

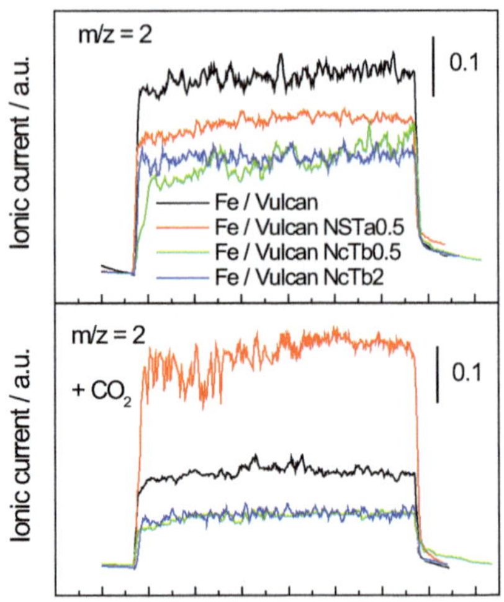

Figure 9.4 MSTs for CO_2-free (upper panel) and CO_2 saturated (bottom panel) solution in 0.5 M H_2SO_4 on Fe/C catalysts ($E_i = -0.2$ V, $E_f = -1.8$ *vs.* Ag/AgCl/Sat).
Reproduced from S. Pérez-Rodriguez, G. Garcia, L. Calvillo, V. Celorrio, E. Pastor and M. J. Lázaro, *International Journal of Electrochemistry*, 2011, 249804,[24] http://dx.doi.org/10.4061/2011/249804.

On the basis of these results, a reaction mechanism for the CO_2 electroreduction reaction was proposed and two main pathways were suggested:

(i) Formation of adsorbed CO. This path is favoured when the metal surface is highly covered by adsorbed hydrogen and interacts preferentially with adsorbed carbon dioxide instead to react with each other forming molecular hydrogen. Hence, the H_2 evolution will be inhibited. This path appears to be governed by Fe/Vulcan, Fe/Vulcan NcTb0.5 and Fe/Vulcan NcTb2, in which the mass current signal associated with hydrogen production decreased in the presence of CO_2. From $CO_{(ad)}$ species different hydrocarbons and alcohols may be obtained.

(ii) $COOH_{(ad)}$ production: this pathway is favoured on surfaces with preference to produce H_2. Hence, HER is usually favoured in parallel to this reaction. Fe/Vulcan NSTa0.5 follows principally this reaction, since this catalyst enhanced the HER. This path gives mainly formic acid, although the possibility of the formation of $CO_{(ad)}$ and hence of other products cannot be discarded.

In *a posteriori* work[23] it has been demonstrated that the formation of the mass fragments $m/z = 29$ and $m/z = 45$ is not observed on GDEs based on Pt due to a strong adsorption of species from CO_2 reduction. These adsorbates act as a self-poison inhibiting further reduction. Thus, the use of iron-based electrodes seems to favour the adsorption of species from CO_2 reduction and their posterior reduction, while in the case of Pt hydrogen is the main product due to water reduction.

It should be recalled that DEMS is limited to the detection of gaseous and volatile species and therefore adsorbed, liquid or solid species cannot be detected. In this context, during CO_2 reduction, different species may be adsorbed on the electrode surface, acting as an intermediate of the reaction or as self-poisoning inhibiting further hydrogenation. In order to obtain complementary information about these adsorbates, DEMS measurements are often combined with *in situ* FTIR spectroscopy or other *in situ* spectroscopic techniques.

9.3 *In Situ* Spectroscopies

The CO_2 reduction reaction has been also studied using electronic and vibrational spectroscopies. Electronic spectroscopy addresses the transitions, which take place between electronic states of a sample. Such transitions may occur in the ultraviolet (UV), visible (vis) or near infrared region (NIR). Therefore, electronic spectroscopy is usually utilized to elucidate the nature of electronic changes induced by redox processes. On the other hand, vibrational spectroscopy, including infrared (IR) and Raman, provide structural information, resulting useful for the studying of the double layer, soluble redox products and surface adsorbed species.

9.3.1 *In Situ* Fourier Transform Infrared Spectroscopy (FTIRS)

FTIR spectro-electrochemistry has been widely used to investigate the reaction intermediates and the possible pathways during the electrochemical reduction of CO_2 as this method allows the characterization of adsorbed, soluble and gaseous species, as well as the determination of short-lived radical ions. However, the main limitation of this technique is that aqueous and permanent polar electrolytes are strongly IR active, leading to significant background interference, which can be avoided by expensive isotopes species such as heavy water. Furthermore, *in situ* FTIR spectroscopy requires the use of sensitive instrumentation, including special optical window materials, powerful detectors (*e.g.* mercury/cadmium/tellurium, MCT), polarized light sources and/or adequate modulation techniques. Finally, species must be IR active and therefore they must follow the selection rule, *i.e.* only adsorbed species with dynamic dipole moment perpendicular to the surface will be observed in the vibrational spectrum; on the other hand, species with zero net dipole moment are "invisible" to the IR radiation.

In FTIRS, the electrochemical cell is usually coupled to the chamber (under vacuum or dry chamber) of an infrared detector, and the most popular set-up works by external reflection and includes an invisible prism (*e.g.* CaF_2) to IR radiation at the bottom of the spectro-electrochemical cell. In this way, it is possible to analyze the structure and orientation of molecular species, as well as the dynamics and strength of adsorption-desorption processes at or near the electrode–electrolyte interface, under potential or current control conditions.

Although time-resolved spectroscopy is possible, steady-state measurements are more usual. The common procedure is to apply single potential steps from a reference potential, in which no faradaic reaction occurs, to the potential of interest. Spectra are usually depicted as a reflectance (or transmittance) ratio R/R_0, where R and R_0 are the reflectances measured at the sample and the reference potential, respectively. In this way, positive and negative bands represent the loss and gain of species at the sampling potential, respectively. Deep information about *in situ* FTIR spectroscopy set-ups can be found elsewhere.[19,55–67]

The electrochemically modulated infrared spectroscopy (EMIRS) technique was the first approach for infrared studies *in situ* to the electrochemical experiments.[68,69] The technique involved the use of external reflection from bulk metal mirror electrodes. This first approach made use of a dispersive instrument and required modulation of the electrode potential and demodulation of the reflectance signal using sensitive detection in order to amplify the signal. The potential modulated technique was subsequently adapted to Fourier transform infrared interferometers by Pons and coworkers.[62–64] This technique was namely subtractively normalized interfacial Fourier transform infrared spectroscopy (SNIFTIRS). The introduction of Fourier transform to infrared interferometers improved significantly the technique since high signal-to-noise ratios are obtained owing to averaging of numerous scans and, hence, modulation of the electrode potential is not required. Additionally, the IR spectra are quickly recorded, increasing the efficiency of the technique.

In situ FTIRS has been the most used method to study the operating mechanism and to detect the reaction intermediates during the electrochemical reduction of CO_2.[38–42,70–86] Several catalysts, including polycrystalline and single crystalline electrodes of diverse materials such as Cu, Pb, Au, Ni, Pt and Pd, have been studied for the CO_2 reduction reaction by *in situ* FTIRS. Among them, metals with low overvoltage toward the HER, such as Pt, Pd[38–42,72,73,75,80,82] and Cu[76–80,83] have attracted special attention. In addition, molecular catalysts[74,85] and semiconductor cathodes[81] have been also characterized by this technique.

Highly relevant is the work of Beden *et al.*,[42] in which the CO_2 reduction reaction to adsorbed CO on a Pt electrode was revealed for the first time by EMIRS. Linearly bonded adsorbed CO (band at 2060 cm^{-1}) and an important amount of CO species adsorbed at higher coordination surface sites (band at 1865 cm^{-1}) on the metal was found in 1 M H_2SO_4, as can be seen in Figure 9.5. They also observed an increment of the HER due to catalyst poisoning by strongly adsorbed CO, which inhibits further reduction of CO_2.

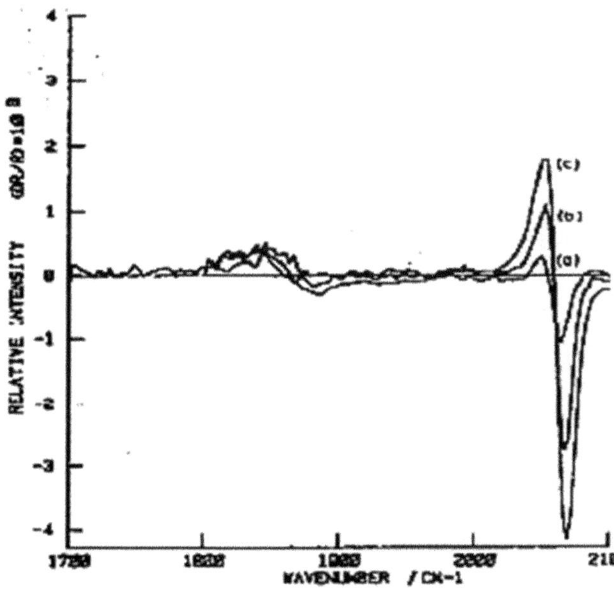

Figure 9.5 EMIRS spectra on a Pt electrode in 0.1 M H_2SO_4 in the presence of dissolved CO_2 for modulation between (a) +50 mV and +250 mV, (b) +50 mV and +350 mV, (c) +50 mV and +450 mV.
Reproduced from *Journal of Electroanalytical Chemistry*, **139**, B. Beden, A. Bewick, M. Razaq and J. Weber, On the nature of reduced CO_2 An IR spectroscopic investigation, 203–206,[42] Copyright 1982, with permission from Elsevier.

Later, the group of Iwasita studied the CO_2 reduction on the three basal planes of Pt single-crystal electrodes in acid media and significant achievements were obtained.[38–40] Interestingly, the activity and selectivity toward the CO_2 reduction strongly depend on the orientation of the Pt surfaces being Pt(110), the most active for this reaction, followed by Pt(100) and Pt(111) sites. An important presence of adsorbed species at the Pt(110) and Pt(100) surfaces was confirmed after holding the potential at 0.10 V in a 0.1 M $HClO_4$ solution saturated with CO_2 (Figures 9.6 and 9.7), while a small amount of adsorbate was found at Pt(111) sites (Figure 9.8).

Linearly bonded CO and traces of multibonded CO, with bands at around 2056 and 1817 cm^{-1}, were obtained as the major products at Pt(110), as can be observed in Figure 9.9. Additionally, the presence of adsorbed carbonate species at potentials above 0.5 V (band at 1430 cm^{-1}) was confirmed by FTIR.

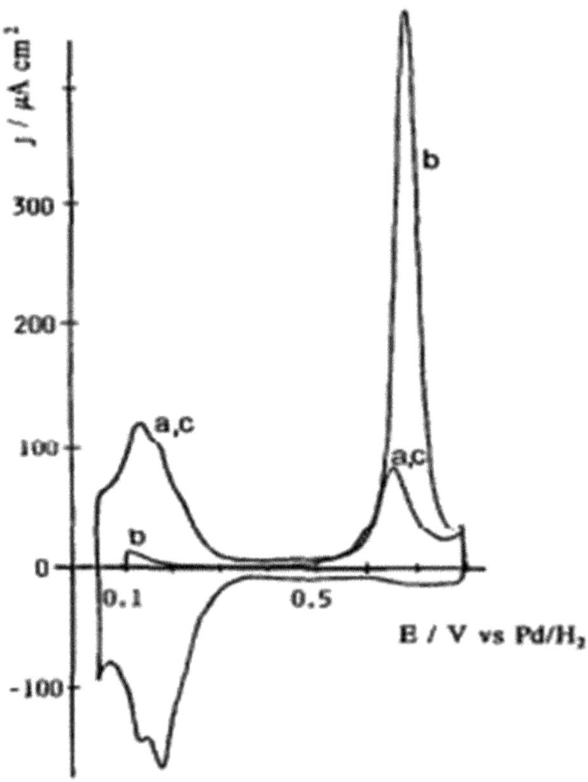

Figure 9.6 Voltammograms for the Pt(110) electrode in a 0.1 M $HClO_4$ solution saturated with CO_2 ($\nu = 0.10$ V s^{-1}): (a) stationary voltammogram; (b) stripping of the adsorbate after holding the potential at 0.1 V for 2 min; (c) potential cycle recorded just after stripping the adsorbate. Reproduced from *Journal of Electroanalytical Chemistry*, **369**, A. Rodes, E. Pastor and T. Iwasita, Structural effects on CO_2 reduction at Pt single crystal electrodes Part 1. The Pt(110) surface, 183–191,[38] Copyright 1994, with permission from Elsevier.

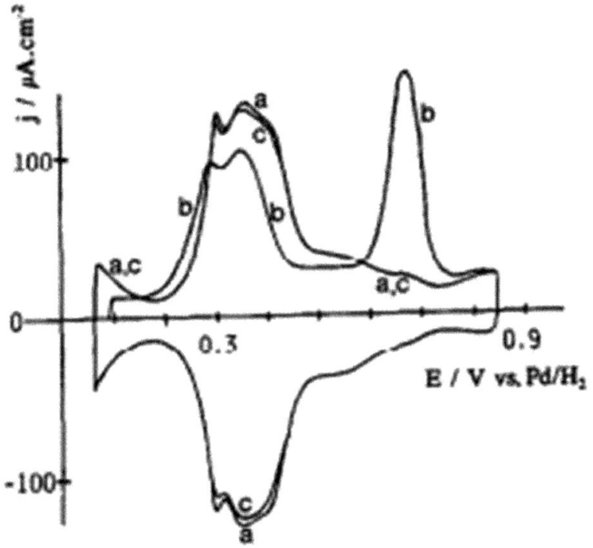

Figure 9.7 Voltammograms for the Pt(100) electrode in a 0.1 M $HClO_4$ solution saturated with CO_2 ($\nu = 0.10\ V\,s^{-1}$): (a) stationary voltammogram; (b) stripping of the adsorbate after holding the potential at 0.08 V for 15 min; (c) potential cycle recorded just after stripping the adsorbate. Reproduced from *Journal of Electroanalytical Chemistry*, **377**, A. Rodes, E. Pastor and T. Iwasita, Structural effects on CO_2 reduction at Pt single crystal electrodes Part 3. Pt(100) and related surfaces, 215–225,[39] Copyright 1994, with permission from Elsevier.

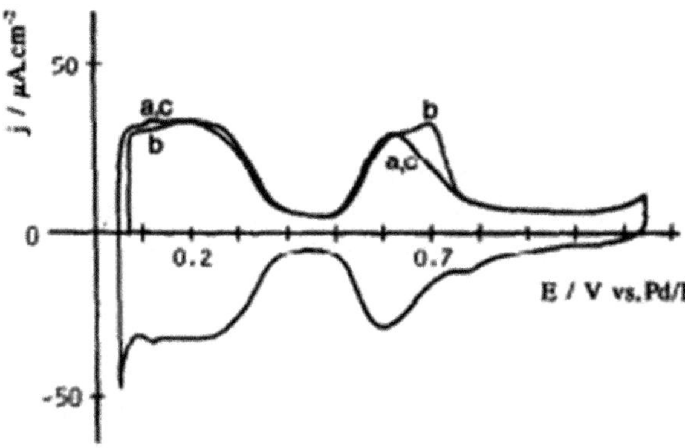

Figure 9.8 Voltammograms for the Pt(111) electrode in a 0.1 M $HClO_4$ solution saturated with CO_2 ($\nu = 0.05\ V\,s^{-1}$): (a) stationary voltammogram; (b) stripping of the adsorbate after holding the potential at 0.07 V for 5 min; (c) potential step after (b).
Reproduced from *Journal of Electroanalytical Chemistry*, **373**, A. Rodes, E. Pastor and T. Iwasita, Structural effects on CO_2 reduction at Pt single crystal electrodes Part 2. Pt(111) and vicinal surfaces in the [011] zone, 167–175,[40] Copyright 1994, with permission from Elsevier.

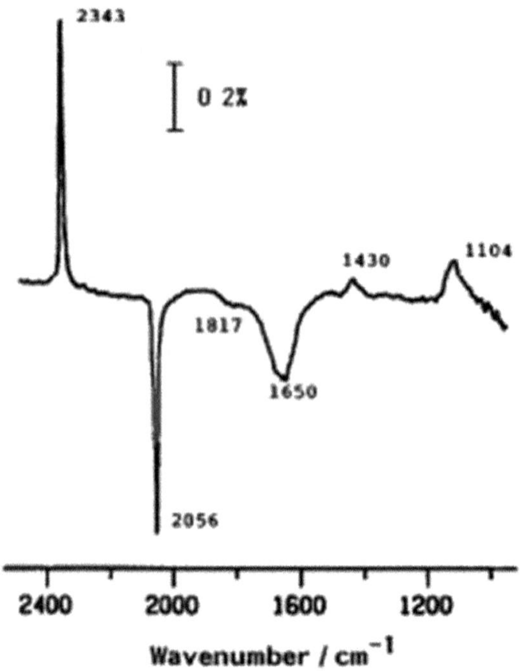

Figure 9.9 FTIR reflectance spectrum from a Pt(110) surface. The reference spectrum was taken at 0.50 V in a CO_2 saturated 0.1 M $HClO_4$ solution. The sample spectrum was obtained after applying a potential step to 0.10 V. Positive-going features represent loss and negative-going features represent gain of species at 0.10 V.
Reproduced from *Journal of Electroanalytical Chemistry*, **369**, A. Rodes, E. Pastor and T. Iwasita, Structural effects on CO_2 reduction at Pt single crystal electrodes Part 1. The Pt(110) surface, 183–191,[38] Copyright 1994, with permission from Elsevier.

In contrast, multibonded CO was mainly found at a well-ordered Pt(100) electrode surface (Figure 9.10). Finally, the spectrum of Pt(111) did not show bands in the frequency region 2100–1800 cm^{-1} related to the formation of CO-like species.

On the other hand, Taguchi and co-workers[41] studied the electrochemical reduction of CO_2 on polycrystalline Pt and Pd electrodes in neutral solution. A higher adsorption affinity of "reduced CO_2" was found on Pt than on Pd electrodes. They suggested that adsorbate species from the CO_2 reduction reaction acts as a catalytic self-poisoning and reaction intermediate on Pt and Pd, respectively.

The CO_2 reduction reaction at single-crystal or polycrystalline copper-based electrodes has been also investigated in detail, and there is consensus that the production of hydrocarbons occurs *via* adsorbed CO as reaction intermediate on these copper-based materials.[76–79,83] Relevant is the work of Hori *et al.*,[79] in which it was shown that the spectroscopic features of

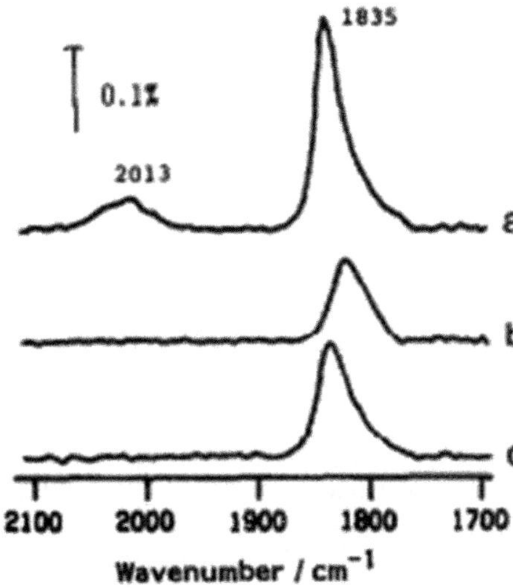

Figure 9.10 *In situ* FTIR reflectance spectrum from a Pt(100) surface in a CO_2 saturated 0.1 M $HClO_4$ solution. The sample spectrum was obtained after applying a potential step to 0.80 V. The reference spectrum was taken at: (a) 0.08 V; (b) 0.25 V and (c) 0.40 V. Positive-going features represent gain of species at 0.80 V.
Reproduced from *Journal of Electroanalytical Chemistry*, **377**, A. Rodes, E. Pastor and T. Iwasita, Structural effects on CO_2 reduction at Pt single crystal electrodes Part 3. Pt(100) and related surfaces, 215–225,[39] Copyright 1994, with permission from Elsevier.

reduced CO_2 were identical to those of adsorbed CO. In *posterior* works,[76–78] they found that specifically adsorbed anions from the electrolyte remained on the electrode surface below the potential of zero charge (pzc), while at more negative potentials CO is adsorbed, displacing these adsorbed anions.

On the other hand, the group of Ito[83] studied the adsorption of CO and the electrochemical reduction of CO_2 in polycrystalline Cu and Ag in 0.05 M Na_2SO_4 and 3.5 M KCl. Two kinds of adsorbed linear CO were observed on polycrystalline copper after their adsorption at −0.6 V and *posterior* oxidation in 0.05 M, as can be observed in Figure 9.11. They assigned the bands at *ca.* 2000 and *ca.* 2100 cm^{-1} to CO adsorbed on terrace and adatom defect Cu atoms, respectively. In contrast, only an absorption band at *ca.* 2000 cm^{-1} was obtained for linearly adsorbed CO on Ag.

Spectroscopic features of CO_2 reduction also showed the presence of linearly adsorbed CO on both Cu and Ag electrodes (Figure 9.12). They also observed desorption of CO on an Ag surface by the positive potential sweep from −0.7 to −0.5 V, while when Cu was used as the electrode CO desorption did not occur. Additionally, several bands appeared in the frequency region 1500–1200 cm^{-1} at the surface of this last metal, which were assigned to the vibration of CO_3^{2-} species. Similar results were obtained by using *in situ* surface-enhanced Raman spectroscopy (see later in Section 9.3.2).

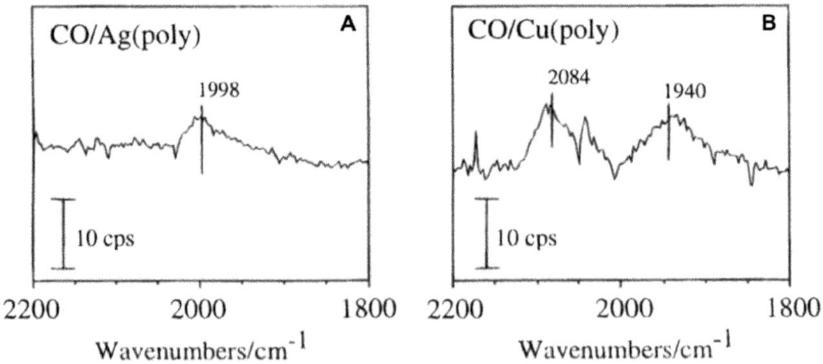

Figure 9.11 IRAS spectra of CO in a 3.5 M KCl solution: (a) on an Ag electrode; (b) on a Cu electrode. Electrode potential $= -0.6$ V/SHE.
Reprinted with permission from I. Oda, H. Ogasawara and M. Ito, *Langmuir*, 1996, **12**, 1094–1097.[83] Copyright 1996 American Chemical Society.

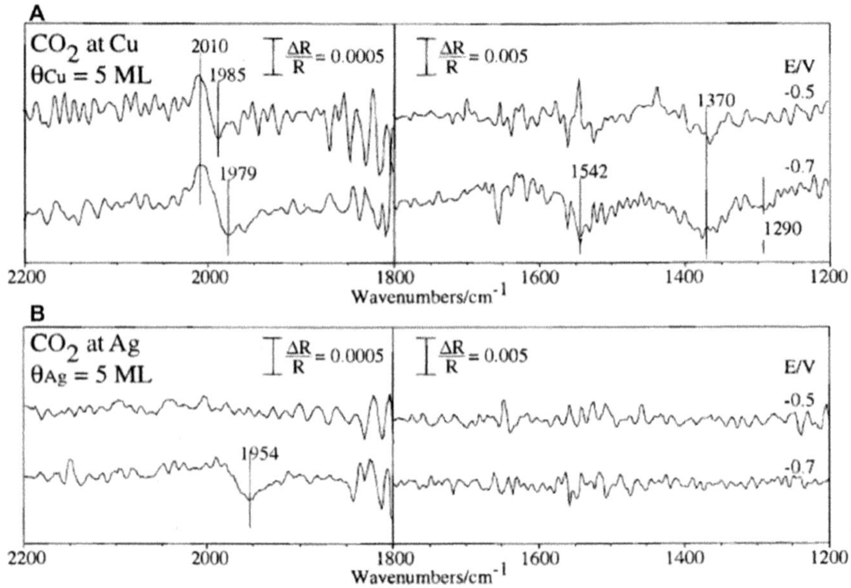

Figure 9.12 FTIRS spectra of CO_2 reduction in a 0.05 M Na_2SO_4 solution: (a) on a Cu electrode; (b) on a Ag electrode. Reference potential $= -0.3$ V/SHE; $\theta = 5$ ML.
Reprinted with permission from I. Oda, H. Ogasawara and M. Ito, *Langmuir*, 1996, **12**, 1094–1097.[83] Copyright 1996 American Chemical Society.

However, Ortiz and coworkers[80] did not obtain so promising results for the electrochemical reduction of CO_2 on various metal electrodes, including Sn, Cu, Au, In, Ni, Ru and Pt, in methanol. Dissolved CO_2 increased the cathodic current at potentials below −1.3 V *vs.* Ag/AgCl at Sn, Cu, Au, In and Ni electrodes, while on Pt and Ru surfaces the current was lower in the presence of CO_2. However, it was concluded from the FTIR spectra that there was no CO_2 reduction on any of the studied metals in methanol and that the only reduction product, *i.e.* $CO_3{}^{2-}$, was formed by reaction of CO_2 with hydroxyl anions produced in the electroreduction of residual water. On the other hand, Innocent *et al.* studied the electroreduction of CO_2 on a lead electrode by FTIR spectroscopy,[70,71] in which only bands associated to formate in aqueous solution and oxalate in aprotic media were achieved.

9.3.2 *In Situ* Raman Spectroscopy

In situ surface-enhanced Raman spectroscopy appears as an important tool to characterize the electrochemical interface during CO_2 reduction and provides attractive features, which address several limitations of other spectro-electrochemical techniques, such as IR. In this sense, Raman can be easily employed to study electrochemical aqueous systems due to the weak scattering in aqueous systems of this technique. The accessibility to study the low frequency region makes Raman an ideal and complementary technique to FTIR spectroscopy. Another important advantage of Raman is that there is no requirement of special optical window materials, since excitation is often conducted in the visible region. However, the use of high power monochromatic light sources (laser) and sensitive detectors are required, due to the poor sensibility of light scattering methods. Thus, developments in Raman instrumentation and methodology are necessary in order to expand and make more fruitful this powerful technique.

Therefore, Raman appears as an important vibrational spectro-electrochemistry technique, which can be coupled to electrochemical systems. This method is a suitable tool for the study of electrode–electrolyte interfaces and surfaces at the molecular level, being one of the most promising methods for the investigation of redox reactions. It can determine the bonding strength or orientation of adsorbed species. Raman spectro-electrochemistry measurements involve the study of scattered light along the time under potential control. Time-resolved spectra are also possible, although they are not as common.

Raman spectroscopy is usually employed for easily polarized molecules. This offers an important advantage of Raman *vs.* IR since permanent polar solutions (such as water) present a weak signal, allowing the *in situ* study of electrochemical reactions in aqueous media without the important background interference, which is one of the main limitations of the IR technique. However, Raman scatter is a weak phenomenon and consequently several enhancement approaches have been proposed to increase the sensibility. Van Duyne and coworkers were the first to demonstrate the potential of *in situ* resonance Raman (RR) to electrochemical systems.[87–92] A further improvement of Raman spectroscopy was achieved by Fleischmann *et al.*[93] using surface enhanced Raman scattering (SERS). This surface-sensitive technique enhances Raman scattering by molecules adsorbed on metal surfaces or by nanostructures such as plasmonic-magnetic silica nanotubes, allowing detecting single molecules. More extensive information about Raman spectro-electrochemistry can be found elsewhere.[19,65,66,94,95]

In situ Raman spectro-electrochemistry is an adequate tool to detect adsorbed species such as reaction intermediates and to estimate the variation of adsorbate coverage under potential control, and hence Raman spectro-electrochemistry has been used to study the electrochemical reduction of CO_2 on diverse metallic electrodes.[49,83,96–100] Some examples include the work by McQuillan *et al.*,[96] in which adsorbed CO_2^- and carboxy species were obtained on Ag electrodes during the cathodic reduction of CO_2 in 0.1 M $NaHCO_3$.

Further investigations related to CO_2 reduction were carried out by SERS.[49,83,97–100] For instance, Batista and co-workers[49] studied CO_2 electro-reduction on a Cu electrode in K_2SO_4 aqueous solutions at different pH ranges. In acidic solutions (pH $\approx$ 2.5) the presence of strong bands in the region 1580–1620 cm^{-1} (Figure 9.13) indicated the formation of adsorbed ethylene on the copper surface at −0.2 V *vs.* Ag/AgCl. This result was really noticeable since ethylene was spectroscopically identified for the first time at Cu during CO_2 electroreduction. However, CO bands were not detected, suggesting the reaction does not occur *via* adsorbed carbon monoxide.

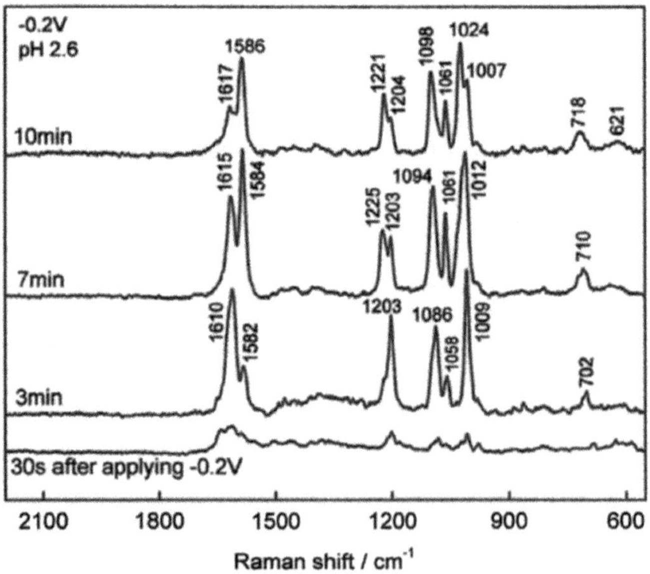

Figure 9.13 SERS spectra as a function of time collected on a Cu electrode at a fixed potential of −0.2 V in a CO_2-saturated 0.1 M K_2SO_4 solution, pH $\approx$ 2.6. CO_2 was bubbled into the solution at 0 V *vs.* Ag/AgCl.
Reproduced from *Journal of Electroanalytical Chemistry*, **629**, E. A. Batista and M. L. A. Temperini, Spectroscopic evidences of the presence of hydrogenated species on the surface of copper during CO_2 electroreduction at low cathodic potentials, 158–163,[49] Copyright 2009, with permission from Elsevier.

In contrast, at higher pH (pH ≈ 6.7) bands associated to adsorbed CO (2080 cm^{-1}, 360 cm^{-1} and 280 cm^{-1}, Figure 9.14) were obtained at potentials more negative than −0.5 V. In addition, spectral features indicated the presence of a complex structure containing a double C=C bond different to ethylene (band at 1645 cm^{-1} in Figure 9.14), a carboxyl group (band at 1390 cm^{-1}) and C–H bonds (several bands in the region 800–1500 cm^{-1}) at −0.5 V *vs.* Ag/AgCl. At more negative potentials the bands in the region between 800 and 1700 cm^{-1} tended to diminish and disappear, indicating a greater affinity toward the formation of adsorbed CO.

In solutions with pH = 9.5, the relative intensity for adsorbed CO increased, showing that the formation of this intermediate is favoured in basic medium, as can be seen in Figure 9.15. Therefore, the main conclusion

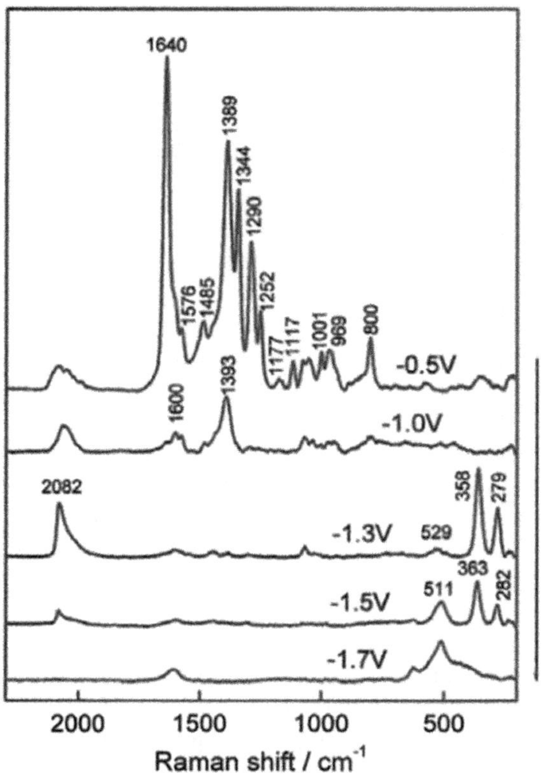

Figure 9.14 SERS spectra as a function of potential collected on a Cu electrode in a CO_2-saturated 0.1 M K_2SO_4 solution, pH ≈ 6.7. CO_2 was bubbled into the solution at 0 V *vs.* Ag/AgCl, a step to −0.5 V was applied, and then the subsequent potential steps followed.
Reproduced from *Journal of Electroanalytical Chemistry*, **629**, E. A. Batista and M. L. A. Temperini, Spectroscopic evidences of the presence of hydrogenated species on the surface of copper during CO_2 electroreduction at low cathodic potentials, 158–163,[49] Copyright 2009, with permission from Elsevier.

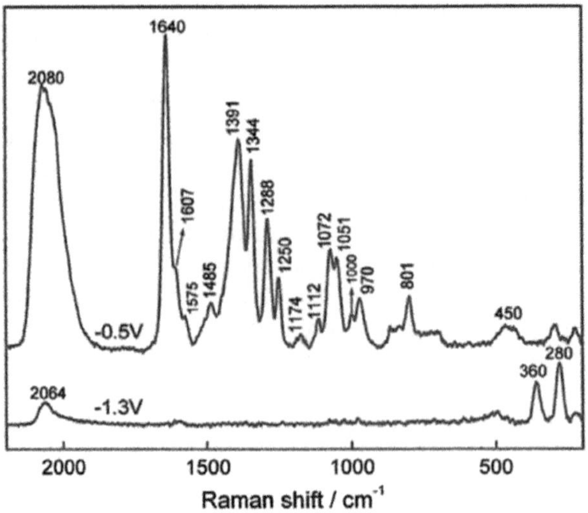

Figure 9.15 SERS spectra of a Cu electrode collected at −0.5 and −1.3 V in a CO_2-saturated 0.1 M K_2SO_4 solution, pH ~ 9.5.
Reproduced from *Journal of Electroanalytical Chemistry*, **629**, E. A. Batista and M. L. A. Temperini, Spectroscopic evidences of the presence of hydrogenated species on the surface of copper during CO_2 electroreduction at low cathodic potentials, 158–163,[49] Copyright 2009, with permission from Elsevier.

suggested is a change of the operating mechanism during the CO_2 reduction to hydrocarbon species formation with the pH of the media, since adsorbed CO is only detected in alkaline media.

As was mentioned above (Section 9.3.1), the group of Ito[83] studied the electrochemical reduction of CO_2 in polycrystalline Cu and Ag by FTIR and SERS. Both spectro-electrochemical techniques showed the formation of linearly adsorbed CO on Cu and Ag electrodes. CO_3^{2-} species were also observed at a Cu electrode.

Recently, Santos *et al.*[99] reported the formation of different types of adsorbed CO during CO_2 electroreduction on a Cu electrode in 1-*n*-butyl-3methyl imidazolium tetrafluoroborate (BMI · BF_4) by SERS. Interestingly, the presence of a thin film of Cu_2O anticipated the reduction of CO_2 to CO, in other words a lower overpotential was required for the formation of this compound. Besides the adsorbed CO, other species were also detected by SERS, including the BMI carbene and BMI-CO_2 adduct.

On the other hand, the group of Dryfe performed an extensive study of the cation influence on the electroreduction of CO_2 on Au and Pt electrodes.[97] Tetrabutylammonium tetrafluoroborate ($TBABF_4$) and its alkali metals (lithium, sodium and rubidium) analogues, $LiBF_4$, $NaBF_4$ and $RbBF_4$ were used as electrolytes in *N*-methylpyrrolidone solution. The main conclusion indicates that products from the electroreduction of CO_2 are formed at lower overpotentials using metallic rather than non-metallic cations. However, the resultant alteration of the surface environment was found to deactivate the electrode to further CO_2 transformation.

Remarkable is the research work of Kenis, Gewirth and coworkers,[98,100] on the electrochemical reduction of CO_2 on Ag and Au electrodes using 1 M KOH solution as electrolyte with N-containing additives such as benzotriazole or ethanolamin. The SERS spectra recorded during CO_2 reduction on Au only exhibited a peak associated with CO when N-containing additives were present in the electrolyte. These additives did not affect the CO_2 reduction activity or product distribution of the Au electrodes. In contrast, on Ag electrodes the N-containing additives led to enhanced CO production. They explained this behaviour by a different stabilization degree of the adsorbed CO-species depending on the active metal. On Ag, CO formed was only weakly adsorbed on the active surface and that weaker adsorption was encouraged by the addition of an additive. Thus, on Ag, the role of the additive is to further destabilize the adsorbed CO, to be easily removed and to enhance the CO_2 reduction activity to volatile CO. On the other hand, CO adsorption on Au is much weaker than that on Ag (*ca.* 0.28 eV *vs. ca.* 0.40 eV, respectively), and consequently, the CO product is already destabilized and the additives have no effect on CO_2 reduction.

9.3.3 *In Situ* UV–vis Spectroscopy

UV–vis spectro-electrochemistry is a suitable tool to study electron transfer reaction pathways and fundamental molecular states at the interface. One of the main advantages includes the high amount of "invisible" electrolytes in the UV–vis wavelength range. However, *in situ* UV–vis spectroscopy is less popular than other spectro-electrochemical methods, such as FTIRS and Raman spectroscopy and, hence, there are more difficulties with the equipment and background. Another important limitation in this methodology is that spectra features are broad and observing changes during the electrochemical reactions is quite tricky, since they may derive from different sources, such as modifications at the valence band of metal ions, and dispersion of supported catalysts, reactants, products and reaction intermediates.

In situ UV–vis spectroscopy detects absorbing species in this frequency region and therefore is able to identify reaction intermediates, reactants and products. Similar to IR spectroscopy, UV–vis measurements can be performed by reflectance or transmittance along the time under potential control or by time-resolved spectra. More extensive information about this technique can be found elsewhere.[19,65,66,101–103] Briefly, the beginning of *in situ* UV–vis studies was based on modulated specular reflectance spectroscopy, which involved external reflection from massive electrodes and potential modulation of the working electrode.[104] Further improvement of UV–vis measurements was achieved by the application of diode array detectors and consequently a large number of spectra can be recorded, minimizing the measurement time and improving the signal quality.[105,106] In

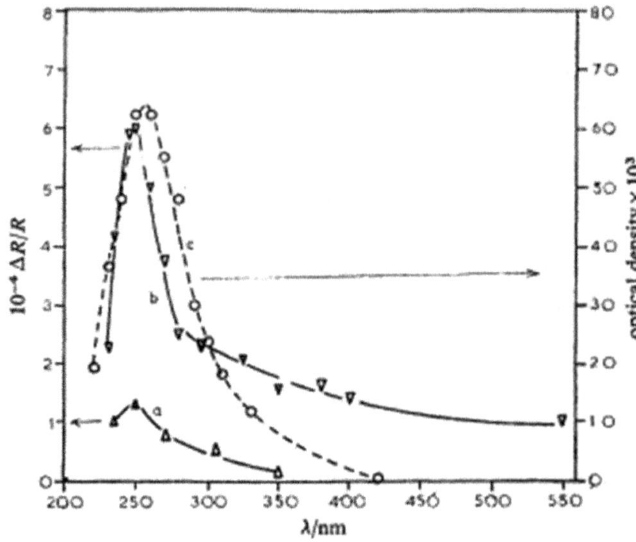

Figure 9.16 Spectral dependence of the optical response on pulsing a lead cathode in aqueous 0.1 M tetramethylammonium chloride solution with CO_2 from (a) −1.0 to −1.6 V, (b) −1.0 to −1.8 V, at 30 Hz. Curve (c) is a plot of the spectrum for $CO_2^{\bullet-}$.
Reproduced from ref. 104 with permission from The Royal Society of Chemistry.

the last few decades the technique has been improved due to the advances in electronic, optics and the use of "invisible" materials such as indium tin oxide (ITO) electrodes, in which the internal configuration can be adopted. *In situ* UV–vis spectro-electroscopy has been also employed to detect and identify adsorbed and soluble reaction intermediates generated during the reduction of CO_2.[104,107–110]

In this context, Aylmer-Kelly *et al.* suggested the presence of $CO_2^{\bullet-}$ as a reaction intermediate during the CO_2 electroreduction on Pb in aqueous and non-aqueous solutions.[104] Figure 9.16 shows the results obtained in aqueous 0.1 M tetramethylammonium chloride solution in the presence of CO_2 at a Pb cathode after stepping the potential from −1.0 to −1.6 V (curve a) or to −1.8 V (curve b). As can be seen, both curves are quite similar to the spectrum of $CO_2^{\bullet-}$ (curve c), presenting a single peak at 250 nm.

An evident red shift of the band associated with the formation of the $CO_2^{\bullet-}$ intermediate was found in a non-aqueous media, from 250 nm in water (hydrogen bonding solvent) to 285 nm in propylene carbonate (non-hydrogen bonding solvent), as can be seen in Figure 9.17. In addition, a second peak at longer wavelengths appeared in non-aqueous media, indicative of the presence of another intermediate, $(CO_2)_2^{\bullet-}$, formed from the reaction of CO_2 and $CO_2^{\bullet-}$ radical anion, according to the eqn (9.15).

$$CH_2 + CO_2^{\bullet-} \rightarrow (CH_2)_2^{\bullet-}. \quad (9.15)$$

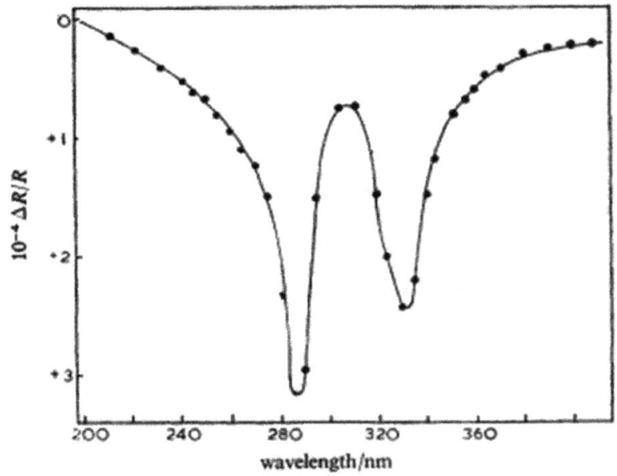

Figure 9.17 Spectral dependence of the optical response on pulsing a lead cathode in 0.4 M tetramethylammonium perchlorate in propylene carbonate saturated with CO_2 from 0.1 to 0.2 V at 30 Hz. Reference electrode: Li/0.5 M Li^+.
Reproduced from ref. 104 with permission from The Royal Society of Chemistry.

On the other hand, molecular catalysts have been widely characterized with this technique since they develop several features in the UV–vis region, which makes it unique for performing kinetic and mechanistic studies.[107–110] For example, Isaacs *et al.*[107] studied the electrochemical reduction of CO_2 on a modified glassy carbon by polymeric M-aminophthalocyanines (M = Co, Ni, Fe) in an aqueous electrolyte. They observed the great importance of the central ion, which strongly influences the CO_2 reduction reaction, *e.g.*, the Co-based electrode needs higher overpotentials than the Ni-based catalyst to overcome the reaction. The reaction products were analyzed by gas chromatography and a different distribution was found depending on the electrode: formic acid was the main product at the Co-based electrode, while formic acid and formaldehyde were formed at the Ni-polymer material. Finally, formaldehyde and H_2 were mostly detected at the Fe-based catalyst. In addition to phtalocyanines, porphyrins and pyridine-based catalysts are usually employed to study the CO_2 reduction reaction.[108–110] In this context, Abruña and coworkers suggested the formation of a *sesqui*-coordinated bipyridine intermediate during the cathodic CO_2 reduction on a rhenium bipyridine carbonyl electrode.[108,109]

9.4 Summary

In this chapter, a review of the most promising results for the electrochemical reduction of CO_2 using *in situ* spectroscopy and spectrometry techniques has been addressed. Several catalysts, including polycrystalline and single crystalline electrodes, of diverse metals such as Cu, Pt, Pd, Au, Ni and Pb have been used for the CO_2 reduction reaction by *in situ* techniques. Among them, metals with low overvoltage toward the hydrogen evolution reaction (such as Pt and Pd) and Cu-based electrodes have attracted special attention due to their affinity for adsorbing hydrogen or their high production rate to the formation of hydrocarbons and alcohols, respectively. The spectro-electrochemical characterization of supported-catalysts, mesoporous active materials and molecular catalysts has been also successfully achieved over the last few decades.

High-added-value products, such as acid formic, methanol, methane, carbon monoxide and ethane, which may be used as raw materials in other chemical processes, or as fuels, have been detected by means of the *in situ* techniques. From these results, different reaction pathways have been suggested in order to further understand the mechanism of this complex process.

Despite these results, *in situ* techniques have not been extensively used as these methods are not commercial and some of them have recently been introduced into the market and, hence, set-ups have to be adapted to the required application. One of the main disadvantages of the process is the strong hydrogen formation by water reduction, which occurs as a competitive reaction in aqueous electrolytes. The huge amount of gaseous hydrogen facilitates the depletion of the active phase, and consequently the configuration has to be carefully designed to solve this issue. On the other hand, the application of the different techniques will depend on the electrode involved in the process or the operating conditions. For example, DEMS appears as a powerful tool for the identification of gaseous and volatile products/intermediates of the reaction, while the detection of adsorbed species can be studied by *in situ* spectroscopy techniques. On the other hand, Raman can be easily employed to study electrochemical aqueous systems due to the weak scattering of water, while aqueous and permanent polar electrolytes are strongly IR active, leading to significant background interference. Thus, *in situ* techniques may offer complementary information.

In summary, *in situ* spectroscopy and spectrometry techniques are robust methods for research on the electrochemical reduction of CO_2. However, further efforts are necessary for gaining a deeper knowledge of the reaction mechanism and the intermediates.

Acknowledgements

The authors gratefully acknowledge financial support given by Spanish MINECO (*CTQ2011-28913-C02-01 and 02*). GG acknowledges the Viera y Clavijo program (*ACIISI & ULL*) for financial support.

References

1. W. Paik, T. N. Andersen and H. Eyring, *Electrochim. Acta*, 1969, **14**, 1217–1232.

2. J. Jordan and P. T. Smith, Free-radical intermediate in the electroreduction of carbon dioxide, in *Proceedings of the Chemical Society*, 1960, pp. 246–247.
3. Y. B. Vassiliev, V. S. Bagotsky, N. V. Osetrova, O. A. Khazova and N. A. Mayorova, *J. Electroanal. Chem. Interfacial Electrochem.*, 1985, **189**, 271–294.
4. Y. B. Vassiliev, V. S. Bagotzky, N. V. Osetrova and A. A. Mikhailova, *J. Electroanal. Chem.*, 1985, **189**, 311–324.
5. H. De Jesús-Cardona, C. Del Moral and C. R. Cabrera, *J. Electroanal. Chem.*, 2001, **513**, 45–51.
6. Y. Hori, H. Wakebe, T. Tsukamoto and O. Koga, *Electrochim. Acta*, 1994, **39**, 1833–1839.
7. Y. Hori, K. Kikuchi and S. Suzuki, *Chem. Lett.*, 1985, 1695–1698.
8. H. Noda, S. Ikeda, Y. Oda, K. Imai, M. Maeda and K. Ito, *Bull. Chem. Soc. Jpn.*, 1990, **63**, 2459–2462.
9. M. Azuma, K. Hashimoto, M. Hiramoto, M. Watanabe and T. Sakata, *J. Electrochem. Soc.*, 1990, **137**, 1772–1778.
10. K. Ohta, M. Kawamoto, T. Mizuno and D. A. Lowy, *J. Appl. Electrochem.*, 1998, **28**, 717–724.
11. S. Kaneco, N. H. Hiei, Y. Xing, H. Katsumata, H. Ohnishi, T. Suzuki and K. Ohta, *Electrochim. Acta*, 2002, **48**, 51–55.
12. S. Kaneco, N. H. Hiei, Y. Xing, H. Katsumata, H. Ohnishi, T. Suzuki and K. Ohta, *J. Solid State Electrochem.*, 2003, **7**, 152–156.
13. M. Azuma, K. Hashimoto, M. Hiramoto, M. Watanabe and T. Sakata, *J. Electroanal. Chem.*, 1989, **260**, 441–445.
14. K. Hara, A. Kudo and T. Sakata, *J. Electroanal. Chem.*, 1995, **391**, 141–147.
15. S. Nakagawa, A. Kudo, M. Azuma and T. Sakata, *J. Electroanal. Chem.*, 1991, **308**, 339–343.
16. C. Cremers and D. Bayer, Differential electrochemical mass spectrometry (DEMS) technique for direct alcohol fuel cell characterization, in *Polymer Electrolyte Membrane and Direct Methanol Fuel Cell Technology*, ed. C. Hartnig and C. Roth, Woodhead Publishing, 2012, pp. 65–86.
17. H. Baltruschat, *J. Am. Soc. Mass Spectrom*, 2004, **15**, 1693–1706.
18. H. Wang, E. Rus and H. D. Abruña, *Anal. Chem.*, 2010, **82**, 4319–4324.
19. A. J. Bard and L. R. Faulkner, *Electrochemical Methods: Fundamentals and Applications*, Wiley, New York, 2000.
20. D. Kolbe and W. Vielstich, *Electrochim. Acta*, 1996, **41**, 2457–2460.
21. G. M. Brisard, A. P. M. Camargo, F. C. Nart and T. Iwasita, *Electrochem. Commun.*, 2001, **3**, 603–607.
22. M. E. Martins, Electrochemical behaviour of carbon dioxide on platinum and rhodium, in *Anales de la Asociación Química Argentina*, 2005.
23. S. Pérez-Rodríguez, F. Barreras, E. Pastor and M. J. Lazaro, *Int. J. Hydrogen Energy*, 2016, **41**, 19756–19765.
24. S. Pérez-Rodríguez, G. García, L. Calvillo, V. Celorrio, E. Pastor and M. J. Lázaro, *Int. J. Electrochem.*, 2011, **2011**, 249804.

25. S. Pérez-Rodríguez, D. Sebastián, M. J. Lazaro and E. Pastor, *J. Catal.*, 2017, **355**, 156–166.
26. S. Pérez-Rodríguez, N. Rillo, M. J. Lázaro and E. Pastor, *Appl. Catal., B*, 2015, **163**, 83–95.
27. D. Plana, J. Florez-Montano, V. Celorrio, E. Pastor and D. J. Fermin, *Chem. Commun.*, 2013, **49**, 10962–10964.
28. P. Dubé and G. M. Brisard, *J. Electroanal. Chem.*, 2005, **582**, 230–240.
29. S. Wasmus, E. Cattaneo and W. Vielstich, *Electrochim. Acta*, 1990, **35**, 771–775.
30. T. Smolinka, M. Heinen, Y. X. Chen, Z. Jusys, W. Lehnert and R. J. Behm, *Electrochim. Acta*, 2005, **50**, 5189–5199.
31. J. Shen, R. Kortlever, R. Kas, Y. Y. Birdja, O. Diaz-Morales, Y. Kwon, I. Ledezma-Yanez, K. J. P. Schouten, G. Mul and M. T. M. Koper, *Nat. Commun.*, 2015, **6**, 8177.
32. J. Monzo, Y. Malewski, R. Kortlever, F. J. Vidal-Iglesias, J. Solla-Gullon, M. T. M. Koper and P. Rodriguez, *J. Mater. Chem. A*, 2015, **3**, 23690–23698.
33. R. Reske, M. Duca, M. Oezaslan, K. J. P. Schouten, M. T. M. Koper and P. Strasser, *J. Phys. Chem. Lett.*, 2013, **4**, 2410–2413.
34. R. Kortlever, K. H. Tan, Y. Kwon and M. T. M. Koper, *J. Solid State Electrochem.*, 2013, **17**, 1843–1849.
35. R. Kas, R. Kortlever, A. Milbrat, M. T. M. Koper, G. Mul and J. Baltrusaitis, *Phys. Chem. Chem. Phys.*, 2014, **16**, 12194–12201.
36. R. Kortlever, C. Balemans, Y. Kwon and M. T. M. Koper, *Catal. Today*, 2015, **244**, 58–62.
37. S. Pérez-Rodríguez, CO_2 valorization by electrochemical reduction using catalysts supported on nanostructured carbon materials, Thesis, University of Zaragoza, 2015.
38. A. Rodes, E. Pastor and T. Iwasita, *J. Electroanal. Chem.*, 1994, **369**, 183–191.
39. A. Rodes, E. Pastor and T. Iwasita, *J. Electroanal. Chem.*, 1994, **377**, 215–225.
40. A. Rodes, E. Pastor and T. Iwasita, *J. Electroanal. Chem.*, 1994, **373**, 167–175.
41. S. Taguchi, A. Aramata and M. Enyo, *J. Electroanal. Chem.*, 1994, **372**, 161–169.
42. B. Beden, A. Bewick, M. Razaq and J. Weber, *J. Electroanal. Chem.*, 1982, **139**, 203–206.
43. J. Giner, *Electrochim. Acta*, 1963, **8**, 857–865.
44. O. Guillén-Villafuerte, G. Garcia, M. C. Arevalo, J. L. Rodriguez and E. Pastor, *Electrochem. Commun.*, 2016, **63**, 48–51.
45. M. Gattrell, N. Gupta and A. Co, *J. Electroanal. Chem.*, 2006, **594**, 1–19.
46. S. Kaneco, K. Iiba, H. Katsumata, T. Suzuki and K. Ohta, *Electrochim. Acta*, 2006, **51**, 4880–4885.
47. S. Kaneco, K. Iiba, K. Ohta and T. Mizuno, *J. Solid State Electrochem.*, 1999, **3**, 424–428.

48. S. Kaneco, Y. Ueno, H. Katsumata, T. Suzuki and K. Ohta, *Chem. Eng. J.*, 2006, **119**, 107–112.
49. E. A. Batista and M. L. A. Temperini, *J. Electroanal. Chem.*, 2009, **629**, 158–163.
50. Y. Hori, H. Konishi, T. Futamura, A. Murata, O. Koga, H. Sakurai and K. Oguma, *Electrochim. Acta*, 2005, **50**, 5354–5369.
51. B. Jermann and J. Augustynski, *Electrochim. Acta*, 1994, **39**, 1891–1896.
52. R. Kortlever, J. Shen, K. J. P. Schouten, F. Calle-Vallejo and M. T. M. Koper, *J. Phys. Chem. Lett.*, 2015, **6**, 4073–4082.
53. D. Tegtmeyer, J. Heitbaum and A. Heindrichs, *Ber. Bunsenges. Phys. Chem.*, 1989, **93**, 201–206.
54. S. Pérez-Rodríguez, M. Corengia, G. García, C. F. Zinola, M. J. Lázaro and E. Pastor, *Int. J. Hydrogen Energy*, 2012, **37**, 7141–7151.
55. C. Korzeniewski, IR Spectroelectrochemistry. Instrumentation and Applications of External Reflection, ATR, and Transmission Sampling, in *In-situ Spectroscopic Studies of Adsorption at the Electrode andElectrocatalysis*, ed. S.-G. Sun, P. A. Christensen and A. Wieckowski, Elsevier, Amsterdam, 2007, pp. 179–208.
56. K. Ashley and S. Pons, *Trends Anal. Chem.*, 1985, **4**, 142–145.
57. K. Ashley, *Talanta*, 1991, **38**, 1209–1218.
58. K. Ashley and S. Pons, *Chem. Rev.*, 1988, **88**, 673–695.
59. C. Korzeniewski and M. W. Severson, *Spectrochim. Acta, Part A*, 1995, **51**, 499–518.
60. T. Davidson, B. S. Pons, A. Bewick and P. P. Schmidt, *J. Electroanal. Chem.*, 1981, **125**, 237–241.
61. S. Pons, *J. Electron Spectrosc. Relat. Phenom.*, 1987, **45**, 303–322.
62. S. Pons, T. Davidson and A. Bewick, *J. Electroanal. Chem.*, 1984, **160**, 63–71.
63. S. Pons, T. Davidson and A. Bewick, *J. Am. Chem. Soc.*, 1983, **105**, 1802–1805.
64. S. Pons, T. Davidson and A. Bewick, *J. Electroanal. Chem.*, 1982, **140**, 211–216.
65. T. E. Keyes and R. J. Forster, Spectroelectrochemistry, in *Handbook of Electrochemistry*, Elsevier, Amsterdam, 2007, pp. 591–635.
66. J. A. Crayston, Spectroelectrochemistry, in *Comprehensive Coordination Chemistry II*, Elsevier, Amsterdam, 2004, pp. 775–789.
67. S. M. Stole, Infrared spectroelectrochemistry: A probe of the molecular architecture of the electrochemical interface, in *Electrochemical Interfaces: Modern Techniques for In Situ Interface Characterization*, ed. H. D. Abruña, VCH, New York, 1991, pp. 339–410.
68. A. Bewick and K. Kunimatsu, *Surf. Sci.*, 1980, **101**, 131–138.
69. A. Bewick, K. Kunimatsu and B. S. Pons, *Electrochim. Acta*, 1980, **25**, 465–468.
70. B. Innocent, D. Pasquier, F. Ropital, F. Hahn, J. M. Léger and K. B. Kokoh, *Appl. Catal., B*, 2010, **94**, 219–224.
71. B. Eneau-Innocent, D. Pasquier, F. Ropital, J. M. Léger and K. B. Kokoh, *Appl. Catal., B*, 2010, **98**, 65–71.

72. P. A. Christensen, A. Hamnett, A. V. G. Muir and N. A. Freeman, *J. Electroanal. Chem.*, 1990, **288**, 197–215.
73. J. Desilvestro and S. Pons, *J. Electroanal. Chem. Interfacial Electrochem.*, 1989, **267**, 207–220.
74. E. R. Pérez, J. R. Garcia, D. R. Cardoso, B. R. McGarvey, E. A. Batista, U. P. Rodrigues-Filho, W. Vielstich and D. W. Franco, *J. Electroanal. Chem.*, 2005, **578**, 87–94.
75. M. C. Arévalo, C. Gomis-Bas, F. Hahn, B. Beden, A. Arévalo and A. J. Arvia, *Electrochim. Acta*, 1994, **39**, 793–799.
76. Y. Hori, O. Koga, Y. Watanabe and T. Matsuo, *Electrochim. Acta*, 1998, **44**, 1389–1395.
77. O. Koga, S. Teruya, K. Matsuda, M. Minami, N. Hoshi and Y. Hori, *Electrochim. Acta*, 2005, **50**, 2475–2485.
78. O. Koga, Y. Watanabe, M. Tanizaki and Y. Hori, *Electrochim. Acta*, 2001, **46**, 3083–3090.
79. Y. Hori, O. Koga, H. Yamazaki and T. Matsuo, *Electrochim. Acta*, 1995, **40**, 2617–2622.
80. R. Ortiz, O. P. Márquez, J. Márquez and C. Gutiérrez, *J. Electroanal. Chem.*, 1995, **390**, 99–107.
81. B. Aurian-Blajeni, M. Ahsan Habib, I. Taniguchi and J. O. M. Bockris, *J. Electroanal. Chem. Interfacial Electrochem.*, 1983, **157**, 399–404.
82. K. Chandrasekaran and J. O. M. Bockris, *Surf. Sci.*, 1987, **185**, 495–514.
83. I. Oda, H. Ogasawara and M. Ito, *Langmuir*, 1996, **12**, 1094–1097.
84. P. A. Christensen, A. Hamnett and A. V. G. Muir, *J. Electroanal. Chem. Interfacial Electrochem.*, 1988, **241**, 361–371.
85. M. Nakayama, M. Iino and K. Ogura, *J. Electroanal. Chem.*, 1997, **440**, 251–257.
86. A. V. Rudnev, U. E. Zhumaev, A. Kuzume, S. Vesztergom, J. Furrer, P. Broekmann and T. Wandlowski, *Electrochim. Acta*, 2016, **189**, 38–44.
87. D. L. Jeanmaire, M. R. Suchanski and R. P. Van Duyne, *J. Am. Chem. Soc.*, 1975, **97**, 1699–1707.
88. D. L. Jeanmaire and R. P. Van Duyne, *J. Am. Chem. Soc.*, 1976, **98**, 4034–4039.
89. D. L. Jeanmaire and R. P. Van Duyne, *J. Am. Chem. Soc.*, 1976, **98**, 4029–4033.
90. D. L. Jeanmaire and R. P. Van Duyne, *J. Electroanal. Chem.*, 1975, **66**, 235–247.
91. M. R. Suchanski and R. P. Van Duyne, *J. Am. Chem. Soc.*, 1976, **98**, 250–252.
92. R. P. Van Duyne, M. R. Suchanski, J. M. Lakovits, A. R. Siedle, K. D. Parks and T. M. Cotton, *J. Am. Chem. Soc.*, 1979, **101**, 2832–2837.
93. M. Fleischmann, P. J. Hendra and A. J. McQuillan, *Chem. Commun.*, 1973, 80–81.
94. J. E. Pemberton, Surface Enhanced Raman, in *Electrochemical Interfaces: Modern Techniques for In Situ Interface Characterization*, ed. H. D. Abruña, VCH, New York, 1991, pp. 193–263.

95. Z.-Q. Tian and B. Ren, Raman spectroscopy of electrode surfaces, in *Encyclopedia of Electrochemistry*, Wiley-VCH Verlag GmbH & Co. KGaA, New York, 2007.
96. A. J. McQuillan, P. J. Hendra and M. Fleischmann, *J. Electroanal. Chem.*, 1975, **65**, 933–944.
97. B. M. Setterfield-Price and R. A. W. Dryfe, *J. Electroanal. Chem.*, 2014, **730**, 48–58.
98. J. L. Oberst, H. R. Jhong, P. J. A. Kenis and A. A. Gewirth, *J. Solid State Electrochem.*, 2016, **20**, 1149–1154.
99. V. O. Santos, I. R. Leite, A. G. Brolo and J. C. Rubim, *J. Raman Spectrosc.*, 2016, **47**, 674–680.
100. C. E. Tornow, M. R. Thorson, S. Ma, A. A. Gewirth and P. J. A. Kenis, *J. Am. Chem. Soc.*, 2012, **134**, 19520–19523.
101. A. Petr, L. Dunsch and A. Neudeck, *J. Electroanal. Chem.*, 1996, **412**, 153–158.
102. R. J. Mortimer, Spectroelectrochemistry, Methods and Instrumentation, in *Encyclopedia of Spectroscopy and Spectrometry*, ed. J. C. Lindon, Academic Press, Oxford, 2nd edn, 1999, pp. 2633–2639.
103. A. Neudeck, F. Marken, R. G. Compton, UV/Vis/NIR spectroelectrochemistry, in *Electroanalytical Methods: Guide to Experiments and Applications*, Springer, Berlin, 2010, pp. 179–200.
104. A. W. B. Aylmer-Kelly, A. Bewick, P. R. Cantrill and A. M. Tuxford, *Faraday Discuss.*, 1973, **56**, 96–107.
105. A. Neudeck and L. Dunsch, *Electrochim. Acta*, 1995, **40**, 1427–1434.
106. A. Neudeck and L. Dunsch, *J. Electroanal. Chem.*, 1995, **386**, 135–148.
107. M. Isaacs, F. Armijo, G. Ramírez, E. Trollund, S. R. Biaggio, J. Costamagna and M. J. Aguirre, *J. Mol. Catal. A: Chem.*, 2005, **229**, 249–257.
108. A. I. Breikss and H. D. Abruña, *J. Electroanal. Chem.*, 1986, **201**, 347–358.
109. C. R. Cabrera and H. D. Abruña, *J. Electroanal. Chem.*, 1986, **209**, 101–107.
110. K. A. Grice and C. P. Kubiak, Recent studies of rhenium and manganese bipyridine carbonyl catalysts for the electrochemical reduction of CO_2, in *Advances in Inorganic Chemistry*, ed. A. Michele and R. van Elsen, Academic Press, New York, 2014, pp. 163–188.

CHAPTER 10

Surface-selective and Time-resolved Spectro-electrochemical Studies of CO_2 Reduction Mechanisms

ALEXANDER J. COWAN

Department of Chemistry and the Stephenson Institute for Renewable Energy, University of Liverpool, Liverpool, UK
Email: A.J.Cowan@Liverpool.ac.uk

10.1 Introduction to Detecting Short-lived Intermediates at Surfaces and in Solution

Interrogation of intermediates, electrode surfaces and of the wider electric double layer (EDL) *in situ* is a non-trivial but essential task if the mechanisms of molecular catalysed and metal electrode mediated electrochemical CO_2 reduction are to be elucidated. Surface species are important—although the detection of electrochemically generated intermediates in the bulk can provide insight into chemical mechanism it does not inform the experimentalist on the balance of interactions between the electrode, molecular intermediate, solvent and electrolyte ions that are key in determining activity. In Chapter 9 the use of Raman and IR spectroscopy to study electrode interfaces is covered and several of the studies discussed highlight that advances in conventional FTIR instrumentation now makes it possible to identify even sub-monolayer concentrations of species through

Energy and Environment Series No. 21
Electrochemical Reduction of Carbon Dioxide: Overcoming the Limitations of Photosynthesis
Edited by Frank Marken and David Fermin

Published by the Royal Society of Chemistry, www.rsc.org

standard IR absorption-reflectivity approaches (IRAS). Here we build upon this section and explore specialist techniques that are particularly relevant to either the detection of short-lived or low concentration surface intermediates. Whilst IRAS and in particular SNIFTIRS (subtractively normalized interfacial Fourier transform IR spectroscopy) experiments represent a powerful tool[1–3] they often require hundreds or even of thousands of individual spectra to be averaged to obtain sufficient levels of sensitivity leading to data acquisition times of minutes to hours which in many cases is incompatible with short short-lived surface intermediates and the dynamics of the EDL.[4] The need for extensive averaging arises from the difficulties in detecting and discriminating small concentrations of catalytically relevant species at the surface from species in the bulk which may have similar spectral features to those of interest. Therefore we focus on approaches where surface selectivity is achieved either through intrinsic selection rules or through surface enhanced phenomena greatly decreasing data acquisition times, making time-resolved experiments viable.

10.2 Surface Enhanced Raman Spectroscopy

Raman scattering arises due to the excitation of a molecule to a virtual (or real in the case of resonance Raman) excited state, which is followed by its subsequent relaxation back to a different vibrational state in the ground electronic state.[5] Typically scattered photons are of lower energy than the incident probe light due to Stokes scattering that corresponds to molecules being excited from the ground to first vibrational state levels ($\Delta\nu = +1$), with anti-Stokes ($\Delta\nu = -1$) being far weaker. As Raman spectroscopy typically uses a visible or nIR laser as the excitation source, it has allowed the acquisition of spectral data from electrode surfaces even through thick electrolyte layers, greatly simplifying the design of spectro-electrochemical cells. However, it can be challenging to obtain sufficient sensitivity; it is a weak effect with $\sim$1 in 10^7 typically being Raman scattered. Fortunately enhancement factors of the order of $>10^5$ have been obtained for molecules on many nanostructured metals, including those commonly used in electrochemical CO_2 reduction (*e.g.* Cu, Ag, Au),[6] which has allowed detection of even single molecules at surfaces.[7] Surface enhanced Raman spectroscopy (SERS) has been used extensively to study CO_2 reduction and is in part been covered elsewhere in this book (see Chapter 9), where examples of specific studies are given. Given the importance of the technique to the field, and to provide context to the other closely related techniques that are discussed in greater detail, here a brief background into the mechanism of SERS is given and we focus on the potential advantages and disadvantages of the technique with respect to the study of electrochemical CO_2 reduction and explore potential future research directions.

The mechanism of enhancement occurring in SERS has been heavily debated in the literature, with at least two mechanisms being proposed,[7–9] and extensive reviews of SERS to study electrochemical processes exist elsewhere.[10,11] Briefly, chemical enhancement of the Raman signal through charge transfer between the molecule and the metal surface may be present, however the electromagnetic enhancement mechanism is thought to be the most signficiant.[8] Here excitation of a localized surface plasmon in the nanostructured metal by an external oscillating field occurs when the frequency of the external light source matches that resonant frequency of the surface plasmon, which for small nanostructures of the coinage metals typically occurs in the visible part of the spectrum. The localized surface plasmon gives rise to a strong local electric field leading to enhancement of the Raman signal. As the electric field strength decays rapidly as the distance from the surface is increased SERS is essentially a surface/EDL specific technique (Figure 10.1).

During electrochemical CO_2 reduction SERS has been used to study a range of electrode materials, including Cu, Ag, and Au, in both aqueous and non-aqueous solvents,[2,12–16] with studies reporting the potential dependent detection of surface adsorbed CO_2, CO and hydrocarbons, which correlated with the electrochemically measured potentials of catalysis. Notably, in some studies SERS allowed the detection of intermediate oxygenated surface species despite them only being a minor product during catalysis, highlighting the sensitivity of the technique.[14] Although to the best of our knowledge SERS has not been applied extensively to the study of molecular electrocatalysts at surfaces it has proved to be a powerful tool for rationalizing the role of additives,[12,13,16] such as ionic liquids and amine additives during heterogeneous electrocatalysis, showing how molecular species can promote CO production at sites of weaker adsorption aiding catalysis.

Although a powerful technique, it has been highlighted that CO_2 reduction activity can depend strongly on the nanostructure with recent

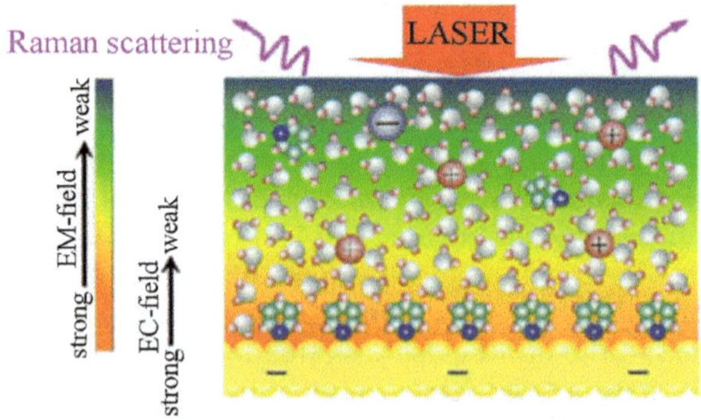

Figure 10.1 Scheme of an electrode surface held negative to the potential of zero charge that also exhibits SERS. The SERS enhancement rapidly decreases as distance increases away from the electrode surface, making SERS a powerful probe of surface and EDL processes and structure. Adapted from ref. 10 with permission from the Royal Society of Chemistry.

examples of greatly modified selectivities and activities on Ag and Au electrodes with angular nanostructures where large electric fields may be formed having high levels of activity, making it feasible that the SERS active substrates may not be an ideal model of more commonly employed electrodes.[17,18] The formation of a strong electric field can also generate hot electrons and cause localized heat rises[7,19] on the SERS substrate, which may also modify (electro-)catalytic activity. An additional consideration is that detailed studies of the SERS mechanism have highlighted that the strongest electric fields arise at a small number of "hot-spots" on the SERS substrate, which typically occur at junctions and necking points between nanoparticles.[8] As the SERS enhancement factor scales roughly to the magnitude of the electric field to the fourth power, Raman signals will be dominated by these sites, making it challenging to obtain a comprehensive picture of the entire electrode area.[20] Many of these disadvantages can be alleviated by the use of tip enhanced Raman spectroscopy (TERS), a technique which uses an AFM tip of a suitable SERS active metal. Here the AFM tip can be scanned over the electrode surface ensuring that the surface is well characterized whilst also providing a high spatial resolution along with the SERS enhancement. To the best of my knowledge only one TERS study relevant to CO_2 reduction has been reported to date,[21] where a cobalt porphyrin molecular electrocatalyst on an Au surface was shown to have reduced ambient CO_2 leading to the formation of a CO adduct. Although performing TERS on electrochemical systems with liquid layers present is challenging, recent experiments have demonstrated the feasibility of such an approach,[11,22,23] making it likely that this will become a particularly fruitful tool for the study of CO_2 reduction mechanisms.

10.3 Surface Enhanced Infrared Absorption Spectroscopy (SEIRAS)

Within a few years of the first reports of SERS, it was shown that significant enhancement factors could also be achieved during IR absorption spectroscopy when molecules are absorbed on thin metallic films.[24] Although SEIRAS enhancement factors are of the order of 10^1–10^2 *versus* equivalent IRAS experiments, well below those seen in SERS (with respect to a non-enhanced Raman experiment),[25] the rapid detection of sub-monolayer concentrations is still possible due to the reasonably high starting sensitivity of IRAS. Electrochemical CO_2 SEIRAS experiments are also particularly amenable to the use an ATR geometry (Figure 10.2b), providing a further advantage over traditional IRAS experiments that utilize a reflection geometry (Figure 10.2a). In standard reflective mode, electrolyte absorption of the IR beam leads to the use of very thin (1–2 μm is common) electrolyte layers leading to high cell-time constants and limitations of mass-transport with subsequent local pH changes during CO_2 reduction.[4] In contrast, evaporation of a thin metallic electrode directly onto a suitable prism and

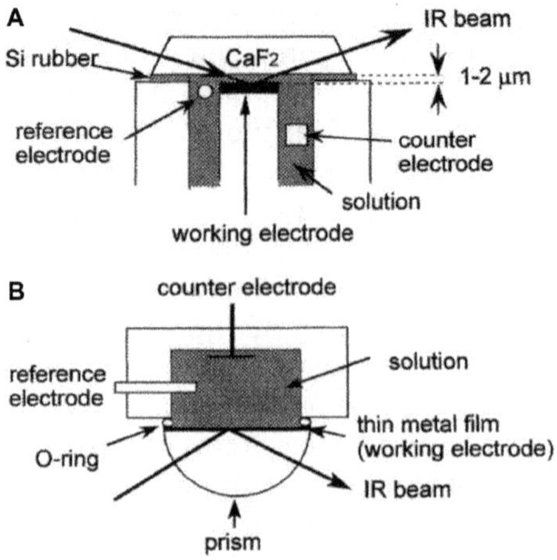

Figure 10.2 Electrochemical cells for IRAS/SEIRAS (a) and SEIRAS using an ATR geometry (b). The use of an ATR geometry minimizes IR absorption by the solvent.
Reproduced with permission from M. Osawa, *Bull. Chem. Soc. Japan.*, 1997, **70**(12), 2861–2880.

subsequent ATR sampling leads to minimal solvent IR losses. This is particularly important for the spectroscopy of CO_2 reduction in aqueous media where the presence of water bending modes (~1600 cm^{-1}) can mask the spectral region where the stretching modes of CO_2 adducts and reduction intermediates (*e.g.* $M–CO_2^{\bullet -}$) would be expected to occur.[26]

The models of the enhancement mechanism of SEIRAS has been developed extensively by Osawa and others and reviewed elsewhere.[4,27–29] Similar to SERS an electromagnetic mechanism is thought to dominate with a potential additional contribution from a charge transfer mechanism, both of which give a strong enhancement to molecules at, or near (<10 nm), the electrode surface.[29] Infrared surface enhancements studies have been reported on a wide range of metals including those used in SERS (Ag, Au, Cu) and many other common metals (*e.g.* Ni, Pd, Pt), which are known to be active for CO_2 reduction. The SEIRAS effect has also been shown to be highly dependent on the parameter used during electrode preparation and active materials are best described not as continuous thin film electrodes, but instead as islands of nanoparticles[4] (Figure 10.3 inset). In contrast to SERS where a laser excitation source of suitable frequency to match the plasmon resonance is used, in SEIRAS the broad absorption of the metal islands across the mid-IR region makes it possible to generate an enhancement, the strength of which is significantly less sensitive to the frequency being probed, leading to less perturbation of band positions.[30,31] Additionally, during SEIRAS experiments the mechanism of enhancement is proposed to be more homogenous than SERS, suggesting that SEIRAS is able to provide a better probe of the aggregate population at an electrode surface.[32]

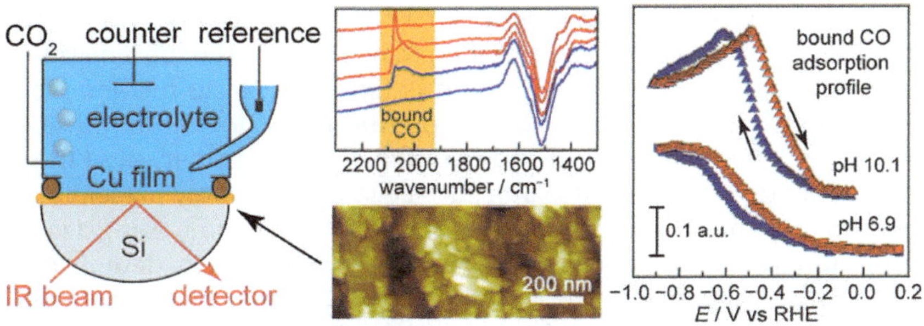

Figure 10.3 SEIRAS studies on nanostructured copper electrodes for CO_2 reduction in water monitor the CO surface population as function of potential and pH, rationalizing the improved activity under alkali conditions. Reprinted with permission from A. Wuttig, C. Liu, Q. Peng, M. Yaguchi, C. H. Hendon, K. Motobayashi, S. Ye, M. Osawa, and Y. Surendranath, *ACS Cent. Sci.*, 2016, **2**, 522–528. Copyright © 2016 American Chemical Society.

To date SEIRAS has been used primarily to study CO_2 reduction at metal electrodes and application to molecular electrocatalysis appears minimal to date. Early works explored CO_2 reduction on Pt electrodes in H_2SO_4, with data showing CO being formed and interacting *via* both linearly bound and multiple bonded configurations.[33] Notably the very high signal:noise ratios that could be achieved within short data acquisition times in these experiments allowed for time-resolved experiments on the seconds timescale, and although none have been reported for CO_2 reduction, time-resolved electrochemical SEIRAS has been reported with microsecond resolution using "step-scan" spectrometers.[34] Recently the mechanism of CO_2 reduction on Au in both organic and aqueous solvents has been reported.[30,35,36] Au is a particularly interesting electrode material due to reported high selectivity for CO production (*vs.* H_2 evolution) even in aqueous solutions.[37] Wuttig and co-workers carried out a SEIRAS study that explored the factors behind the activity and selectivity of Au electrodes and they observed CO produced during CO_2 reduction binding to Au in both a bridged and linear manner, with irreversible CO binding at bridge sites leading to a loss of ~20% of surface area during catalysis.[30] The authors proposed that partial poisoning of the Au electrode was in line with recent reports that catalytic activity may arise from a small fraction of the surface, such as at grain boundaries and step-edges.[38,39] Notably, even under conditions where H_2 evolution occurred, Au–H was not observed and electrokinetic studies suggested that initial proton transfer to the surface followed by electron transfer may be rate limiting during hydrogen evolution limiting activity at higher pH. In contrast, CO_2 reduction is limited by only initial electron transfer making it possible that the ease of proton transfer to the surface may be vital, providing a rationale for the pH dependent selectivity of catalysis. Further SEIRAS studies have since questioned the possible poisoning by CO at bridging sites, suggesting that the low level of Pt deposition on the Au film, arising from the Pt counter electrode, may lead to the observation of irreversibly bound CO.[36] This suggestion is likely to lead to further debate; however, what is most striking about this finding is not the implications for past work[30] but instead the demonstration of the sensitivity of SEIRAS to be able to identify intermediates at a surface site that is only present at low concentrations. In a series of further elegant experiments using isotopic labelling and potential steps Dunwell and co-workers also demonstrated both how CO removal from the Au surface is facilitated by the adsorption of electrolyte cations at negative potentials (Figure 10.4) and how the equilibrium between solvated CO_2 and bicarbonate controls local CO_2 concentration at the electrode surface.[36]

Further studies have also explored the mechanism of electrocatalytic reduction of CO_2 at Cu electrodes[40,41] with adsorbed H and CO both being detected, in contrast to the studies on Au.[30] Significantly H_{ads} is seen to be able to displace CO_{ads} and the concentration of CO_{ads} decreased as H_{ads} increased at the most negative potentials providing a rationale for the loss of

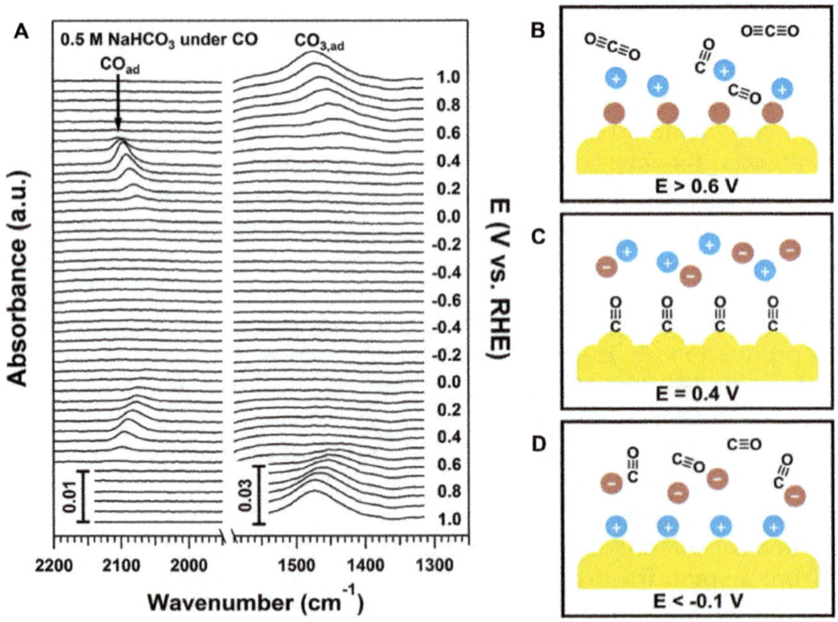

Figure 10.4 *In situ* ATR-SEIRAS studies of surface absorbed CO and carbonate on a gold electrode recorded at 10 mV s^{-1} (a). SEIRAS studies indicated that when the electrode is held at very positive and negative potentials electrolyte anions and cations, respectively, are able to displace surface CO preventing poisoning.
Reprinted with permission from M. Dunwell, Q. Lu, J. M. Heyes, J. Rosen, J. G. G. Chen, Y. Yan, F. Jiao, and B. J. Xu, *J. Am. Chem. Soc.*, 2017, **139**, 3774–3783. Copyright 2017 American Chemical Society.

selectivity towards CO_2 reduction at very negative potentials. Intriguingly, despite Cu being known to be an effective electrocatalyst for the production of hydrocarbon products,[42,43] no evidence for hydrogenated carbon products was observed, despite such species having previously been reported in analogous SERS studies,[14] which may be speculated to suggest that hydrocarbon generation only occurs at localized sites. Currently the number of papers that have reported the use of SEIRAS to study CO_2 reduction is relatively small. It is, however, feasible that many experiments that do not explicitly identify with the classification may also be benefitting from enhancement effects; for example a recent FTIR study on the mechanism of CO_2 reduction on Ag reported intermediates such as Ag–COO^- and Ag–COOH utilized a thin layer of Ag (50 nm), consisting of 10–20 nm crystallites, deposited on a Ge ATR element, with only 64 scans per spectrum required to obtain very high quality data.[44] It is therefore clear that surface enhanced vibrational spectroscopy is increasingly being deployed as a powerful tool for the study of low concentrations of reactive intermediates. We anticipate rapid future growth in this field with the ability to carry out fast time-resolved experiments using commercially available apparatus being an exciting opportunity.

10.4 Sum-frequency Generation Spectroscopy

In the final two sections of this chapter we will examine vibrational spectroscopic techniques that have utilized short pulsed lasers to interrogate either short-lived intermediates of relevance to electrocatalytic CO_2 reduction or to study electrode surfaces in surface specific spectroscopic studies during CO_2 catalysis. In contrast to surface enhanced techniques such as SERS and SEIRAS the use of a spectroscopic method that is intrinsically surface specific to discriminate between bulk and surface species can avoid the need to prepare roughened metal surfaces, which, as discussed above, may not be representative of a typical electrode used during catalysis.

IR–vis sum frequency generation (IR–vis SFG) spectroscopy is a second order non-linear spectroscopy that is intrinsically surface selective, providing the spectra of molecules at interfaces. Using SFG spectroscopy submonolayer sensitivities are routinely reported and polarization control can yield details on the geometry of molecules at a surface.[45,46] The theory of SFG spectroscopy has been covered in detail in several recent reviews and only a brief overview will be provided here.[45,47–49] During an SFG experiment short IR and visible laser pulses with frequencies ω_{vis} and ω_{IR} are overlapped onto the sample of interest and the generated sum frequency photon ($\omega_{SFG} = \omega_{vis} + \omega_{IR}$) is detected (see Figure 10.5). When ω_{IR} matches that of a surface vibrational mode a resonant enhancement of the SFG signal occurs and further enhancements are also possible if ω_{vis} is resonant with the interfacial molecule electronic state. During an experiment the ω_{vis} is usually kept fixed at a single value and ω_{IR} is either tuned across the IR region of interest (if a spectrally narrow (typically <10 cm^{-1}) picosecond (ps) source is

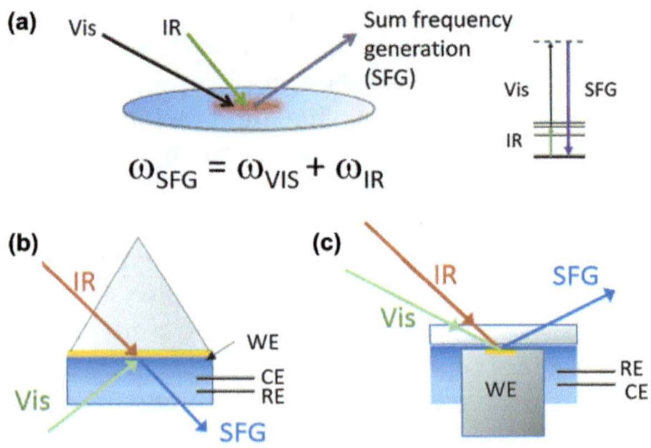

Figure 10.5 Principle of IR–vis SFG spectroscopy (a), commonly spectro-electrochemical cells for SFG spectroscopy, using (b) an ATR and (c) reflective geometries. Using an Au electrode a plasmon resonance enhancement has been reported.[52] WE, CE, RE refers to the working, counter and reference electrode respectively.

employed) or a very short broadband femtosecond IR pulse (when bandwidths of 400–500 cm^{-1} are commonly reported), is employed.[45,50] The selectivity of SFG to surface/interfacial species arises as due to interference effects the light from the centro-symmetric bulk media is completely extinguished, leaving the readily detectable sum frequency signal from the surface, which occurs in the visible, enabling high detection sensitivities. This selectivity is sometimes discussed within the dipole approximation—for a mode to be SFG active it must be both Raman and IR active, ruling out materials with a centre of symmetry.

Rey and Dlott have recently highlighted the benefits of SFG for the study of buried electrochemical interfaces and highlighted that within the experimental model of Figure 10.6 the SFG signal will nearly entirely arise from species within the EDL, with only a minimal contribution from the top-window electrolyte interface.[50] Although the SFG signal is generated in the visible the presence of IR losses from the bulk electrolyte are still an important consideration, and in some cases they can lead to well defined, potential "phantom transitions" in the SFG spectrum that do not actually arise from the interfacial species.[50,51] IR losses can however be overcome in many studies by the use of an ATR sampling geometry (see Figure 10.5), which has been shown to also benefit from surface plasmon enhancement of the SFG signal when noble metal electrodes are employed.[52]

In an important study in 2011 it was reported that CO_2 reduction occurred with a very low overpotential on Ag electrodes in mixed water/1-ethyl-3-methylimidazolium tetrafluoroborate (EMIM-BF_4) electrolytes, with stabilization of the initially reduced CO_2 by $EMIM^+$ being proposed to occur.[53] This initial result has since been followed up by dozens of further reports

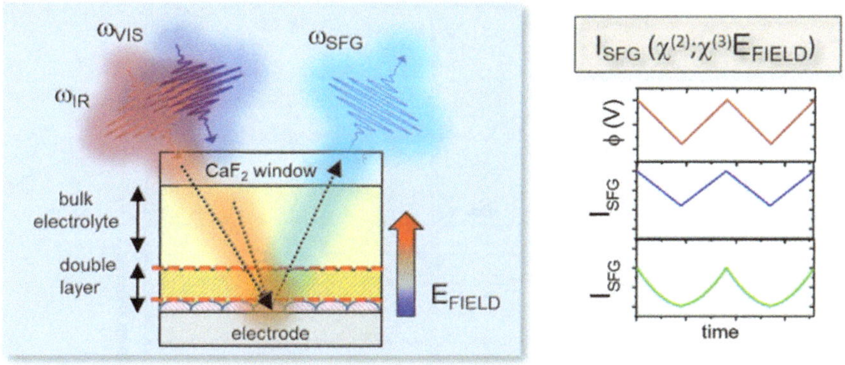

Figure 10.6 IR–vis SFG spectroscopy can directly address the chemistry of the electrode surface through analysis of the potential dependence of vibrational frequencies and the EDL through analysis, particularly of the non-resonant (NR) response, of the potential dependence of the SFG intensity due to third order effects.
Reprinted from *Electroanal. Chem.*, **800**, N. G. Rey and D. D. Dlott, Studies of electrochemical interfaces by broadband sum frequency generation, 114–125, Copyright 2017, with permission from Elsevier.

identifying ionic liquid mediated CO_2 reduction and the mechanism of such processes have now been studied by SFG spectroscopy by the group of Dlott.[54–57] We believe that the studies on EMIM-BF_4 represent the first to explore a real, complex, CO_2 reduction system *in situ* by SFG spectroscopy. On polycrystalline platinum electrodes it was shown that EMIM-BF_4 is retained at the electrode surface and a potential dependent SFG feature at 2340 cm^{-1} was also reported. On the basis of the intrinsic surface selectivity and the observation of CO_2 (a centro-symmetric molecule) by SFG, the band was assigned to a surface CO_2-EMIM intermediate.[57] A later SFG study also explored this system in more detail and importantly confirmed the presence and potential dependence of the proposed CO_2-EMIM adduct (Figure 10.7).[56] This is a significant set of results confirming that the ionic liquids studied are able to act in a catalytic manner to stabilize the otherwise high energy $CO_2^{\bullet -}$ radical anion.

On many electrodes NR SFG occurs in addition to the resonant signals arising from molecules at the interface. Although the NR background can provide a phase reference aiding the determination of molecular orientation at metal electrodes such as Ag and Au the magnitude of the NR can swamp the resonant signals so some studies deliberately suppress NR responses by the inclusion of small temporal delay between the fs IR and a fs derived time-asymmetric visible laser pulse.[58] However, recent SFG studies of Ag electrodes in EMIM-BF_4 examined both the potential dependent NR and resonant SFG signals providing important insights into the related electro-catalytic studies.[54,55] Here it was shown that the potential dependence of the NR SFG arises from the total EDL and not just the electrode surface.

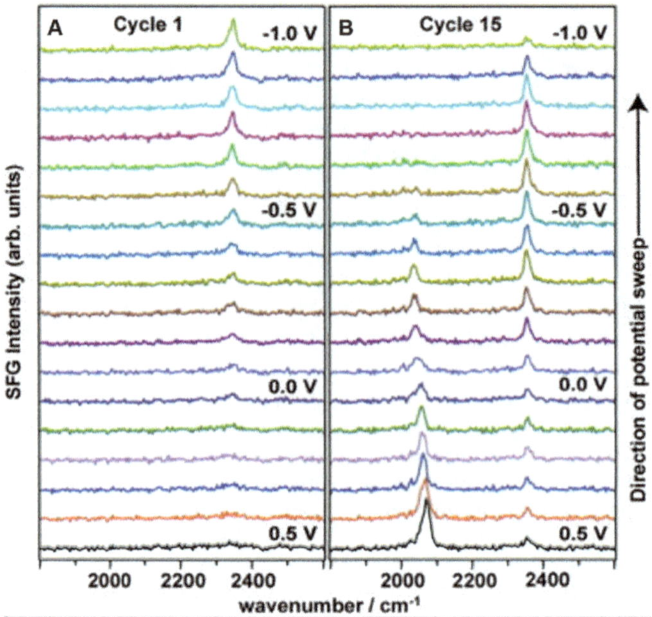

Figure 10.7 Potential dependent resonant SFG response of a Pt electrode in EMIM-BF_4 during CO_2 reduction. The feature at *ca.* 2340 cm^{-1} has been assigned to a CO_2-EMIM adduct at the electrode surface.
Reprinted from *J. Electroanal. Chem.*, **800**, B. Braunschweig, P. Mukherjee, J. L. Haan and D. D. Dlott, Vibrational sum-frequency generation study of the CO_2 electrochemical reduction at Pt/EMIM-BF4 solid/liquid interfaces, 144–150, Copyright 2017 with permission from Elsevier.

A detailed discussion of the mechanisms of the potential dependence of SFG signals at electrified interfaces has recently been published[50] and the SFG studies[59,60] build upon the work of Eisenthal and others on the potential dependence of second harmonic generation spectroscopy.[60,61] Briefly, with some electrode/electrolyte combinations very large DC electric fields can be generated, making third order terms a significant contributor to the SFG polarization. Under many conditions the second order terms still dominate and a linear dependence of SFG intensity on applied potential is observed. However, under the extreme where the third order terms dominate, the NR SFG response is expected to be quadratic with respect to potential with the potential dependent SFG intensity plot appearing as a series of overlaid parabola, the shape of which provides insights into the EDL.[50] Previous SFG studies[62] have indicated that the electrified interface consisted of a Helmholtz layer that was ~1 ion thick with Pt/EMIM-BF_4 making it feasible that large fields can also be generated with Ag/EMIM-BF_4 and in line with these studies a quadratic dependence of NR SFG intensity on applied potential was reported for EMIM-BF_4 at Ag.[54] Here the NR SFG response had a minimum in the NR SFG intensity at −1.33 V, which was followed by a

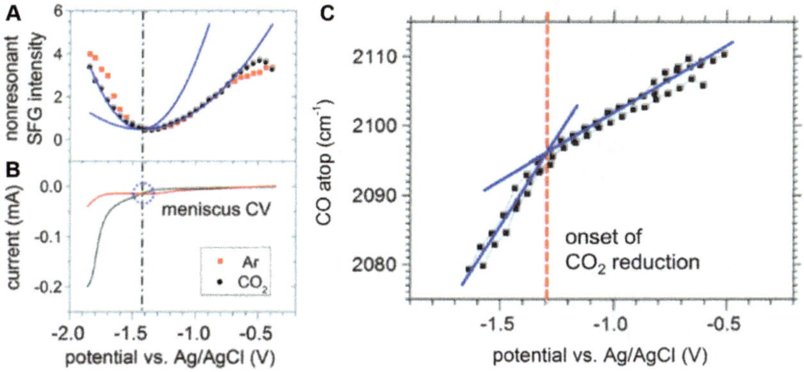

Figure 10.8 Potential dependence of NR SFG intensity during measurements with a Ag electrode in EMIM-BF_4 (left). The behaviour can be modelled using two parabolas with the change in curvature at the potential where CO_2 catalysis onsets indicating a change in the EDL structure at this point. The resonant SFG ν(CO) mode also shows a large change in Stark shift as catalysis onsets, again indicating change in the EDL structure (right). Adapted from *J. Electroanal. Chem.*, **800**, N. García Rey and D. D. Dlott, Studies of electrochemical interfaces by broadband sum frequency generation, 114–125, Copyright 2017, with permission from Elsevier.

change in gradient at more negative potentials. Importantly, −1.33 V is the potential for the onset of electrocatalytic CO_2 reduction, leading to the conclusion that the EDL undergoes a major transition at the point at which CO_2 reduction initiates, likely due to a structural reorganization of the ionic liquid within the EDL (Figure 10.8). The change in the EDL at the onset potential is also detectable by a sudden change in the CO Stark tuning rate (Figure 10.8) indicating a change in the electric field strength.[54] Significantly this structural rearrangement appears both in the presence and absence of CO_2 and its potential varies with water content, in line with the onset potential for CO_2 reduction,[55] suggesting that the structural transition within the EDL is controlling the onset potential for catalysis and is not a result of the onset of catalysis. These first works studying the electrochemical reduction of CO_2 by SFG are important as they both (i) demonstrate the capability of the technique to probe both the electrode surface and the EDL and (ii) detail approaches to analyze the sometimes complex potential dependent SFG responses that can be recorded.[50]

The ability of IR–vis SFG spectroscopy to discriminate surface species has also been used to examine molecular systems of relevance to electrocatalytic CO_2 reduction. Re(bpy)$(CO)_3$Cl is a well-known electrocatalyst[63] and photocatalyst.[64] Past electrochemical studies on Re(bpy)$(CO)_3$Cl immobilized on TiO_2[65] have led to interest into the binding of derivatives of this complex to TiO_2 surfaces with combined SFG and DFT studies exploring the binding orientation of catalysts derivatized with carboxylic acid groups to aid adhesion to TiO_2, highlighting the ability of the technique to resolve molecular geometries of molecules at surfaces.[66–68] Whilst IR pump–SFG probe experiments of similar systems recorded the rate of vibrational relaxation of the bound complexes providing insights into the surface specific interactions occurring,[69] a conclusion expanded upon in later 2D-SFG studies that reported a structural heterogeneity of bound catalysts due to the range of surface environments.[70] SFG spectroscopy has also been used to explore a range of Re(bpy)$(CO)_3$Cl derivatives at Au electrodes, although in the absence of an applied potential providing details of surface orientations,[71–74] vibrational cooling rates[73] and of the substrate–catalyst interactions themselves. Indeed, until very recently SFG studies of molecular electrocatalysts for CO_2 reduction had only been carried out in the absence of applied potentials. A concern had been that the quadratic nature of the intensity of the SFG intensity with number density[75,76] would make it challenging to measure the expected small and transient populations of homogeneous electrocatalysts at electrode surfaces. However, very recently some of us reported[77] an *in situ* SFG study exploring the role of electrode–catalyst interactions in enabling efficient CO_2 reduction with Mo(bpy)$(CO)_4$ on gold electrodes. Here we explored the previously proposed strong interaction between the metal carbonyl complex and the gold electrode.[78] The strong interaction between the electrode and the catalyst allowed us to obtain clear SFG signals for the catalyst and solvent at the electrode surface, even though they were only transiently present, and we were even able to identify the formation of a catalytic active species, $[Mo(bpy)(CO)_3]^{2-}$, at potentials 0.6 V positive of the solution reduction potential for this species, in line with predictions from past catalytic studies.[78] Critically, the SFG experiments were also able to measure Stark tuning rates for the metal carbonyl complexes, providing confirmation that true surface species were being probed, and through potential dependent studies of the SFG response of solvent molecules at the surface (see Figure 10.9) it was also possible to identify that the presence of a strong interaction between the gold electrode and a pre-catalytic intermediate, $[Mo(bpy)(CO)_4]^{\bullet-}$, was critical in determining the onset of CO_2 catalysis. This initial study on molecular electrocatalysis coupled to the work of Dlott, in particular on metal electrodes for CO_2 reduction, demonstrates the ability of SFG spectroscopy to derive a remarkable level of detail on processes within the EDL. It is anticipated that the increased uptake of heterodyned SFG spectroscopy (that gives a linear dependence of I_{SFG} on number density[75]) and the improved availability of robust ps/fs laser systems will lead to a range of reports on molecular electrocatalysts for CO_2 reduction in the near future.

10.5 Pulse Radiolysis and Time-resolved Spectroscopy

Although not in itself a direct probe of electrochemically generated species, fast time-resolved vibrational spectroscopy coupled to pulse radiolysis offers a novel route to studying short-lived reduced or oxidized intermediates that

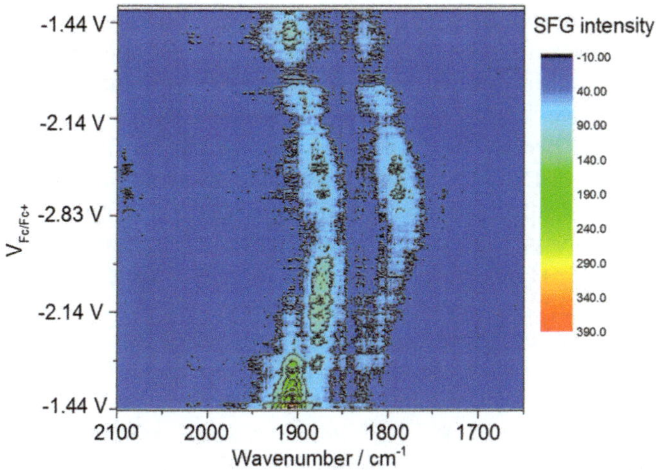

Figure 10.9 *In situ* SFG spectra of the CO_2 reduction catalyst [Mo(bpy)(CO)$_4$] at a gold electrode surface recorded *in situ* during a single CV measurement (5 mV s^{-1}). The spectrum shows the ν(CO) region with the initial complex being reduced to form the active catalyst. The entire spectro-electrochemical survey took less than 5 min to complete.
Reproduced with permission from *J. Am. Chem. Soc.*, 2017, **139**(39), 13791–13797, http://doi.org/10.1021/jacs.7b06898. © 2017 American Chemical Society. Published under the terms of the CC BY 4.0 licence, https://creativecommons.org/licenses/by/4.0/.

are formed relevant to molecular catalytic mediated CO_2 reduction, which may only exist on the nanoseconds timescale. Time-resolved electrochemical measurements using voltage steps coupled to spectroscopic studies are challenging due to the need to employ micro-electrodes to lower cell time constants and traditionally such species have been studied following photochemical generation. Pulse radiolysis is particularly useful for the study of systems that may be challenging to photosensitize, which prevents the use of the more commonly employed UV/vis pump–vibrational spectroscopy probe experiments.[79] A further considerable advantage of the generation of radical species *via* the use of pulse radiolysis is that it is proposed to be a "clean" technique, avoiding the use of sacrificial electron/hole donors that are typically required in photo-initiated systems. The principles of pulse radiolysis have been recently reviewed elsewhere,[79] but briefly the energy from short electron pulse (often ps–ns in length) is deposited into the reaction media generating a mixture of ionized solvent molecules. In the case of acetonitrile initially generated $CH_3CN^{+\bullet}$ can be scavenged by formate leading to the very rapid production of $CO_2^{\bullet-}$, a very strong reducing agent (-1.9 V_{NHE}) that in turn can reduce the catalyst/molecule of interest.[80]

Pulse-radiolysis has a long history of application to the study of electro-catalytic CO_2 reduction mechanisms with studies using UV/vis spectroscopy to probe the rate constants and binding constants of CO_2 and H^+ with

Ni(cyclam)$^+$ and derivatives,[81,82] reduced cobalt porphyrins[83] and cobalt 14-membered tetrazamacrocyles.[84] However, UV/vis spectra are typically relatively broad, making definitive assignments complicated. More recently several studies have begun to use time-resolved IR and Raman spectroscopies to directly probe the molecular structures. Janik and Tripathi reported a spectrum of $CO_2^{-\bullet}$ that has bands at 1298 and 742 cm^{-1} in aqueous solvents, itself an important possible intermediate during CO_2 reduction, using pulse radiolysis and time-resolved resonance Raman spectroscopy (Figure 10.10) with pH dependence spectroscopy allowing for determination of a pK_a of 3.4 ± 0.2.[85]

Grills and co-workers developed a time-resolved infrared (TRIR) pulse radiolysis experiment that has been applied to two of the most widely studied complexes for electrocatalytic CO_2 reduction, [Re(bpy)(CO)$_3$Cl] and derivatives of [Mn(bpy)(CO)$_3$Br].[86–88] Previous FTIR spectro-electrochemical studies on [Mn(bpy)(CO)$_3$Br] had demonstrated that the generation of the catalytic active species, [Mn(bpy)(CO)$_3$]$^-$, occurs following the reduction of a dimeric species [(Mn(bpy)(CO)$_3$)$_2$], although the mechanism of dimer formation was unproven.[89–91] In the pulse radiolysis works the initial one electron reduced species [Mn(4,4′-^{t}Bu$_2$-bpy)(OCHO)(CO)$_3$]$^\bullet$ was observed. Furthermore the kinetics of the formation of the neutral radical, [Mn(4,4′-^{t}Bu$_2$-bpy)(CO)$_3$]$^\bullet$, followed by its subsequent dimerization was measured for the first time.[87] Subsequent pulse-radiolysis TRIR studies have also explored derivatives of this electrocatalyst with bulky substituents on the bipyridine ligand.[88] These modified ligands are thought to have a dual role, containing Lewis base sites to direct the binding of a Brønsted acid, improving the catalytic onset, whilst the steric bulk is included to prevent the previously studied dimerization, to aid the generation of the active catalyst at more positive potentials. Remarkably whilst TRIR

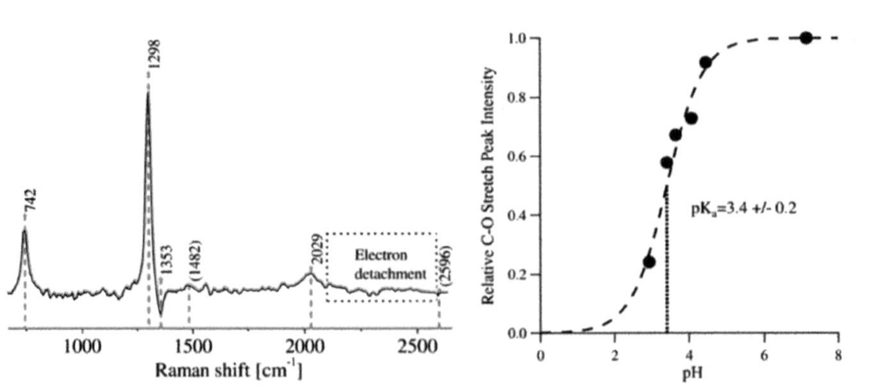

Figure 10.10 Resonance Raman spectrum of the $CO_2^{-\bullet}$ radical anion generated by pulse radiolysis (a) and pH dependence of the C–O stretching mode to determine the pK_a (b).
Figure adapted with permission from *J. Chem. Phys.*, 2016, **144**, 154307, with permission from AIP Publishing.

studies on catalysts with less bulky ligands (^{t}Bu) on the 4,4′ site of bipyridine showed extremely high dimerization rates for the initially generated $[Mn(R_2\text{-bpy})(CO)_3]^\bullet$ radicals ($2k_{dim} \approx 1 \times 10^9\ M^{-1}\,s^{-1}$)[87] the inclusion of the large ($(MeO)_2Ph$) on the 6,6′ position completely turned off dimerization with a Mn radical species persisting throughout the experiment (ms timescale). Although limited in number it is notable that the pulse radiolysis–time-resolved spectroscopies have great potential to allow for the identification of even the shortest lived reactive intermediates.

10.6 Outlook and Summary

Surface sensitive and surface selective vibrational spectroscopies are increasingly being applied to deliver intricate details of intermediates at electrode surfaces offering a route to rational design of new catalysts and electrodes. Remarkable levels of detail, such as the orientation of molecules at the electrode interface, are being reported and of great promise are techniques such as SFG spectroscopy that probe the structure and nature of both the intermediates and the EDL. To date such measurements on electrochemical systems have often focused on relatively slow timescales (often seconds or milliseconds), yet pulse-radiolysis studies in solution indicate that many important intermediates may be extremely short-lived and addressing species at electrode surfaces on the sub-millisecond timescale following a perturbation is an exciting prospect, offering a route to further enhancing our understanding of the mechanistic factors controlling the electrochemical reduction of carbon dioxide.

References

1. E. E. Benson, C. P. Kubiak, A. J. Sathrum and J. M. Smieja, *Chem. Soc. Rev.*, 2009, **38**, 89–99.
2. M. C. Figueiredo, I. Ledezma-Yanez and M. T. M. Koper, *ACS Catal.*, 2016, **6**, 2382–2392.
3. L. Wang, K. Gupta, J. B. M. Goodall, J. A. Darr and K. B. Holt, *Faraday Discuss.*, 2017, **197**, 517–532.
4. M. Osawa, *Bull. Chem. Soc. Jpn.*, 1997, **70**, 2861–2880.
5. J. M. Brown, *Molecular Spectroscopy*, Oxford University Press, 1998.
6. J. Qiao, Y. Liu, F. Hong and J. Zhang, *Chem. Soc. Rev.*, 2013, **43**, 631–675.
7. T. Hartman, C. S. Wondergem, N. Kumar, A. van den Berg and B. M. Weckhuysen, *J. Phys. Chem. Lett.*, 2016, **7**, 1570–1584.
8. B. Sharma, R. R. Frontiera, A.-I. Henry, E. Ringe and R. P. Van Duyne, *Mater. Today*, 2012, **15**, 16–25.
9. G. C. Schatz and R. P. Van Duyne, in *Handbook of Vibrational Spectroscopy*, John Wiley & Sons, Ltd, 2006.
10. D.-Y. Wu, J.-F. Li, B. Ren and Z.-Q. Tian, *Chem. Soc. Rev.*, 2008, **37**, 1025–1041.

11. S. Zaleski, A. J. Wilson, M. Mattei, X. Chen, G. Goubert, M. F. Cardinal, K. A. Willets and R. P. Van Duyne, *Acc. Chem. Res.*, 2016, **49**, 2023–2030.
12. K. G. Schmitt and A. A. Gewirth, *J. Phys. Chem. C*, 2014, **118**, 17567–17576.
13. V. O. Santos, I. R. Leite, A. G. Brolo and J. C. Rubim, *J. Raman Spectrosc.*, 2016, **47**, 674–680.
14. E. A. Batista and M. L. A. Temperini, *J. Electroanal. Chem.*, 2009, **629**, 158–163.
15. I. Oda, H. Ogasawara and M. Ito, *Langmuir*, 1996, **12**, 1094–1097.
16. J. L. Oberst, H.-R. Molly Jhong, P. J. A. Kenis and A. A. Gewirth, *J. Solid State Electrochem.*, 2016, **20**, 1149–1154.
17. S. Liu, H. Tao, L. Zeng, Q. Liu, Z. Xu, Q. Liu and J.-L. Luo, *J. Am. Chem. Soc.*, 2017, **139**, 2160–2163.
18. M. Liu, Y. Pang, B. Zhang, P. De Luna, O. Voznyy, J. Xu, X. Zheng, C. T. Dinh, F. Fan, C. Cao, F. P. G. de Arquer, T. S. Safaei, A. Mepham, A. Klinkova, E. Kumacheva, T. Filleter, D. Sinton, S. O. Kelley and E. H. Sargent, *Nature*, 2016, **537**, 382–386.
19. R. Long, Y. Li, L. Song and Y. Xiong, *Small*, 2015, **11**, 3873–3889.
20. P. L. Stiles, J. A. Dieringer, N. C. Shah and R. P. Van Duyne, *Annu. Rev. Anal. Chem.*, 2008, **1**, 601–626.
21. K. F. Domke and B. Pettinger, *ChemPhysChem*, 2009, **10**, 1794–1798.
22. T. Schmid, B.-S. Yeo, G. Leong, J. Stadler and R. Zenobi, *J. Raman Spectrosc.*, 2009, **40**, 1392–1399.
23. Z.-C. Zeng, S.-C. Huang, D.-Y. Wu, L.-Y. Meng, M.-H. Li, T.-X. Huang, J.-H. Zhong, X. Wang, Z.-L. Yang and B. Ren, *J. Am. Chem. Soc.*, 2015, **137**, 11928–11931.
24. A. Hartstein, *Phys. Rev. Lett.*, 1980, **45**, 201–204.
25. A. Fujishima, K. Kohayakawa and K. Honda, *Bull. Chem. Soc. Jpn.*, 1975, **48**, 1041–1042.
26. M. Aresta, A. Dibenedetto and E. Quaranta, *SpringerLink*, 2016, 35–69.
27. F. Neubrech, C. Huck, K. Weber, A. Pucci and H. Giessen, *Chem. Rev.*, 2017, **117**, 5110–5145.
28. M. Osawa, K. Ataka, K. Yoshii and T. Yotsuyanagi, *J. Electron Spectrosc. Relat. Phenom.*, 1993, **64**, 371–379.
29. *Near-Field Optics and Surface Plasmon Polaritons*, ed. S. Kawata, Springer, 2001.
30. A. Wuttig, M. Yaguchi, K. Motobayashi, M. Osawa and Y. Surendranath, *Proc. Natl. Acad. Sci. U. S. A.*, 2016, **113**, E4585–E4593.
31. L. W. H. Leung, A. Wieckowski and M. J. Weaver, *J. Phys. Chem.*, 1988, **92**, 6985–6990.
32. A. Miki, S. Ye and M. Osawa, *Chem. Commun.*, 2002, 1500–1501.
33. T. Smolinka, M. Heinen, Y. X. Chen, Z. Jusys, W. Lehnert and R. J. Behm, *Electrochim. Acta*, 2005, **50**, 5189–5199.
34. M. Osawa, *Langmuir*, 1994, **10**, 640–642.
35. A. V. Rudnev, U. E. Zhumaev, A. Kuzume, S. Vesztergom, J. Furrer, P. Broekmann and T. Wandlowski, *Electrochim. Acta*, 2016, **189**, 38–44.

36. M. Dunwell, Q. Lu, J. M. Heyes, J. Rosen, J. G. Chen, Y. Yan, F. Jiao and B. Xu, *J. Am. Chem. Soc.*, 2017, **139**, 3774–3783.
37. Y. Hori, A. Murata, K. Kikuchi and S. Suzuki, *J. Chem. Soc., Chem. Commun.*, 1987, 728–729.
38. Y. Chen, C. W. Li and M. W. Kanan, *J. Am. Chem. Soc.*, 2012, **134**, 19969–19972.
39. W. Zhu, R. Michalsky, Ö. Metin, H. Lv, S. Guo, C. J. Wright, X. Sun, A. A. Peterson and S. Sun, *J. Am. Chem. Soc.*, 2013, **135**, 16833–16836.
40. A. Wuttig, C. Liu, Q. Peng, M. Yaguchi, C. H. Hendon, K. Motobayashi, S. Ye, M. Osawa and Y. Surendranath, *ACS Cent. Sci.*, 2016, **2**, 522–528.
41. J. Heyes, M. Dunwell and B. Xu, *J. Phys. Chem. C*, 2016, **120**, 17334–17341.
42. K. P. Kuhl, E. R. Cave, D. N. Abram and T. F. Jaramillo, *Energy Environ. Sci.*, 2012, **5**, 7050–7059.
43. Y. Hori, A. Murata and R. Takahashi, *J. Chem. Soc., Faraday Trans. 1*, 1989, **85**, 2309–2326.
44. N. J. Firet and W. A. Smith, *ACS Catal.*, 2017, 7, 606–612.
45. H.-F. Wang, *Prog. Surf. Sci.*, 2016, **91**, 155–182.
46. Y. R. Shen, *Nature*, 1989, **337**, 519–525.
47. S. Ye and M. Osawa, *Chem. Lett.*, 2009, **38**, 386–391.
48. M. Buck and M. Himmelhaus, *J. Vac. Sci. Technol., A*, 2001, **19**, 2717–2736.
49. Y. R. Shen, *J. Phys. Chem. C*, 2012, **116**, 15505–15509.
50. N. G. Rey and D. D. Dlott, *J. Electroanal. Chem.*, 2017, **800**, 114–125.
51. B. G. Nicolau, N. García-Rey, B. Dryzhakov and D. D. Dlott, *J. Phys. Chem. C*, 2015, **119**, 10227–10233.
52. W.-T. Liu and Y. R. Shen, *Proc. Natl. Acad. Sci. U. S. A.*, 2014, **111**, 1293–1297.
53. B. A. Rosen, A. Salehi-Khojin, M. R. Thorson, W. Zhu, D. T. Whipple, P. J. A. Kenis and R. I. Masel, *Science*, 2011, **334**, 643–644.
54. N. García Rey and D. D. Dlott, *J. Phys. Chem. C*, 2015, **119**, 20892–20899.
55. N. G. Rey and D. D. Dlott, *Phys. Chem. Chem. Phys.*, 2017, **19**, 10491–10501.
56. B. Braunschweig, P. Mukherjee, J. L. Haan and D. D. Dlott, *J. Electroanal. Chem.*, 2017, **800**, 144–150.
57. B. A. Rosen, J. L. Haan, P. Mukherjee, B. Braunschweig, W. Zhu, A. Salehi-Khojin, D. D. Dlott and R. I. Masel, *J. Phys. Chem. C*, 2012, **116**, 15307–15312.
58. A. Lagutchev, A. Lozano, P. Mukherjee, S. A. Hambir and D. D. Dlott, *Spectrochim. Acta, Part A*, 2010, **75**, 1289–1296.
59. K. C. Jena, P. A. Covert and D. K. Hore, *J. Phys. Chem. Lett.*, 2011, **2**, 1056–1061.
60. F. M. Geiger, *Annu. Rev. Phys. Chem.*, 2009, **60**, 61–83.
61. K. B. Eisenthal, *Chem. Rev.*, 2006, **106**, 1462–1477.
62. S. Baldelli, *J. Phys. Chem. B*, 2005, **109**, 13049–13051.

63. J. Hawecker, J.-M. Lehn and R. Ziessel, *J. Chem. Soc., Chem. Commun.*, 1984, 328–330.
64. J. Hawecker, J.-M. Lehn and R. Ziessel, *J. Chem. Soc., Chem. Commun.*, 1983, 536–538.
65. F. Cecchet, M. Alebbi, C. A. Bignozzi and F. Paolucci, *Inorg. Chim. Acta*, 2006, **359**, 3871–3874.
66. C. L. Anfuso, R. C. Snoeberger, A. M. Ricks, W. Liu, D. Xiao, V. S. Batista and T. Lian, *J. Am. Chem. Soc.*, 2011, **133**, 6922–6925.
67. C. L. Anfuso, D. Xiao, A. M. Ricks, C. F. A. Negre, V. S. Batista and T. Lian, *J. Phys. Chem. C*, 2012, **116**, 24107–24114.
68. A. Ge, B. Rudshteyn, B. T. Psciuk, D. Xiao, J. Song, C. L. Anfuso, A. M. Ricks, V. S. Batista and T. Lian, *J. Phys. Chem. C*, 2016, **120**, 20970–20977.
69. C. Calabrese, H. Vanselous and P. B. Petersen, *J. Phys. Chem. C*, 2016, **120**, 1515–1522.
70. H. Vanselous, A. M. Stingel and P. B. Petersen, *J. Phys. Chem. Lett.*, 2017, **8**, 825–830.
71. M. L. Clark, B. Rudshteyn, A. Ge, S. A. Chabolla, C. W. Machan, B. T. Psciuk, J. Song, G. Canzi, T. Lian, V. S. Batista and C. P. Kubiak, *J. Phys. Chem. C*, 2016, **120**, 1657–1665.
72. J. Wang, M. L. Clark, Y. Li, C. L. Kaslan, C. P. Kubiak and W. Xiong, *J. Phys. Chem. Lett.*, 2015, **6**, 4204–4209.
73. C. L. Anfuso, A. M. Ricks, W. Rodríguez-Córdoba and T. Lian, *J. Phys. Chem. C*, 2012, **116**, 26377–26384.
74. Z. Li, J. Wang, Y. Li and W. Xiong, *J. Phys. Chem. C*, 2016, **120**, 20239–20246.
75. I. V. Stiopkin, H. D. Jayathilake, A. N. Bordenyuk and A. V. Benderskii, *J. Am. Chem. Soc.*, 2008, **130**, 2271–2275.
76. H. Vanselous and P. B. Petersen, *J. Phys. Chem. C*, 2016, **120**, 8175–8184.
77. G. Neri, P. M. Donaldson and A. J. Cowan, *J. Am. Chem. Soc.*, 2017, **139**(39), 13791–13797.
78. J. Tory, B. Setterfield-Price, R. A. W. Dryfe and F. Hartl, *ChemElectroChem*, 2015, **2**, 213–217.
79. D. C. Grills, J. A. Farrington, B. H. Layne, J. M. Preses, H. J. Bernstein and J. F. Wishart, *Rev. Sci. Instrum.*, 2015, **86**, 044102.
80. D. C. Grills, J. A. Farrington, B. H. Layne, S. V. Lymar, B. A. Mello, J. M. Preses and J. F. Wishart, *J. Am. Chem. Soc.*, 2014, **136**, 5563–5566.
81. J. Schneider, H. Jia, K. Kobiro, D. E. Cabelli, J. T. Muckerman and E. Fujita, *Energy Environ. Sci.*, 2012, **5**, 9502–9510.
82. C. A. Kelly, E. L. Blinn, N. Camaioni, M. D'Angelantonio and Q. G. Mulazzani, *Inorg. Chem.*, 1999, **38**, 2756.
83. D. Behar, T. Dhanasekaran, P. Neta, C. M. Hosten, D. Ejeh, P. Hambright and E. Fujita, *J. Phys. Chem. A*, 1998, **102**, 2870–2877.
84. E. Fujita and R. van Eldik, *Inorg. Chem.*, 1998, **37**, 360–362.
85. I. Janik and G. N. R. Tripathi, *J. Chem. Phys.*, 2016, **144**, 154307.
86. D. C. Grills, A. R. Cook, E. Fujita, M. W. George, J. M. Presses and J. F. Wishart, *Appl. Spectrosc.*, 2010, **64**, 563.

87. D. C. Grills, J. A. Farrington, B. H. Layne, S. V. Lymar, B. A. Mello, J. M. Preses and J. F. Wishart, *J. Am. Chem. Soc.*, 2014, **136**, 5563–5566.
88. K. T. Ngo, M. McKinnon, B. Mahanti, R. Narayanan, D. C. Grills, M. Z. Ertem and J. Rochford, *J. Am. Chem. Soc.*, 2017, **139**, 2604–2618.
89. M. Bourrez, F. Molton, S. Chardon-Noblat and A. Deronzier, *Angew. Chem., Int. Ed. Engl.*, 2011, **50**, 9903–9906.
90. J. M. Smieja, M. D. Sampson, K. A. Grice, E. E. Benson, J. D. Froehlich and C. P. Kubiak, *Inorg. Chem.*, 2013, **52**, 2484–2491.
91. B. D. Rossenaar, F. Hartl, D. J. Stufkens, C. Amatore, E. Maisonhaute and J.-N. Verpeaux, *Organometallics*, 1997, **16**, 4675–4685.

Subject Index

Page numbers in *italic* refer to figures; page numbers with suffix T refer to tables.